Applications of Mathematics

Stochastic Modelling and Applied Probability

46

T0206022

Springer

Applications of Mathematics

1 Fleming/Rishel, **Deterministic and Stochastic Optimal Control** (1975)
2 Marchuk, **Methods of Numerical Mathematics,** Second Ed. (1982)
3 Balakrishnan, **Applied Functional Analysis,** Second Ed. (1981)
4 Borovkov, **Stochastic Processes in Queueing Theory** (1976)
5 Liptser/Shiryayev, **Statistics of Random Processes I: General Theory,** Second Ed. (1977)
6 Liptser/Shiryayev, **Statistics of Random Processes II: Applications,** Second Ed. (1978)
7 Vorob'ev, **Game Theory: Lectures for Economists and Systems Scientists** (1977)
8 Shiryayev, **Optimal Stopping Rules** (1978)
9 Ibragimov/Rozanov, **Gaussian Random Processes** (1978)
10 Wonham, **Linear Multivariable Control: A Geometric Approach,** Third Ed. (1985)
11 Hida, **Brownian Motion** (1980)
12 Hestenes, **Conjugate Direction Methods in Optimization** (1980)
13 Kallianpur, **Stochastic Filtering Theory** (1980)
14 Krylov, **Controlled Diffusion Processes** (1980)
15 Prabhu, **Stochastic Storage Processes: Queues, Insurance Risk, Dams, and Data Communication,** Second Ed. (1998)
16 Ibragimov/Has'minskii, **Statistical Estimation: Asymptotic Theory** (1981)
17 Cesari, **Optimization: Theory and Applications** (1982)
18 Elliott, **Stochastic Calculus and Applications** (1982)
19 Marchuk/Shaidourov, **Difference Methods and Their Extrapolations** (1983)
20 Hijab, **Stabilization of Control Systems** (1986)
21 Protter, **Stochastic Integration and Differential Equations** (1990)
22 Benveniste/Métivier/Priouret, **Adaptive Algorithms and Stochastic Approximations** (1990)
23 Kloeden/Platen, **Numerical Solution of Stochastic Differential Equations** (1992)
24 Kushner/Dupuis, **Numerical Methods for Stochastic Control Problems in Continuous Time,** Second Ed. (2001)
25 Fleming/Soner, **Controlled Markov Processes and Viscosity Solutions** (1993)
26 Baccelli/Brémaud, **Elements of Queueing Theory** (1994)
27 Winkler, **Image Analysis, Random Fields, and Dynamic Monte Carlo Methods: An Introduction to Mathematical Aspects** (1994)
28 Kalpazidou, **Cycle Representations of Markov Processes** (1995)
29 Elliott/Aggoun/Moore, **Hidden Markov Models: Estimation and Control** (1995)
30 Hernández-Lerma/Lasserre, **Discrete-Time Markov Control Processes: Basic Optimality Criteria** (1996)
31 Devroye/Györfi/Lugosi, **A Probabilistic Theory of Pattern Recognition** (1996)
32 Maitra/Sudderth, **Discrete Gambling and Stochastic Games** (1996)
33 Embrechts/Klüppelberg/Mikosch, **Modelling Extremal Events** (1997)
34 Duflo, **Random Iterative Models** (1997)
35 Kushner/Yin, **Stochastic Approximation Algorithms and Applications** (1997)
36 Musiela/Rutkowski, **Martingale Methods in Financial Modeling: Theory and Application** (1997)

(continued after index)

Hong Chen   David D. Yao

Fundamentals of Queueing Networks

Performance, Asymptotics, and Optimization

Springer

Hong Chen
Faculty of Commerce and
  Business Administration
University of British Columbia
Vancouver, British Columbia V6T 1Z2
Canada
chen@hong.commerce.ubc.ca
http://hong.commerce.ubc.ca/

David D. Yao
Department of Operations Research and
  Industrial Engineering
Columbia University
New York, NY 10027
USA
yao@ieor.columbia.edu
http://www.ieor.columbia.edu/~yao/

Managing Editors

I. Karatzas
Departments of Mathematics and Statistics
Columbia University
New York, NY 10027, USA

M. Yor
CNRS, Laboratoire de Probabilités
Université Pierre et Marie Curie
4, Place Jussieu, Tour 56
F-75252 Paris Cedex 05, France

Mathematics Subject Classification (2000): 60K25, 90-02, 60G55

Library of Congress Cataloging-in-Publication Data
Chen, Hong.
    Fundamentals of queueing networks: performance, asymptotics, and optimization /
Hong Chen, David D. Yao.
        p. cm. – (Applications of mathematics; 46)
    Includes bibliographical references and index.
    ISBN 978-1-4419-2896-2
    1. Queuing theory.   2. Stochastic analysis.   I. Yao, David D., 1950–   II. Title.   III. Series.
QA274.8 .C44 2001
519.8′2–dc21   00-052280

Printed on acid-free paper.

Production managed by A. Orrantia; manufacturing supervised by Jeffrey Taub.
Electronically imposed from PDF output of LaTeX files created by the authors and formatted by
David Kramer.

Printed in the United States of America.

9 8 7 6 5 4 3 2 1

Springer-Verlag   New York Berlin Heidelberg
A member of BertelsmannSpringer Science+Business Media GmbH

Learning is but an adjunct to ourself,
And where we are our learning likewise is.

— *Love's Labour's Lost*

Preface

The objective of this book is to collect in a single volume the essentials of stochastic networks, from the classical product-form theory to the more recent developments such as diffusion and fluid limits, stochastic comparisons, stability, control (dynamic scheduling) and optimization. The selection of materials inevitably is a reflection upon our bias and preference, but it is also driven to a large extent by our desire to provide a graduate-level text that is well balanced in breadth and depth, suitable for the classroom. Given the wide-ranging applications of stochastic networks in recent years, from supply chains to telecommunications, it is also our hope that the book will serve as a useful reference for researchers and students alike in these diverse fields.

The book consists of three parts. The first part, Chapters 1 through 4, covers (continuous-time) Markov-chain models, including the classical Jackson and Kelly networks, the notion of quasi-reversible queues, and stochastic comparisons. The second part, Chapters 5 through 10, focuses on Brownian models, including limit theorems for generalized Jackson networks and multiclass feedforward networks, an in-depth examination of stability in a Kumar–Seidman network, and Brownian approximations for general multiclass networks with a mixture of priority and first-in-first-out disciplines. The third part, Chapters 11 and 12, discusses scheduling in both queueing (stochastic) and fluid (deterministic) networks, along with topics such as conservation laws, polymatroid optimization, and linear programming.

In terms of teaching, there are several ways to use the book, depending on the background of the class. A basic and balanced coverage will include

Chapters 1, 2, and 4 from the first part, Chapters 5 through 8 from the second part, while leaving the other chapters for independent reading (with a term paper as part of the evaluation, for instance). It is also possible to focus on either the first part or the second part, supplemented by one or both chapters of the third part. The first option requires only a modest background in Markov-chain models, whereas the second option will suit students who have more preparation in advanced mathematics, real analysis in particular.

The book is based on our class notes accumulated over many years of graduate teaching on stochastic networks at the University of British Columbia and Columbia University. Also, in the fall of 1998, while both of us were in Hong Kong, we cotaught a graduate class at the Chinese University, based on a preliminary version of the book. For their careful reading of the manuscript and able assistance at various stages, we want to thank many of our students and associates, in particular, Yingdong Lu, Alan Sheller-Wolf, Xinyang Shen, Hengqing Ye, Hanqin Zhang, and Li Zhang. In addition, we are very grateful to Ward Whitt for offering many insightful comments to several chapters, and for sharing his encyclopedic knowledge of the literature. Last but not least, Hong Chen wishes to express his deepest gratitude to Avi Mandelbaum for mentoring and guiding him through the early years of his research career.

In the course of writing this book, which spans the last four or five years, we have been supported by several grants from NSERC and NSF. During our sojourn in Hong Kong, we were supported by grants from the RGC, and by the University of Science and Technology and the Chinese University. We thank these funding agencies and institutions for their sponsorship.

Hong Chen
David Yao
July 2000

Contents

Preface   **vii**

List of Figures   **xv**

List of Tables   **xvii**

1 Birth–Death Queues   **1**
   1.1 Basics 1
   1.2 Time Reversibility 4
   1.3 Stochastic Orders 8
   1.4 Notes 11
   1.5 Exercises 11

References   **13**

2 Jackson Networks   **15**
   2.1 Open Network 16
   2.2 Closed Network 19
   2.3 Semiopen Network 21
   2.4 Throughput Function 23
   2.5 Throughput Computation 25
      2.5.1 Convolution Algorithm 26
      2.5.2 Mean Value Analysis 27
   2.6 Time Reversal 28

2.7 Notes 32
2.8 Exercises 33

References   **35**

3 Stochastic Comparisons   **37**
3.1 Monotonicity 38
   3.1.1 PF_2 Property 38
   3.1.2 Likelihood Ratio Ordering 39
   3.1.3 Shifted Likelihood Ratio Ordering 42
3.2 Concavity and Convexity 46
3.3 Multiple Servers 49
3.4 Resource Sharing 52
   3.4.1 Aggregation of Servers 52
   3.4.2 Aggregation of Nodes 53
3.5 Arrangement and Majorization 56
3.6 Notes 61
3.7 Exercises 63

References   **65**

4 Kelly Networks   **69**
4.1 Quasi-Reversible Queues 70
4.2 Symmetric Queues 76
   4.2.1 Phase-Type Distributions and Processes . . . 76
   4.2.2 Multiclass $M/PH/1$ Queue 77
4.3 A Multiclass Network 81
4.4 Poisson Flows 87
4.5 Arrival Theorems 89
4.6 Notes 91
4.7 Exercises 93

References   **95**

5 Technical Desiderata   **97**
5.1 Convergence and Limits 98
5.2 Some Useful Theorems 100
5.3 Brownian Motion 102
5.4 Two Fundamental Processes 106
5.5 Limit Theorems for the Two Fundamental Processes 108
   5.5.1 Functional Strong Law of Large Numbers . . . 109
   5.5.2 Functional Central Limit Theorem 110
   5.5.3 Functional Law of the Iterated Logarithm . . 111
   5.5.4 Functional Strong Approximation 112
5.6 Exponential Rate of Convergence 114

5.7   Notes 116
5.8   Exercises 117

References   **123**

6   Single-Station Queues   **125**
6.1   Queue-Length and Workload Processes 126
6.2   One-Dimensional Reflection Mapping 127
6.3   Fluid Limit (FSLLN) 135
6.4   Diffusion Limit (FCLT) 138
6.5   Approximating the G/G/1 Queue 142
6.6   Functional Law of the Iterated Logarithm 144
6.7   Strong Approximation 146
6.8   Exponential Rate of Convergence 151
6.9   Notes 153
6.10 Exercises 154

References   **157**

7   Generalized Jackson Networks   **159**
7.1   The Queueing Network Model 160
7.2   Oblique Reflection Mapping 164
7.3   A Homogeneous Fluid Network 167
7.4   A Reflected Brownian Motion 172
7.5   Approximating the Network 175
7.6   Functional Law of the Iterated Logarithm 179
7.7   Strong Approximation 183
7.8   Fluid Limit (FSLLN) 186
7.9   Diffusion Limit (FCLT) 188
7.10 Closed Networks 196
      7.10.1 Reflection Mapping, Fluid Model, and RBM 196
      7.10.2 Approximating a Generalized Closed Network 200
      7.10.3 Fluid Limit (FSLLN) 201
      7.10.4 Diffusion Limit (FCLT) 202
7.11 Notes 204
7.12 Exercises 206

References   **213**

8   A Two-Station Multiclass Network   **215**
8.1   Model Description and Motivation 215
8.2   A Fluid Model with Priorities 220
8.3   Stability 227
8.4   Fluid Limit 233
8.5   Diffusion Limit 236

8.6   More Examples 243
8.7   Notes 247
8.8   Exercises 250

References   **255**

9   Feedforward Networks   **259**
9.1   The Single Station Model 260
9.2   FLIL: The Single Station Case 263
9.3   Strong Approximation: The Single Station Case 267
9.4   The Feedforward Network Model 274
      9.4.1   Primitive Data and Assumptions 274
      9.4.2   Performance Measures and Dynamics 277
9.5   FLIL: The Network Case 278
9.6   Strong Approximation: The Network Case 279
9.7   Performance Analysis and Approximations 285
9.8   Numerical Examples 287
      9.8.1   A Single Station with Two Job Classes 287
      9.8.2   Two Stations in Tandem 289
9.9   Notes 291
9.10 Exercises 294

References   **297**

10 Brownian Approximations   **299**
10.1 The Queueing Network Model 300
      10.1.1 Notation and Conventions 300
      10.1.2 The Primitive Processes 302
      10.1.3 The Derived Processes 303
10.2 Two-Moment Characterization of Primitive Processes . . . 305
10.3 The SRBM Approximation 308
10.4 Discussions and Variations 313
      10.4.1 Issues Surrounding the SRBM 313
      10.4.2 Kumar–Seidman Network 315
      10.4.3 State-Space Collapse 316
10.5 Stationary Distribution of the SRBM 318
10.6 Special Cases and Numerical Results 320
      10.6.1 A Single-Class Queue with Breakdown 320
      10.6.2 A Variation of the Bramson Network 321
      10.6.3 A More Complex Multiclass Network 325
10.7 Notes 327
10.8 Exercises 331

References   **333**

11 Conservation Laws   **337**
11.1 Polymatroid 338
   11.1.1 Definitions and Properties 338
   11.1.2 Optimization 340
11.2 Conservation Laws 342
   11.2.1 Polymatroid Structure 342
   11.2.2 Examples 344
   11.2.3 Optimal Scheduling 347
11.3 Generalized Conservation Laws 349
   11.3.1 Definition 349
   11.3.2 Examples 351
11.4 Extended Polymatroid 353
   11.4.1 Equivalent Definitions 353
   11.4.2 Connections to Sub/Supermodularity 355
11.5 Optimization over EP 360
11.6 Notes 364
11.7 Exercises 366

References   **369**

12 Scheduling of Fluid Networks   **375**
12.1 Problem Formulation and Global Optimality 375
12.2 The Myopic Procedure via LP 378
12.3 Properties of the Myopic Procedure 380
   12.3.1 Termination Rules 380
   12.3.2 Stability and Clearing Time 382
12.4 The Single-Station Model 385
   12.4.1 Description of the Solution Procedure 385
   12.4.2 Summary of the Algorithm 388
   12.4.3 Remarks 390
12.5 Proof of Global Optimality: Single-Station Model 391
12.6 Notes 395
12.7 Exercises 396

References   **399**

Index   **401**

List of Figures

3.1 A closed cyclic network 40

4.1 Kumar–Seidman network 94

5.1 Four sample paths of a unit Poisson process 109
5.2 Four sample paths of a unit Poisson process: An illustration of the functional strong law of large numbers 110
5.3 Four sample paths of a unit Poisson process: An illustration of the functional central limit theorem 111

6.1 Four sample paths of a reflected Brownian motion 130

7.1 A three-station generalized Jackson network 161

8.1 Kumar–Seidman network 216
8.2 Queue lengths by class in a Kumar–Seidman network with Poisson arrivals and exponential services 217
8.3 Queue length by station in a Kumar–Seidman network with Poisson arrivals and exponential services 218
8.4 Kumar–Seidman network with deterministic arrival and service times 219
8.5 Kumar–Seidman network with deterministic arrival and service times 220
8.6 A three-station variation of a Kumar–Seidman network . . 243

8.7 Total queue length of the three station variation of Kumar–
     Seidman network with Poisson arrivals and exponential ser-
     vices 243
8.8 A six-class version of the Kumar–Seidman network 244
8.9 Queue length by station in a six-class variation of a Kumar–
     Seidman network with $m = (0.35, 0.25, 0.35, 0.25, 0.3, 0.3)$,
     Poisson arrivals, and exponential services 244
8.10 Queue length by station in a six-class variation of a Kumar–
     Seidman network with $m = (0.3, 0.0.3, 0.3, 0.3, 0.3, 0.3)$, Pois-
     son arrivals and exponential services 245
8.11 A variation of the Bramson network 246
8.12 Queue length by station in a variation of Bramson network
     with Poisson arrivals and exponential services 247
8.13 Queue length by station in a variation of Bramson network
     with deterministic arrivals and services 248
8.14 A Three-Station Variation of Bramson Network 248
8.15 Total queue length in a three-station variation of the Bram-
     son network with Poisson arrivals and exponential service
     times 249

9.1 A multiclass single-station queue 260
9.2 A two-station six-class feedforward network 275
9.3 A two-class single-station network 287
9.4 A two-station tandem queue 289

10.1 A two-station multiclass network 325
10.2 Cumulative stationary distributions of waiting time in the
     complex queueing network: group 1 jobs 328
10.3 Cumulative stationary distributions of waiting time in the
     complex queueing network: group 3 jobs 329

List of Tables

7.1 Parameters of the generalized Jackson queueing network . . 180
7.2 Average sojourn time of the generalized Jackson queueing
     network 181

9.1 Average queue length in the single-station queue shown in
     Figure 9.3 290
9.2 System specifications of a two-station tandem queue 291
9.3 Average queue length of network 1 292
9.4 Average queue length of network 2 293

10.1 Parameters of a single-station queue with breakdown 321
10.2 Average sojourn time in the single-class single-server queue
     with breakdown 322
10.3 Average waiting times in the Bramson network 323
10.4 Average sojourn time of each job class in the Bramson network324
10.5 Mean service times of three cases of the complex queueing
     network 325
10.6 Average sojourn time of each job class in the complex queue-
     ing network 326

1

Birth–Death Queues

This chapter introduces three preliminary topics that will serve as the basis for much of the first part of this book (the first four chapters), which focuses on continuous-time Markov chain models. The three topics are (a) the birth–death process and related queueing models in Section 1.1, (b) the notion of time reversibility of a Markov chain in Section 1.2, and (c) likelihood ratio ordering and stochastic ordering in Section 1.3.

1.1 Basics

Let $\{Y(t), t \geq 0\}$ denote a birth–death process, on the state space $\mathcal{Z}_+ := \{0, 1, 2, \ldots\}$. Both the birth rates and the death rates are state-dependent, denoted by $\lambda(n)$ and $\mu(n)$, given $Y(t) = n$. Assume $\lambda(n) > 0$ for all $n \in \mathcal{Z}_+$; and $\mu(0) = 0$, $\mu(n) > 0$ for all $n \geq 1$. Note that in the queueing context, "birth" corresponds to arrivals and "death" corresponds to service completions. Hence, $\mu(0) = 0$ signifies the suspension of the service facility when there is no job in the system. In some later applications, we shall allow $\lambda(n) = 0$ for $n \geq N$, with N being a given positive integer. In this case, the state space of Y is truncated to $\{0, 1, 2, \ldots, N\}$.

The above is a quite general model, which includes the following as special cases:

- the single-server model with a constant service rate $\mu > 0$, i.e., $\mu(n) \equiv \mu$, for all $n \geq 1$;

- the multiple-server model with c servers and each following a constant service rate $\mu > 0$, i.e., $\mu(n) = \min\{c, n\} \cdot \mu$, for all $n \geq 1$.

If, in addition to the above, the birth rate is also a constant, $\lambda(n) \equiv \lambda > 0$ for all n, then the above two cases correspond to the $M/M/1$ and $M/M/c$ queues. Also note that the general model is sometimes denoted by $M(n)/M(n)/1$.

Let $\pi := \{\pi(n) \ n \in \mathcal{Z}_+\}$ denote the equilibrium distribution of the birth–death process. Then, from standard Markov chain theory, we know that π satisfies the following system of linear equations:

$$\lambda(0)\pi(0) = \mu(1)\pi(1),$$
$$[\lambda(1) + \mu(1)]\pi(1) = \lambda(0)\pi(0) + \mu(2)\pi(2),$$
$$\cdots$$
$$[\lambda(n) + \mu(n)]\pi(n) = \lambda(n-1)\pi(n-1) + \mu(n+1)\pi(n+1),$$
$$\cdots$$

The above is known as "flow-balance equations" (or "equilibrium balance equations"): They balance the (probability) flow out of each state with the flow into the same state. Substituting the first equation into the second one simplifies the latter to

$$\lambda(1)\pi(1) = \mu(2)\pi(2).$$

Continuing this procedure simplifies the system of balance equations to the following:

$$\lambda(n)\pi(n) = \mu(n+1)\pi(n+1), \quad \forall n \in \mathcal{Z}_+. \tag{1.1}$$

Hence, we can derive, recursively,

$$\pi(n) = \pi(0) \cdot \frac{\lambda(0) \cdots \lambda(n-1)}{\mu(1) \cdots \mu(n)}, \quad n \geq 1. \tag{1.2}$$

Assuming

$$\sum_{n=1}^{\infty} \frac{\lambda(0) \cdots \lambda(n-1)}{\mu(1) \cdots \mu(n)} < \infty, \tag{1.3}$$

and applying the normalizing condition $\sum_{n=0}^{\infty} \pi(n) = 1$ to (1.2), we have

$$\pi(0) = \left[1 + \sum_{n=1}^{\infty} \frac{\lambda(0) \cdots \lambda(n-1)}{\mu(1) \cdots \mu(n)}\right]^{-1}. \tag{1.4}$$

In the special case of the $M/M/1$ queue, the condition in (1.3) reduces to the convergence of a geometric series: $\sum_{0}^{\infty} \rho^n < \infty$ (where $\rho := \lambda/\mu$), which is equivalent to $\rho < 1$. And the equilibrium distribution in (1.2) is reduced to a geometric distribution:

$$\pi(n) = (1 - \rho)\rho^n, \quad \forall n \in \mathcal{Z}_+;$$

in particular, $\pi(0) = 1 - \rho$.

The sojourn time of the birth–death process $\{Y(t)\}$ in each state n follows an exponential distribution with parameter (rate) $\lambda(n) + \mu(n)$. When the sojourn time expires, the transitions are either upward to $n + 1$ or downward to $n - 1$, respectively with probabilities

$$\frac{\lambda(n)}{\lambda(n) + \mu(n)}$$

and

$$\frac{\mu(n)}{\lambda(n) + \mu(n)}.$$

A technique called "uniformization" can be used to convert $\{Y(t)\}$ to an equivalent process $\{\tilde{Y}(t)\}$ (equivalent in the sense that the two processes follow the same probability law). To this end, assume that the birth and death rates are bounded,

$$\lambda(n) \leq \eta_1 \quad \text{and} \quad \mu(n) \leq \eta_2,$$

for some positive constants η_1 and η_2; and write $\eta := \eta_1 + \eta_2$. Let

$$0 < \tau_1 < \tau_2 < \cdots < \tau_k < \cdots$$

denote a sequence of event epochs of a Poisson process with rate η. Let $\{U_k, k = 1, 2, \ldots\}$ be a sequence of i.i.d. random variables following a uniform distribution on the interval $[-\eta_2, \eta_1]$. Construct the process $\{\tilde{Y}(t)\}$ recursively as follows: Let $\tilde{Y}(0) = Y(0)$. Suppose $\tilde{Y}(\tau_{k-1})$ is constructed. Continue the construction as follows: Let

$$\tilde{Y}(t) = \tilde{Y}(\tau_{k-1}) := y, \quad t \in [\tau_{k-1}, \tau_k).$$

At $t = \tau_k$, let

$$\tilde{Y}(\tau_k) = y - \mathbf{1}\{U_k \in [-\mu(y), 0]\} + \mathbf{1}\{U_k \in (0, \lambda(y)]\}, \tag{1.5}$$

where $\mathbf{1}\{\cdot\}$ denotes the indicator function. In other words, in between the two consecutive Poisson event epochs τ_{k-1} and τ_k, $\tilde{Y}$ has no change. At τ_k, there is a state transition if $U_k \in [-\mu(y), \lambda(y)]$, which happens with probability

$$p_y := \frac{\lambda(y) + \mu(y)}{\eta},$$

whereas with probability $1 - p_y$, there is no transition, i.e., the process will remain in state y, the same as at τ_{k-1}. Therefore, the sojourn time of $\tilde{Y}$ in state y is a random summation: $\sum_{i=1}^{N} T_i$, where T_i's are i.i.d. exponential variables with mean $1/\eta$ (interevent times of the Poisson process that generates the $\{\tau_k\}$ sequence), and N, independent of the T_i, is a geometric random variable with success probability p_y. It is easy to verify that this random summation also follows an exponential distribution, with rate $\eta p_y = \lambda(y) + \mu(y)$. Hence, this sojourn time distribution is exactly the same as in the original birth–death process $\{Y(t)\}$. From (1.5), it is also clear that if a transition takes place in state y, it is an upward transition with probability $\frac{\lambda(y)}{\lambda(y)+\mu(y)}$, and a downward transition with probability $\frac{\mu(y)}{\lambda(y)+\mu(y)}$. These are also the same as in $\{Y(t)\}$. Hence, starting from the same initial state, the two Markov chains $\{Y(t)\}$ and $\{\tilde{Y}(t)\}$ must have the same probability law.

1.2   Time Reversibility

Here we assume that the time index t belongs to the entire real line (instead of just the nonnegative half). Also, $\overset{d}{=}$ denotes equal in distribution.

Definition 1.1 A stochastic process $\{X(t)\}$ is *time-reversible* (or, *reversible*, for short) if

$$(X(t_1), \ldots, X(t_n)) \overset{d}{=} (X(\tau - t_1), \ldots, X(\tau - t_n))$$

for all $t_1, \ldots, t_n$, all n, and all τ.

Lemma 1.2 If $\{X(t)\}$ is reversible, then $\{X(t)\}$ is stationary. That is,

$$(X(t_1 + \tau), \ldots, X(t_n + \tau)) \overset{d}{=} (X(t_1), \ldots, X(t_n)),$$

for all $t_1, \ldots, t_n$, all n, and all τ.

Proof. Setting $\tau = 0$ in Definition 1.1, we have

$$(X(t_1), \ldots, X(t_n)) \overset{d}{=} (X(-t_1), \ldots, X(-t_n)).$$

Next, replacing t_i by $t_i + \tau$ in Definition 1.1 for all $i = 1, \ldots, n$, we have

$$(X(t_1 + \tau), \ldots, X(t_n + \tau)) \overset{d}{=} (X(-t_1), \ldots, X(-t_n)).$$

Stationarity then follows from equating the left-hand sides of the above two equations.   □

Therefore, below we focus on stationary processes, in particular stationary Markov chains (or Markov chains in *equilibrium*, or in *steady state*).

Let $\{X(t)\}$ be a stationary Markov chain, with state space $\mathbf{S}$. For ease of discussion, assume ergodicity, i.e., the (continuous-time) Markov chain is irreducible and positive recurrent. Recall that $\{X(t)\}$ is completely characterized by its rate matrix Q, whose entries are

$$Q_{ij} = q(i,j), \quad i \neq j,$$

$$-Q_{ii} = \sum_{j \neq i} q(i,j) := q(i),$$

where

$$q(i,j) = \lim_{h \to 0} \frac{1}{h} \mathsf{P}[X(h) = j | X(0) = i] := \lim_{h \to 0} P_h(i,j)/h, \qquad (1.6)$$

and we assume $q(i) < \infty$ for all i. The stationary (or invariant) distribution, $\pi = (\pi(i))_{i \in \mathbf{S}}$ is a vector of positive numbers (that sum to unity) satisfying

$$\pi(i) \sum_{j \neq i} q(i,j) := \pi(i) q(i) = \sum_{j \neq i} \pi(j) q(j,i),$$

or $\pi' Q = 0$ in matrix form. Note that under ergodicity, not only π is the limiting distribution of $X(t)$ as $t \to \infty$, $\pi(i)$ is also the long-run average proportion of time that the Markov chain is in state i, for all i.

Now, even if $\{X(t)\}$ is *not* (necessarily) reversible, we can still define its *time-reversal*, $\{\tilde{X}(t)\}$, by letting $\tilde{X}(t) = X(\tau - t)$ for all t and for some τ. Since $\{X(t)\}$ is stationary, we can pick $\tau = 0$, for instance, without loss of generality. Note that while $\{X(t)\}$ evolves toward the right of the real line (the time axis), its time-reversal $\{\tilde{X}(t)\}$ evolves in the opposite direction, toward the left of the time line. It is easy to verify the Markov property of $\{\tilde{X}(t)\}$, as well as stationarity. It turns out that the Markov chain $\{X(t)\}$ and its time-reversal have some interesting relations.

Lemma 1.3 Let $\{X(t)\}$ be a stationary Markov chain with state space $\mathbf{S}$, rate matrix Q, and stationary distribution π. Then, the time-reversal $\{\tilde{X}(t)\}$ is also a Markov chain, governed by a rate matrix $\tilde{Q}$, which is defined componentwise as follows:

$$\pi(i)\tilde{q}(i,j) = \pi(j)q(j,i), \quad \forall i,j \in \mathbf{S}, \ i \neq j; \qquad (1.7)$$

Proof. First, observe that π is also the stationary distribution of $\{\tilde{X}(t)\}$, since

$$\mathsf{P}[\tilde{X}(t) = i] = \mathsf{P}[X(\tau - t) = i] = \pi(i)$$

for all t, where the second equality follows from the stationarity of $\{X(t)\}$. For $h > 0$, letting $\tau = h$, we have

$$
\begin{aligned}
\pi(i)P_h(i,j) &= P[X(0) = i, X(h) = j] \\
&= P[\tilde{X}(0) = j, \tilde{X}(h) = i] = \pi(j)\tilde{P}_h(j,i),
\end{aligned}
$$

for any i, j. Dividing both sides by h and letting $h \to 0$ yields the desired relation in (1.7) [cf. (1.6)]. $\square$

The above lemma implies that should $\tilde{Q} = Q$, then the time reversal $\tilde{X}$ has the same probability law as the original process X; hence, X is *reversible*. In fact, the converse is also true (by mimicking the proof of the lemma); i.e., if X is reversible, then (1.7) holds with $\tilde{q} = q$.

Theorem 1.4 A stationary Markov chain $\{X(t)\}$ with state space $\mathbf{S}$ and rate matrix Q is reversible if and only if there exists a probability distribution on $\mathbf{S}$ satisfying

$$
\pi(i)q(i,j) = \pi(j)q(j,i), \quad \forall i, j \in \mathbf{S}, \ i \neq j; \tag{1.8}
$$

in which case π is the invariant distribution of $\{X(t)\}$.

Remark 1.5 The equations in (1.8) are called *detailed balance* equations, as opposed to the *full balance* equations that define the invariant distribution:

$$
\pi(i) \sum_{j \neq i} q(i,j) = \sum_{j \neq i} \pi(j)q(j,i), \quad \forall i \in \mathbf{S}.
$$

(Note that the above are simply a row-by-row display of $\pi Q = 0$.) Obviously, detailed balance is stronger than full balance: Taking summation on both sides of (1.8) over $j \neq i$ yields the full balance equations.

Intuitively, full balance requires that the probability flow coming out of any given state, say i, to all other states—the *outflow*—be equal to the probability flow from all those other states going into the same state i—the *inflow*. In contrast, detailed balance insists that this balance be achieved at a more microscopic level: Outflow equals inflow between each pair of states $i \neq j$.

A (continuous-time) Markov chain (with rate matrix Q and state space $\mathbf{S}$) is known to have a graphical representation: Let each state i be a node, and let the (directed) edge from i to j represent the transition rate $q(i,j)$ if it is positive. If the transition rate satisfies

$$
q(i,j) > 0 \Rightarrow q(j,i) > 0, \quad \forall i, j \in \mathbf{S}, \tag{1.9}
$$

then there is an (undirected) edge between i and j if and only if $q(i,j) > 0$.

Proposition 1.6 Suppose the stationary Markov chain $\{X(t)\}$ has transition rates that satisfy (1.9). Then, it is reversible if the associated graph is a tree.

Proof. Pick any pair of nodes (i, j). If $q(i, j) = 0$, then $q(j, i) = 0$, following (1.9), and the detailed balance equation (1.8) is trivially satisfied. If $q(i, j) > 0$, then there is an edge linking the pair. Since the graph is a tree, the only probability flow between i and j is through this edge. In other words, the full balance equation for state i (or for state j) reduces to the detailed balance equation between i and j. That is, (1.8) is satisfied.   □

Example 1.7 A special case of the above is the birth–death process. The associated graph is a line, or a single-branch tree. Here, $q(i, j) = \lambda(i)$ or $\mu(i)$, respectively for $j = i + 1$ and $j = i - 1$, the birth and the death rates; and the familiar relation [cf. (1.1)]

$$\lambda(i)\pi(i) = \mu(i + 1)\pi(i + 1)$$

is nothing but the detailed balance equation.

Below are some quick implications of Theorem 1.4, with the proofs left as exercise problems.

Corollary 1.8 Suppose $\{X(t)\}$ is a reversible Markov chain on state space **S**. Suppose we truncate the state space to $\mathbf{S}' \subset \mathbf{S}$ (by letting some transition rates be zero, for instance). Then, the modified Markov chain is still reversible, with invariant distribution

$$\pi'(i) = \pi(i) / \sum_{j \in \mathbf{S}'} \pi(j), \quad \forall i \in \mathbf{S}'.$$

Corollary 1.9 Suppose a stationary Markov chain $\{X(t)\}$ has a finite state space and is irreducible. Then, $\{X(t)\}$ is reversible if and only if its rate matrix Q can be expressed as $Q = AD$ with A being a symmetric matrix and D a diagonal matrix.

Example 1.10 Consider the $M/M(n)/1$ queue, with Poisson arrivals at a constant rate λ and state-dependent service rates $\mu(n)$. Let $\{X(t)\}$ be the state process: $X(t)$ denotes the total number of jobs in the system at time t. This is a special case of the birth–death process in Example 1.7. Hence, in equilibrium $\{X(t)\}$ is reversible.

Now consider the time-reversal $\{\tilde{X}(t)\}$, which represents another birth–death queue, whose departure process is the arrival process of the original queue, i.e., a Poisson process with constant rate λ. But because of reversibility, the original queue and its time-reversal have the same probabilistic behavior; and in particular, the departure processes in the two queues are exactly the same. Hence, we can conclude that the departure process from a stationary $M/M(n)/1$ queue is Poisson with constant rate λ, exactly the same as the arrival process.

Furthermore, in the original queue the number of future arrivals after t is independent of the state—the number of jobs in the system—at t.

This is simply because the Poisson arrival process has a constant rate. By reversibility, this immediately translates into the fact that the number of *past* departures *up to* time t in the time-reversed queue, and hence also in the original queue, is independent of the number of jobs in the system at t. This is a counterintuitive result, since one would tend to think that the number of jobs in the system at time t, e.g., whether there is zero or at least one job to keep the server busy, would depend on the number of past departures up to t.

1.3 Stochastic Orders

Suppose X is a nonnegative, integer-valued random variable. Let $p(n) = P(X = n) > 0$ for all $n \in \mathcal{N} = \{0, 1, \ldots, N\}$, where N is a given integer. We allow N to be infinite.

Definition 1.11 The *equilibrium rate* of X is a real-valued, nonnegative function, $r : \mathcal{N} \mapsto \mathfrak{R}_+ = [0, \infty)$ defined as

$$r(0) = 0, \qquad r(n) = p(n-1)/p(n), \quad n = 1, \ldots, N.$$

It follows that the equilibrium rate of X and its probability mass function (pmf) uniquely define each other. In particular,

$$p(n) = p(0)/[r(1)\cdots r(n)], \qquad n = 1, \ldots, N,$$

and

$$p(0) = \left[1 + \sum_{n=1}^{N} 1/(r(1)\cdots r(n))\right]^{-1}.$$

(When $N = \infty$, convergence of the summation is required.)

Example 1.12 For a birth–death queue in equilibrium, with *unit* arrival rate and state-dependent service rate $\mu(n)$, we have

$$\mu(n) = P[Y = n-1]/P[Y = n] = r_Y(n),$$

where Y denotes the number of jobs in the system, and $r_Y(n)$ its equilibrium rate. That is, the equilibrium rate is simply the service rate.

It turns out that the equilibrium rates are useful in comparing random variables under the *likelihood ratio ordering*.

Definition 1.13 X and Y are two discrete random variables. Suppose their pmfs have a common support set $\mathcal{N}$. Let r_X and r_Y denote their equilibrium rates. Then, X dominates Y in *likelihood ratio ordering*, denoted by $X \geq_{\ell r} Y$, if and only if

$$P[X = n]P[Y = n - 1] \geq P[X = n - 1]P[Y = n], \qquad \forall n \in \mathcal{N},$$

or equivalently, $r_X(n) \leq r_Y(n)$ for all $n \in \mathcal{N}$.

The more standard stochastic ordering $X \geq_{st} Y$ is defined as

$$P[X \geq n] \geq P[Y \geq n], \qquad \forall n \in \mathcal{N}, \tag{1.10}$$

which can be verified as equivalent to $Eh(X) \geq Eh(Y)$ for all increasing function $h(\cdot)$. (Here and below, "increasing" means nondecreasing.) The stochastic ordering is, in fact, *weaker* than the likelihood ratio ordering.

Lemma 1.14 $X \geq_{\ell r} Y \Rightarrow X \geq_{st} Y$.

Proof. Suppose $X, Y \in \mathcal{N} = \{0, 1, \ldots, N\}$. Set

$$s(0) = 1, \quad s(n) = [r_X(1) \cdots r_X(n)]^{-1}, \quad n \geq 1;$$

and write

$$S(n) = \sum_{k=0}^{n} s(k).$$

Define $t(n)$ and $T(n)$ similarly, replacing r_X with r_Y. Then, following Definition 1.13, $X \geq_{\ell r} Y$ implies $s(n)t(m) \geq s(m)t(n)$ for all $n \geq m$, which, in turn, implies

$$[s(n) + \cdots + s(N)][t(0) + \cdots + t(n - 1)]$$
$$\geq [s(0) + \cdots + s(n - 1)][t(n) + \cdots + t(N)].$$

Hence,

$$P[X \geq n] = \frac{s(n) + \cdots + s(N)}{S(N)}$$
$$\geq \frac{t(n) + \cdots + t(N)}{T(N)} = P[Y \geq n],$$

for all $n \leq N$; i.e., $X \geq_{st} Y$. $\qquad \square$

Example 1.15 In Example 1.12, increasing the service rate function $\mu(n)$, for each n, will decrease the number of jobs in the system, in the sense of likelihood ratio ordering, in light of Definition 1.13.

Equivalently, for two random variables $X, Y \in \mathcal{N}$, the likelihood ratio ordering $X \geq_{\ell r} Y$ can be viewed as corresponding to two birth–death queues with the same arrival rate, and with state-dependent service rates $r_X(n)$ and $r_Y(n)$, satisfying $r_Y(n) \geq r_X(n)$, for all n.

Example 1.16 Consider a multiserver queue $M/M/c/N$ (N is the limit on the number of jobs allowed in the system at any time.) Suppose each server serves at a constant rate $1/c$ (hence the maximum output rate is 1). Now, reduce the number of servers to $c' < c$, with each server serving at a constant rate $1/c'$ (hence the maximum output rate is maintained at 1). To illustrate the effect of this change, consider the special case of $c' = 1$. In other words, we aggregate the processing capacity of multiple servers to form a single server with an equal capacity. The obvious advantage is that the aggregated single server always works at full capacity whenever there is at least one job present, while in the original case, full capacity is achieved only when there are at least c jobs in the system.

Let X and X' denote the total number of jobs in the system before and after the aggregation. Then clearly,

$$r_X(n) = \min\{n,c\}/c \leq \min\{n,c'\}/c' = r_{X'}(n), \quad n = 0,1,\ldots,N;$$

hence, $X \geq_{\ell r} X'$. Furthermore, this implies $X \geq_{st} X'$, and hence $P[X = N] \geq P[X' = N]$. That is, the aggregation results in a lower blocking probability, and hence an increased throughput.

We conclude this section by applying the uniformization construction of the birth–death queue in Section 1.1 to get a stochastic comparison result in terms of the stochastic ordering $\leq_{st}$.

Example 1.17 As in Example 1.15, we consider the effect of increasing the service rate in a birth–death queue. Suppose there are two such queues, with the same rate $\lambda(n)$, but different service rates: $\mu_1(n) \leq \mu_2(n)$ for all n. Let $\{Y_1(t)\}$ and $\{Y_2(t)\}$ denote the two processes, i.e., for $i = 1,2$, $Y_i(t)$ is the total number of jobs in the system i. We want to show that the two processes are ordered in the stochastic ordering:

$$\{Y_1(t)\} \geq_{st} \{Y_2(t)\}. \tag{1.11}$$

The above means that for any $t_1 < \cdots < t_n$ and any n,

$$(Y_1(t_1),\ldots,Y_1(t_n)) \geq_{st} (Y_2(t_1),\ldots,Y_2(t_n)),$$

which, in turn, means that

$$\mathsf{E}g(Y_1(t_1),\ldots,Y_1(t_n)) \geq \mathsf{E}g(Y_2(t_1),\ldots,Y_2(t_n)),$$

for all increasing functions $g(x_1,\ldots,x_n)$. Note that although the likelihood ratio ordering concluded in Example 1.15 is stronger than the stochastic ordering, the result there holds only for the stationary distribution (i.e., one point in time), whereas the stochastic ordering in (1.11) is a statement about the process over the entire time horizon.

To establish (1.11), it suffices to construct on the same probability space replicas of Y_1 and Y_2, denoted by $\tilde{Y}_1$ and $\tilde{Y}_2$, such that

$$\tilde{Y}_1(t;\omega) \geq \tilde{Y}_2(t;\omega), \qquad \forall t, \, \forall \omega, \tag{1.12}$$

where ω denotes a sample path. That is, $\tilde{Y}_1$ dominates $\tilde{Y}_2$ (deterministically) at every time epoch and over every sample path.

We use the uniformization construction in Section 1.1 to generate the replicas $\tilde{Y}_1$ and $\tilde{Y}_2$. In particular, the common probability space consists of the Poisson event sequence $\{\tau_k\}$ and the sequence of i.i.d. uniform variables $\{U_k\}$. (Note that here we assume $\eta_2 \geq \mu_2(n) \geq \mu_1(n)$ for all n, and, as before, $\eta_1 \geq \lambda(n)$ for all n, and $\eta = \eta_1 + \eta_2$.) And from (1.5), we have

$$\tilde{Y}_1(\tau_k) = y - 1\{U_k \in [-\mu_1(y), 0]\} + 1\{U_k \in (0, \lambda(y)]\},$$
$$\tilde{Y}_2(\tau_k) = y - 1\{U_k \in [-\mu_2(y), 0]\} + 1\{U_k \in (0, \lambda(y)]\}.$$

Since $\mu_1(y) \geq \mu_2(y)$, the above implies $\tilde{Y}_1(\tau_k) \geq \tilde{Y}_2(\tau_k)$. Hence, inductively from the construction, we know that (1.12) holds.

1.4 Notes

The materials on birth–death queues in the first two sections can be found in most introductory texts on probability models and stochastic processes, e.g., Ross [1]. The likelihood ratio ordering and the stochastic ordering in Section 1.3 are standard stochastic order relations; refer to, e.g., Ross [1] and Stoyan [4]. In particular, a result such as likelihood ratio ordering implying stochastic ordering (see Lemma 1.14) is known, and quite easily proved, for continuous random variables with densities (e.g., [1]). The proof for the discrete case (as in Lemma 1.14) is slightly more difficult, and it illustrates the effective use of the relation between the pmf and the equilibrium rate.

The notion of equilibrium rate was first developed in Shanthikumar and Yao [2, 3], along with various applications, in particular in connecting steady-state queueing results to the likelihood ratio ordering.

The materials in all three sections will serve as preliminaries for the next two chapters, when we discuss Jackson networks.

1.5 Exercises

1. Verify that the random summation $\sum_{i=1}^{N} T_i$ involved in the uniformization construction of Section 1.1 indeed follows an exponential distribution.

2. Prove Corollary 1.8.

3. Prove Corollary 1.9.

4. In Example 1.10, since the departure process of the $M/M(n)/1$ queue is Poisson, the interdeparture times must follow an exponential distribution. Directly verify this by conditioning on the number of jobs in the system at a departure epoch.

5. If the arrival (birth) process in Example 1.10 is state-dependent, how would you modify the conclusions there?

6. Show that $X \geq_{st} Y$ if and only if $Eh(X) \geq Eh(Y)$ for all increasing functions $h(x)$.

7. The likelihood ratio ordering in Definition 1.13 also holds for continuous random variables; i.e., $X \geq_{\ell r} Y$ if $f(x)g(y) \geq f(y)g(x)$ for all $x \geq y$, where f and g denote the density functions of X and Y, respectively. Prove Lemma 1.14 in this case.

8. In Example 1.12, show that the likelihood ratio ordering applies to the number of jobs waiting in the queue (i.e., excluding the job in service) as well. That is, increasing the service rate reduces the number of waiting jobs in the likelihood ratio ordering.

9. Repeat the above problem for Example 1.15. Specifically, does the aggregation of servers (while preserving the maximum output rate) increase or decrease the number of waiting jobs in the queue, in the likelihood ratio ordering?

10. Adapt the uniformization construction of Section 1.1 to the $M/M/c/N$ queue; and use it to show the impact of server aggregation, in terms of stochastic ordering for the birth–death process involved in Example 1.16.

11. In Example 1.16, show that the server aggregation also reduces the blocking probability and hence increases the throughput.

References

[1] Ross, S.M. 1996. *Stochastic Processes* (2nd ed.). Wiley, New York.

[2] Shanthikumar, J.G. and Yao, D.D. 1986. The Preservation of Likelihood Ratio Ordering under Convolution. *Stochastic Processes and Their Applications*, **23**, 259–267.

[3] Shanthikumar, J.G. and Yao, D.D. 1986. The Effect of Increasing Service Rates in Closed Queueing Networks. *Journal of Applied Probability*, **23**, 474–483.

[4] Stoyan, D. 1983. *Comparison Methods for Queues and Other Stochastic Models*. Wiley, New York.

2
Jackson Networks

A Jackson network consists of J nodes (or stations), each with one or several servers. The processing times of jobs at each node are i.i.d., following an exponential distribution with unit mean. The service rate, i.e., the rate by which work is depleted, at each node i can be both node-dependent and state-dependent. Specifically, whenever there are x_i jobs at node i, the processing rate is $\mu_i(x_i)$, where $\mu_i(\cdot)$ is a function $\mathcal{Z}_+ \mapsto \Re_+$, with $\mu_i(0) = 0$ and $\mu_i(x) > 0$ for all $x > 0$. Jobs travel among the nodes following a routing matrix $P := (p_{ij})$, where, for $i, j = 1, \ldots, J$, p_{ij} is the probability that a job leaving node i will go to node j.

Note that in a Jackson network all jobs at each node belong to a single "class": All jobs follow the same service-time distribution and the same routing mechanism; and in this sense, the Jackson network is a single-class model. Consequently, there is no notion of priority in serving the jobs: At each node all jobs are served on a first-come-first-served basis.

According to different specifications of the routing matrix, there are three different variations: the *open*, *closed* and *semiopen* networks, which are the topics of the next three sections, Section 2.1, Section 2.2, and Section 2.3, respectively. The throughput function of the closed network and its computation is the focus of Section 2.4 and Section 2.5. In Section 2.6 we examine the Jackson network under time-reversal, a notion introduced in Section 1.2.

2.1 Open Network

In an open network, jobs arrive from outside following a Poisson process with rate $\alpha > 0$. Each arrival is independently routed to node j with probability $p_{0j} \geq 0$, and $\sum_{j=1}^{J} p_{0j} = 1$. Equivalently, this can be viewed as each node j having an independent external Poisson stream of arrivals with rate αp_{0j}. (Note that the probabilities p_{0j} are allowed to be zero.) Upon service completion at node i, a job may go to another node j with probability p_{ij} (which is the (i,j)th element in the routing matrix P as specified earlier, or leave the network with probability $p_{i0} = 1 - \sum_{j=1}^{J} p_{ij}$.

Let λ_i be the overall arrival rate to node i, including both external arrivals and internal transitions. We have the following *traffic equation*:

$$\lambda_i = \alpha p_{0i} + \sum_{j=1}^{J} \lambda_j p_{ji}, \quad i = 1, \ldots, J. \tag{2.1}$$

In matrix notation, the above can be expressed as

$$\lambda = a + P'\lambda,$$

with $\lambda := (\lambda_i)$, $a := (\alpha p_{0i})$. We shall assume that the routing matrix P is *substochastic* (i.e., at least one of the row sums is strictly less than one), with a spectral radius less than unity. Consequently, the matrix $I - P'$ belongs to the class of M-matrices: $I - P'$ has positive diagonal elements and nonpositive off-diagonal elements, and each of its principal minors has an inverse, which is a nonnegative matrix. (Refer to Lemma 2 in Chapter 7 for various properties of an M-matrix.) Hence, we have the following solution to the traffic equation:

$$\lambda = (I - P')^{-1}a. \tag{2.2}$$

Let $X_i(t)$ denote the number of jobs at node i at time t. From the specifications above, clearly, $\{(X_i(t))_{i=1}^{J}; \ t \geq 0\}$ is a continuous-time Markov chain. Consider its stationary distribution, and omit the time index t for simplicity. Set $\mathbf{X} = (X_i)_{i=1}^{J}$ and $\mathbf{x} = (x_i)_{i=1}^{J}$. The Markov chain is governed by the following transition rates, denoted by $q(\cdot, \cdot)$:

$$q(\mathbf{x}, \mathbf{x} + \mathbf{e}_i) = \alpha p_{0i},$$
$$q(\mathbf{x}, \mathbf{x} - \mathbf{e}_i) = \mu_i(x_i)p_{i0},$$
$$q(\mathbf{x}, \mathbf{x} - \mathbf{e}_i + \mathbf{e}_j) = \mu_i(x_i)p_{ij},$$

where $\mathbf{e}_i$ and $\mathbf{e}_j$ denote the ith and the jth unit vector (both of dimension J). Let $\pi(\mathbf{x}) = \mathsf{P}[\mathbf{X} = \mathbf{x}]$ denote the equilibrium (steady-state) distribution. As with the birth–death process analyzed in Section 1.1, $\pi(\mathbf{x})$ is determined by the following system of (full) balance equations, which equates the "probability flow" out of each state $\mathbf{x}$ with the flow into the same state:

$$\pi(\mathbf{x}) \sum_{i=1}^{J} [\alpha p_{0i} + \mu_i(x_i)(1 - p_{ii})]$$

$$= \sum_{i=1}^{J} [\pi(\mathbf{x} - \mathbf{e}_i)\alpha p_{0i} + \pi(\mathbf{x} + \mathbf{e}_i)\mu_i(x_i + 1)p_{i0}]$$

$$+ \sum_{i=1}^{J} \sum_{j \neq i} \pi(\mathbf{x} + \mathbf{e}_i - \mathbf{e}_j)\mu_i(x_i + 1)p_{ij}. \tag{2.3}$$

The above holds for all $\mathbf{x} \in \mathcal{Z}_+^J$, the entire state space. Also, observe that the above is simply a row-by-row display of the matrix equation $\pi'Q = 0$, with π denoting $\{\pi(\mathbf{x}), \mathbf{x} \in \mathcal{Z}_+^J\}$, and Q, the rate matrix, whose entries $q(x, y)$ correspond to the transition rates between each pair of states x and y specified above.

The main result below, Theorem 2.1, relates the distribution of $\mathbf{X} = (X_1, \ldots, X_J)$ to a vector of *independent* random variables $(Y_1, \ldots, Y_J)$, with each Y_i having a probability mass function (pmf) as follows:

$$P[Y_i = n] = P[Y_i = 0] \cdot \frac{\lambda_i^n}{M_i(n)}, \tag{2.4}$$

where

$$M_i(n) = \mu_i(1) \cdots \mu_i(n), \quad n = 1, 2, \ldots,$$

and

$$\sum_{n=1}^{\infty} \frac{\lambda_i^n}{M_i(n)} < \infty, \tag{2.5}$$

so that $P[Y_i = 0]$ is well-defined, namely,

$$P[Y_i = 0] = \left[1 + \sum_{n=1}^{\infty} \frac{\lambda_i^n}{M_i(n)} \right]^{-1}.$$

Clearly, from Chapter 1 we recognize Y_i as the number of jobs in a birth–death queue in equilibrium, with a constant arrival (birth) rate λ_i and state-dependent service (death) rates $\mu_i(n)$.

Theorem 2.1 Provided that the condition in (2.5) is satisfied for all $i = 1, \ldots, J$, the equilibrium distribution of the open Jackson network has the following product form:

$$\pi(\mathbf{x}) = \prod_{i=1}^{J} P[Y_i = x_i],$$

for all $\mathbf{x} \in \mathcal{Z}_+^J$, where Y_i follows the distribution in (2.4).

Proof. It suffices to verify that the equilibrium balance equations in (2.3) are satisfied. From the given product form of $\pi(\mathbf{x})$, taking into account (2.4), we have

$$\begin{aligned} \pi(\mathbf{x}) &= \pi(\mathbf{x} + \mathbf{e}_i)\mu_i(x_i + 1)/\lambda_i \\ &= \pi(\mathbf{x} + \mathbf{e}_i - \mathbf{e}_j)\mu_i(x_i + 1)\lambda_j/[\lambda_i\mu_j(x_j)]. \end{aligned}$$

Substituting these into the right side of (2.3) and canceling out $\pi(\mathbf{x})$ from both side, we have

$$\sum_{i=1}^{J}[\alpha p_{0i} + \mu_i(x_i)(1 - p_{ii})]$$

$$= \sum_{i=1}^{J}\left[\frac{\alpha p_{0i}}{\lambda_i}\mu_i(x_i) + \lambda_i p_{i0}\right] + \sum_{i=1}^{J}\sum_{j\neq i}\frac{\lambda_i}{\lambda_j}p_{ij}\mu_j(x_j). \qquad (2.6)$$

Making use of (2.1), we have

$$\sum_{i=1}^{J}\sum_{j\neq i}\frac{\lambda_i}{\lambda_j}p_{ij}\mu_j(x_j). = \sum_{j=1}^{J}\left[\sum_{i\neq j}\frac{\lambda_i}{\lambda_j}p_{ij}\right]\mu_j(x_j)$$

$$= \sum_{j=1}^{J}\left[1 - p_{jj} - \frac{\alpha p_{0j}}{\lambda_j}\right]\mu_j(x_j).$$

Substituting the above into (2.6) simplifies the latter to

$$\sum_{i=1}^{J}\alpha p_{0i} = \sum_{i=1}^{J}\lambda_i p_{i0},$$

which is nothing but the balance of total input (rate) to the network with the total output. It follows from taking summation on both sides of (2.1) over i:

$$\sum_{i=1}^{J}\alpha p_{0i} = \sum_{i=1}^{J}\lambda_i - \sum_{i=1}^{J}\sum_{j=1}^{J}\lambda_j p_{ji}$$

$$= \sum_{i=1}^{J}\lambda_i - \sum_{j=1}^{J}\lambda_j(1 - p_{j0})$$

$$= \sum_{i=1}^{J}\lambda_i p_{i0}.$$

$\square$

Remark 2.2 Theorem 2.1 reveals that in equilibrium, the J nodes in the network behave independently, each following the equilibrium behavior of a birth–death queue.

The condition in (2.5) is satisfied if for each i there exists a positive constant $\rho_i < 1$ such that

$$\lambda_i^n / [\mu_i(1) \cdots \mu_i(n)] \leq \rho_i^n$$

whenever $n \geq K_i$ for some K_i. Two special cases are of particular interest:

(i) Constant service rate: $\mu_i(x) = \mu_i > 0$ for all $x > 0$. Then, $\rho_i = \lambda_i / \mu_i < 1$ becomes the required condition. In this case, Y_i follows a geometric distribution: $P[Y_i = n] = (1 - \rho_i)\rho_i^n$ for $n = 0, 1, 2, \ldots$.

(ii) Multiple parallel servers: There are c_i servers at node i, each with a constant rate μ_i. That is, $\mu_i(x) = \min\{x, c_i\} \cdot \mu_i$. In this case, the required condition becomes $\rho_i = \lambda_i / (c_i \mu_i) < 1$.

2.2   Closed Network

In many applications the total number of jobs in the network is maintained at a constant level, say N. Once a job completes all of its processing requirements and leaves the network, a new job is immediately released into the network. (Of course, implicit here is the assumption that there is an infinite source of new jobs, ready to be released into the network at any time.) Conceptually, this type of operation can also be viewed as having a fixed number of jobs circulating in the network, with no job ever leaving the network and no external job entering the network; and in this sense, the network is "closed."

Examples of this mode of operation include production systems that maintain a constant level of WIP (work in process), or follows a base-stock control rule (also known as a one-for-one replenishment policy). Also, it is quite common in many automated manufacturing systems that each job being processed needs to be mounted on a specially designed pallet throughout its circulation in the system; once a job is completed, another new job can be released into the system to occupy the pallet. Hence, the total number of pallets naturally dictates the WIP level. In a multilevel computing system, the constant N corresponds to the level of computing. In communication networks, it is quite common to use a fixed number of "tokens" to achieve flow control.

In a closed model, the routing matrix $(r_{ij})_{i,j}^J$ is *stochastic*, i.e., the row sums are all equal to one. In other words, $p_{i0} = p_{0j} = 0$ for all $i, j = 1, \ldots, J$, using the notation in the open model; i.e., no job enters into the network and no job leaves. Assume $(r_{ij})_{i,j}^J$ to be irreducible; in this case, the closed network is known as an irreducible closed network. (A $J \times J$ stochastic matrix P is irreducible if the rank of matrix $(I - P)$ is $J - 1$. In an irreducible closed network, a job may reach from one station to any other station in a finite step with a positive probability.) Let $(v_i)_{i=1}^J$ be the solution to

$$v_i = \sum_{j=1}^J v_j p_{ji}, \quad i = 1, \ldots, J. \tag{2.7}$$

Since the routing matrix is stochastic, the solution to the above system of equations is unique only up to a constant multiplier: There is one degree of freedom. To make the solution unique, we need to add another equation. For instance, set $\sum_{i=1}^J v_i = v$, for some positive constant v, e.g., $v = 1$. This way, $\{v_i; , i = 1, \ldots, J\}$ is in fact the equilibrium distribution of a discrete-time Markov chain with $(r_{ij})_{i,j}^J$ as the probability transition matrix; and $N v_i$ is the arrival rate to node i, the counterpart of λ_i in the open model. Alternatively, we can set $v_j = 1$, for some node j, as the additional equation. For instance, node j can represent some docking (loading/unloading) station, from which all jobs enter into and depart from the network.

The equilibrium balance equations are similar to those of the open network [cf. (2.3)], except that here $\alpha = 0$ and $p_{0i} = p_{i0} = 0$ for all i. Hence, we have

$$\pi(\mathbf{x}) \sum_{i=1}^J \mu_i(x_i)(1 - p_{ii}) = \sum_{i=1}^J \sum_{j \neq i} \pi(\mathbf{x} + \mathbf{e}_i - \mathbf{e}_j) \mu_i(x_i + 1) p_{ij},$$

for all $\mathbf{x} \in \mathcal{Z}_+^J$ such that $|\mathbf{x}| = N$, where $|\mathbf{x}| := x_1 + \cdots + x_J$. Similarly, define $|\mathbf{X}| := X_1 + \cdots + X_J$ and $|\mathbf{Y}| := Y_1 + \cdots + Y_J$, for $\mathbf{X} = (X_i)_{i=1}^J$ and $\mathbf{Y} := (Y_i)_{i=1}^J$.

Theorem 2.3 The closed Jackson network, with a total of N jobs, has the following equilibrium distribution: For all $\mathbf{x} \in \mathcal{Z}_+^J$ such that $|\mathbf{x}| = N$, we have $|\mathbf{x}| = N$,

$$\pi(\mathbf{x}) = \prod_{i=1}^J P[Y_i = x_i] / P[|\mathbf{Y}| = N],$$

where Y_i follows the distribution in (2.4), with $x_i \leq N$, and λ_i replaced by v_i, the solution to the traffic equations in (2.7).

Proof. Follow the proof of Theorem 2.1. It suffices to verify (2.6), with the terms that involve α or p_{i0} set to zero, and λ_i and λ_j replaced by v_i and v_j:

$$\sum_{i=1}^J \mu_i(x_i)(1 - p_{ii}) = \sum_{i=1}^J \sum_{j \neq i} \frac{v_i}{v_j} p_{ij} \mu_j(x_j).$$

Interchanging the two summations on the right-hand side above yields the left-hand side, taking into account (2.7).   □

Remark 2.4 The denominator in $\pi(\mathbf{x})$ comes from the normalizing condition, since

$$P[|\mathbf{Y}| = N] = \sum_{|\mathbf{x}|=N} \prod_{i=1}^{J} P[Y_i = x_i].$$

(Recall that the Y_i's are independent by definition.)

Note that in contrast to the open model, in the closed model the X_i's are *not* independent: They are constrained by the constant N. But Theorem 2.3 reveals that this turns out to be the only source of dependence. In other words, conditioning upon summing up to N, the X_i's are independent. Also note that in the closed model, the state space is finite; hence, no condition such as the one in (2.5) is needed for the existence of the equilibrium distribution.

2.3 Semiopen Network

The semiopen model unifies the features of both open and closed networks. Specifically, the network is open, following the descriptions of the open model in Section 2.1, with the exception that the total number in the network is limited to a maximum of K jobs at any time. We shall refer to K as the "buffer limit." When this limit is reached, external arrivals will be blocked and lost. Since external arrivals follow a Poisson process, due to the memoryless property of the interarrival times, this blocking mechanism is equivalent to stopping the arrival process as soon as the buffer limit is reached. The arrival process will be resumed when a job next leaves the network, bringing the total number of jobs in the network down to $K - 1$.

It turns out that this semiopen model can be reduced to a closed network, with a constant of K jobs and $J+1$ nodes. The additional node, indexed as node 0, again represents the external world, with routing from and to node 0 following the probabilities p_{0j} and p_{i0}, just as in the open network. (Hence, the routing matrix of this closed network of $J + 1$ nodes is the augmented matrix $[p_{ij}]_{i,j=0}^{J}$.) In addition, node 0 is given a "service" function, with service rate $\mu_0(n) = \alpha$ for all $n \geq 1$, and $\mu_0(0) = 0$. (Recall that α is the rate of the external Poisson arrival process.) This way, node 0 effectively generates the arrival process of the original semiopen network. In particular, $x_0 = 0$ means that there are K jobs in the other J nodes (which constitute the original network), and hence $\mu_0(0) = 0$ correctly captures the blocking of external arrivals when the buffer is full.

Modify the traffic equations in (2.1) as follows: Divide both sides by α, and let $v_0 = 1$, $v_i = \lambda_i/\alpha$ for $i = 1, \ldots, J$. Then, (2.1) becomes

$$v_i = \sum_{i=0}^{J} v_j p_{ji}, \quad i = 0, 1, \ldots, J,$$

which is the traffic equation of the closed model [cf. (2.7)] albeit for a network with $J + 1$ nodes. (Furthermore, the additional equation to make the solution unique is already chosen: $v_0 = 1$.)

Theorem 2.5 The semiopen Jackson network, with an overall buffer capacity of K, has the following equilibrium distribution: For all $\mathbf{x} \in \mathcal{Z}_+^J$ such that $|\mathbf{x}| \leq K$,

$$\pi(\mathbf{x}) = \prod_{i=1}^{J} P[Y_i = x_i]/P[|\mathbf{Y}| \leq K],$$

where Y_i follows the distribution in (2.4), with $x_i \leq K$.

Proof. Consider the closed network with $J + 1$ nodes specified above. Its equilibrium distribution satisfies

$$P[X_0 = x_0, X_1 = x_1, \ldots, X_J = x_J] = P[X_1 = x_1, \ldots, X_J = x_J],$$

with $x_0 = K - |\mathbf{x}|$. Since this closed network is equivalent to the original semiopen network, with $v_0 = 1$, $v_i = \lambda_i/\alpha$, we have

$$\begin{aligned}
\pi(\mathbf{x}) &= P[X_1 = x_1, \cdots, X_J = x_J] \\
&= P[X_0 = x_0, X_1 = x_1, \ldots, X_J = x_J] \\
&= \frac{1}{C} \cdot \frac{(1)^{x_0}(\frac{\lambda_1}{\alpha})^{x_1} \cdots (\frac{\lambda_J}{\alpha})^{x_J}}{\alpha^{x_0} M_1(x_1) \cdots M_J(x_J)} \\
&= \frac{\lambda_1^{x_1} \cdots \lambda_J^{x_J}}{C\alpha^K M_1(x_1) \cdots M_J(x_J)},
\end{aligned}$$

where C denotes the normalizing constant, and $M_i(\cdot)$, $i = 1, \ldots, J$, are the same as in (2.4). Now, the last expression above, compared with (2.4), can be written as $\prod_{i=1}^{J} P[Y_i = x_i]/C'$, with a new normalizing constant C', which is simply determined as

$$C' = \sum_{|\mathbf{x}| \leq K} \prod_{i=1}^{J} P[Y_i = x_i] = P[|\mathbf{Y}| \leq K];$$

hence, the desired expression.   □

Remark 2.6 It is obvious, comparing Theorems 2.1 and 2.5, that letting $K \to \infty$ in the semiopen model reduces it to the open model. (The denominator in Theorem 2.5 becomes 1.) On the other hand, letting $\alpha \to \infty$ in the semiopen model, while maintaining $v_i = \lambda_i/\alpha$ a (positive) constant, recovers the closed model. To see this, note from the proof of Theorem 2.5 that because of the α^{x_0} factor in the denominator, the probability will vanish as $\alpha \to \infty$ unless $x_0 = 0$, or equivalently $|\mathbf{x}| \equiv K$ (instead of $|\mathbf{x}| \leq K$). But this exactly reduces to the equilibrium distribution of the closed model. Intuitively, an infinitely large external arrival maintains the total number of jobs in the network at the buffer limit all the time. It also makes new jobs available all the time to be released into the network as soon as an internal job leaves the network.

A summary of the three models is in order. The open model is a special case of the semiopen model, one with infinite buffer capacity. The closed model is also a special case of the semiopen model, one with an infinite external arrival rate. On the other hand, a semiopen network can be reduced to a closed network by adding one more node. Therefore, on the one hand, the semiopen network is the most general of the three models; on the other hand, its analysis is essentially no different from that of a closed model.

In terms of computation, the product form of the open network and the product form in the numerator of the closed and the semiopen networks are trivial. The difficult part is the normalizing constants (denominators) in the closed and the semiopen networks, since they involve convolutions of pmfs. (These constants also relate intimately to major performance measures, as will become evident below.) Observe that in the semiopen network, the normalizing constant is given by

$$P[|\mathbf{Y}| \leq K] = \sum_{n=1}^{K} P[|\mathbf{Y}| = n],$$

with each term on the right-hand side corresponding to a normalizing constant of a closed network with n jobs. Hence, below we shall focus on the closed model.

2.4 Throughput Function

From the joint distribution $\pi(x)$ in Theorem 2.3, we can derive the marginal distribution for X_i. Let $\mathbf{Y}_{-i} = (Y_1, \ldots, Y_{i-1}, Y_{i+1}, \ldots, Y_J)$ be the random vector that removes the component Y_i from $\mathbf{Y}$. Let $|\mathbf{Y}_{-i}|$ denote the sum of the components of $\mathbf{Y}_{-i}$. We have, for $n = 0, 1, \ldots, N$,

$$P[X_i = n] = P[Y_i = n] \cdot \frac{P[|\mathbf{Y}_{-i}| = N - n]}{P[|\mathbf{Y}| = N]}. \tag{2.8}$$

Based on this, the throughput (i.e., average output rate) from node i can be derived as follows:

$$E[\mu_i(X_i)]$$

$$= \sum_{n=0}^{N} \mu_i(n)P[X_i = n]$$

$$= \sum_{n=1}^{N} \mu_i(n)P[Y_i = n] \cdot \frac{P[|\mathbf{Y}_{-i}| = N - n]}{P[|\mathbf{Y}| = N]}$$

$$= v_i \sum_{n=1}^{N} P[Y_i = n - 1] \cdot \frac{P[|\mathbf{Y}_{-i}| = N - n]}{P[|\mathbf{Y}| = N]}$$

$$= v_i \left\{ \sum_{n=0}^{N-1} P[Y_i = n] \cdot \frac{P[|\mathbf{Y}_{-i}| = N - n - 1]}{P[|\mathbf{Y}| = N - 1]} \right\} \cdot \frac{P[|\mathbf{Y}| = N - 1]}{P[|\mathbf{Y}| = N]}$$

$$= v_i \cdot \frac{P[|\mathbf{Y}| = N - 1]}{P[|\mathbf{Y}| = N]},$$

where the third equality is due to

$$\mu_i(n)P[Y_i = n] = v_i P[Y_i = n - 1], \tag{2.9}$$

the detailed balance equation for the birth–death queue [cf. (2.4), with $\lambda_i = v_i$], while the last equality results from recognizing that the sum in the curly brackets is over the marginal distribution of node i in a closed network with $N - 1$ jobs [cf. (2.8)].

Therefore, the throughput of each node in a closed Jackson network is the visit frequency to that node (v_i) times a node-independent quantity, which is a function of the total number of jobs in the network. This node-independent quantity is often referred to as the throughput of the closed Jackson network:

$$TH(N) = \frac{P[|\mathbf{Y}| = N - 1]}{P[|\mathbf{Y}| = N]}. \tag{2.10}$$

The throughput of each node i is then

$$TH_i(N) = E[\mu_i(X_i)] = v_i \cdot TH(N). \tag{2.11}$$

Therefore, while $\mu_i(n)$, the service rate at each node i, describes the *intrinsic* (or potential) rate of job completion at the node, $TH(N)$ gives the *actual* rate of job completion (if $v_i = 1$; otherwise, weight by v_i). Also note that while $\mu_i(n)$ is a function of the number of jobs locally at the node, $TH(N)$, naturally, is a function of the total number of jobs in the network.

From the joint distribution in Theorem 2.1 for the open network, we have

$$\frac{P[|\mathbf{X}| = n - 1]}{P[|\mathbf{X}| = n]} = \frac{\sum_{|\mathbf{x}|=n-1} \prod_{i=1}^{J} \frac{\lambda_i^{x_i}}{M(x_i)}}{\sum_{|\mathbf{x}|=n} \prod_{i=1}^{J} \frac{\lambda_i^{x_i}}{M(x_i)}}.$$

Dividing both the numerator and the denominator by α^n and letting $v_i = \lambda_i/\alpha$, we have, for $n = 1, 2, \ldots,$

$$\frac{P[|\mathbf{X}| = n - 1]}{P[|\mathbf{X}| = n]} = \frac{P[|\mathbf{Y}| = n - 1]}{\alpha P[|\mathbf{Y}| = n]} = \frac{TH(n)}{\alpha}. \qquad (2.12)$$

The above relation between $P[|\mathbf{X}| = n - 1]$ and $P[|\mathbf{X}| = n]$ indicates that $|\mathbf{X}|$, the total number of jobs in the open network, follows the same distribution as the number of jobs in a birth–death queue with arrival rate α and state-dependent service rates $\mu(n) = TH(n)$. In other words, when the total number of jobs in the open network is n, its overall output rate is equal to the throughput function of a closed network with n jobs.

We can apply the same argument to the semiopen model, based on Theorem 2.5, and conclude that (2.12) also holds; the only difference is that the range of n must observe the buffer limit: $n = 1, 2, \ldots, K$.

Proposition 2.7 (i) The throughput of a closed Jackson network with a constant population of N jobs equals $TH(N)$ in (2.10).
(ii) In equilibrium, the total number of jobs in the open or semiopen Jackson network follows the same distribution as the number of jobs in a birth–death queue with the following specification: constant arrival rate α (same as the external arrival rate to the network), and state-dependent service rates that are equal to the throughput function, $TH(n)$, of the same network operating in a closed fashion with a constant of n jobs (with $n \le K$ in the semiopen model).

2.5 Throughput Computation

It is evident from comparing (2.4) and (2.10) that the factors $P[Y_i = 0]$, $i = 1, \ldots, J$, play no role in computing the throughput function: They are canceled out from the numerator and the denominator; also refer to the derivation of (2.12) above. For $1 \le j \le J$ and $n \le N$, write

$$g_i(n) = v_i^n / M_i(n), \; n \ge 1; \quad g_i(0) = 1;$$

and

$$G(j, n) = \sum_{n_1 + \cdots + n_j = n} \prod_{i=1}^{j} g_i(n_i). \qquad (2.13)$$

Then,

$$TH(N) = \frac{G(J, N-1)}{G(J, N)}. \qquad (2.14)$$

Below we present two algorithms for the computation of the throughput function and related performance measures.

2.5.1 Convolution Algorithm

From (2.14), we know that the computation of $TH(N)$ is essentially that of the $G(\cdot, \cdot)$ function. From (2.13), we observe that $G(J, N)$ is a sum with $\binom{N+J-1}{N}$ terms. So a direct summation is out of question. However, based on the simple recursion

$$G(j, n) = \sum_{\ell=0}^{n} G(j-1, \ell) g_j(n - \ell), \qquad (2.15)$$

the computation can be carried out as follows:

- Boundaries: $G(1, n) = g_1(n)$, for all $n = 0, 1, \ldots, N$; and $G(i, 0) = 1$ for all $i = 1, \ldots, J$.

- For $j = 2, \ldots, J$, do:

  - for $n = 1, \ldots, N$, do: (2.15).

Note that the inner loop above takes $O(N^2)$ steps. Hence, the overall computational effort is $O(JN^2)$.

The above algorithm also applies to the computation of the marginal distributions in (2.8). Write $G(n) = G(J, n)$, and denoting $G_{-i}(n)$ as the G value of a network with n jobs and $J - 1$ nodes, with node i removed. Then, the marginal probability in (2.8) can be expressed as follows:

$$P[X_i = n] = g_i(n) G_{-i}(N - n)/G(N). \qquad (2.16)$$

Note that to compute $G_{-i}(N - n)$, all we need is to renumber the nodes so that node i is the last (Jth) node. Then, in the loop that computes $G(N)$ above, the second last step (i.e., $j = J - 1$) yields the $G_{-i}(N - n)$ values.

For networks with constant service rates, μ_i for node i, we have $g_i(n) = \rho_i g_i(n - 1)$, where $\rho_i = v_i/\mu_i$. In this case, the above algorithm further simplifies. In particular, (2.15) becomes

$$G(j, n) = \sum_{\ell=0}^{n-1} G(j-1, \ell) g_j(n - 1 - \ell)\rho_j + G(j-1, n)$$

$$= \rho_j G(j, n-1) + G(j-1, n).$$

Hence, the overall computational effort in this case is $O(JN)$.

2.5.2   Mean Value Analysis

For simplicity, consider the special case of constant service rates: μ_i at node i, for all $i = 1,\ldots,J$. Let $L_i(N)$ and $W_i(N)$ denote the expected total number of jobs and expected delay (queueing plus service) at node i, with the argument N denoting the job population of the network. In this case, $g_i(n) = \rho_i^n$. From (2.16), we have

$$
\begin{aligned}
L_i(N) &= \sum_{n=0}^{N} n P[X_i = n] \\
&= \sum_{n=1}^{N} n \rho_i^n G_{-i}(N - n)/G(N) \\
&= \sum_{n=0}^{N-1} (n+1) \rho_i^{n+1} G_{-i}(N - 1 - n)/G(N) \\
&= \rho_i \left[\sum_{n=0}^{N-1} n \rho_i^n G_{-i}(N - 1 - n) + \sum_{n=0}^{N-1} \rho_i^n G_{-i}(N - 1 - n) \right] \Big/ G(N) \\
&= \rho_i [L_i(N - 1)G(N - 1) + G(N - 1)]/G(N) \\
&= \rho_i \cdot TH(N) \cdot [L_i(N - 1) + 1],
\end{aligned}
$$

where in the last equality we have used the relation $TH(N) = G(N - 1)/G(N)$ [cf. (2.14)].

On the other hand, we have

$$
\begin{aligned}
W_i(N) &= \sum_{n=0}^{N-1} \frac{n}{\mu_i} P(X_i = n) + \frac{1}{\mu_i} \\
&= \frac{1}{\mu_i}[L_i(N - 1) + 1],
\end{aligned}
\tag{2.17}
$$

where the second equation follows from the first equation in the above derivation of $L_i(N)$. Hence, combining the last two results, we have

$$
L_i(N) = v_i \cdot TH(N) \cdot W_i(N).
\tag{2.18}
$$

Since $\sum_{i=1}^{J} L_i(N) = N$, summing over i on both sides of (2.18), we have

$$
TH(N) = N/[\sum_{i=1}^{J} v_i W_i(N)].
$$

Substituting back into (2.18) yields

$$
L_i(N) = \frac{N v_i W_i(N)}{\sum_{j=1}^{J} v_j W_j(N)}.
\tag{2.19}
$$

Therefore, recursively making use of (2.17) and (2.19), with the boundaries $L_i(0) = 0$ for all i, we can compute the mean values $L_i(N)$ and $W_i(N)$, and also $TH_i(N) = v_i TH(N)$ via (2.18), for all $i = 1, \ldots, J$ and for any desired value of N. Clearly, the computational effort of the algorithm, known as "mean value analysis," is also $O(JN)$, the same as for the convolution algorithm (in the special case of constant service rates).

The advantage of mean value analysis is that the main performance measures (means) are computed directly. Mean value analysis also leads to marginal distributions. For instance, to compute the marginal distribution in (2.16), iterating on the recursive relation, $TH(N) = G(N-1)/G(N)$, we have

$$G(N) = [TH(1) \cdots TH(N)]^{-1},$$

with the throughput values computed via mean value analysis. And the G_{-i} term in (2.16) is similarly computed, via the throughput of a network with node i removed.

2.6 Time Reversal

Here we apply the concept of time reversal of a Markov chain discussed in Section 1.2 to analyze the Jackson network. The basis of much of the analysis that follows is Theorem 2.9 below, which is a strengthening of Lemma 1.3. For easy reference, we restate it here:

Lemma 2.8 Suppose $\{X(t)\}$ is a stationary Markov chain with state space **S**, rate matrix Q and stationary distribution π. Then, its time-reversal, denoted by $\{\tilde{X}(t)\}$, is also a Markov chain, governed by a rate matrix $\tilde{Q}$ that satisfies:

$$\pi(i)\tilde{q}(i,j) = \pi(j)q(j,i), \quad \forall i,j \in \mathbf{S}, \ i \neq j. \tag{2.20}$$

Theorem 2.9 Suppose π is a probability distribution with support set **S**, and Q and $\tilde{Q}$ are two rate matrices that satisfy (2.20) and also the following:

$$q(i) := \sum_{j \neq i} q(i,j) = \sum_{j \neq i} \tilde{q}(i,j) := \tilde{q}(i), \quad \forall i \in \mathbf{S}. \tag{2.21}$$

Then, Q and $\tilde{Q}$ are the rate matrices of a stationary Markov chain and its time-reversal, and π is the equilibrium distribution of both.

Proof. Taking the summation on both sides of (2.20) over $j \neq i$ yields

$$\pi(i) \sum_{j \neq i} \tilde{q}(i,j) = \sum_{j \neq i} \pi(j)q(j,i), \quad \forall i \in \mathbf{S}.$$

Subtracting the above equation from $\pi(i)\tilde{q}(i) = \pi(i)q(i)$, which follows from (2.21), we have $0 = \pi'Q$. Hence, π is indeed the equilibrium distribution of $\{X(t)\}$. The rest was already established in Lemma 2.8.   □

Remark 2.10 Note from Lemma 2.8 that we have

$$\tilde{q}(i) = \sum_{j \neq i} \tilde{q}(i,j)$$

$$= \sum_{j \neq i} \pi(j)q(j,i)/\pi(i)$$

$$= \pi(i)q(i)/\pi(i) = q(i),$$

where the third equality follows from the given condition that π is the equilibrium distribution of $\{X(t)\}$. Since this is *not* assumed in Theorem 2.9, the condition (2.21) is required instead.

Recall from Example 1.10 of Section 1.2 that the $M/M(n)/1$ queue, with Poisson arrivals at a constant rate λ and state-dependent service rates $\mu(n)$, is reversible; and consequently, departures from the queue form a Poisson process with constant rate λ, exactly as in the arrival process. Furthermore, the number of past departures *up to* time t is independent of the number of jobs in system at t.

On the other hand, from Proposition 2.7, we know that in aggregation, the open Jackson network behaves very much like the $M/M(n)/1$ queue. Hence, we want to study whether the results for the $M/M(n)/1$ queue mentioned above also hold for the Jackson network. To be sure, the Jackson network itself—or more precisely, the Markov chain associated with it—in general will not satisfy reversibility. (For it to satisfy reversibility, essentially the Markov chain associated with the routing mechanism needs to be reversible, which can be directly verified; see Corollary 2.13 and Remark 2.14 below). However, we can still examine the time-reversed version of the network. If we can claim that it is also a Jackson type (with different parameters of course), then we know that the aggregated *exit* processes from the original network—i.e., the stream of jobs leaving the *network*, combined from all nodes—form a Poisson process, whose past history is independent of the current state of the network. This is because the exit process from the original network is the (external) arrival process to the time-reversed network, and hence must be a Poisson process, since the time-reversed network is claimed to be of the Jackson type.

Note that a network of Jackson type is well-defined; it means that the associated Markov chain, $\{\mathbf{X}(t)\} = \{(X_i(t))_{i=1}^J\}$, is governed by the following transition rates:

$$q(\mathbf{x}, \mathbf{x} + \mathbf{e}^i) = \alpha_i,$$

$$q(\mathbf{x}, \mathbf{x} - \mathbf{e}^i) = \mu_i(x_i)p_{i0},$$

$$q(\mathbf{x}, \mathbf{x} - \mathbf{e}^i + \mathbf{e}^j) = \mu_i(x_i)p_{ij}.$$

(Note in particular that the first equation above is due to the Poisson external arrivals, at constant rate $\alpha_i = \alpha p_{0i}$ to node i.)

Now, consider the time-reversal of $\{\mathbf{X}(t)\}$, denoted by $\{\tilde{\mathbf{X}}(t)\}$. To specify the transition rates of $\{\tilde{\mathbf{X}}(t)\}$, we shall try to relate $\{\tilde{\mathbf{X}}(t)\}$ to the network it represents: the time-reversal of the original network represented by $\{\mathbf{X}(t)\}$. In particular, the external arrival (respectively exit) processes under time-reversal correspond to the exit (respectively external arrival) processes of the original network, and the internal transitions from node i to node j under time-reversal correspond to internal transitions from node j to node i in the original network. From the traffic equations in (2.1), we have

$$1 = \frac{\alpha_i}{\lambda_i} + \sum_{j=1}^{J} \frac{\lambda_j}{\lambda_i} p_{ji}. \tag{2.22}$$

The terms on the right side above represent a breakdown of the jobs that go through node i in the original network: α_i/λ_i is the proportion of jobs from external arrivals, and $\lambda_j p_{ji}/\lambda_i$ is the proportion of jobs from node j (internal transitions). From these, we derive the routing probabilities for the time reversal:

$$\tilde{r}_{i0} = \alpha_i/\lambda_i; \quad \tilde{r}_{ij} = \lambda_j p_{ji}/\lambda_i. \tag{2.23}$$

Hence, the transition rates in the time-revered network are

$$\begin{aligned} \tilde{q}(\mathbf{x}, \mathbf{x} + \mathbf{e}^i) &= \lambda_i p_{i0}, \\ \tilde{q}(\mathbf{x}, \mathbf{x} - \mathbf{e}^i) &= \mu_i(x_i)\alpha_i/\lambda_i, \\ \tilde{q}(\mathbf{x}, \mathbf{x} - \mathbf{e}^i + \mathbf{e}^j) &= \mu_i(x_i)\lambda_j p_{ji}/\lambda_i. \end{aligned}$$

From the first equation above, we have, in addition to the routing probabilities in (2.23),

$$\tilde{r}_{0i} = \lambda_i p_{i0}/\alpha. \tag{2.24}$$

To verify that $\sum_{i=1}^{J} \tilde{p}_{0i} = 1$, from $p_{i0} = 1 - \sum_{j=1}^{J} p_{ij}$, we have

$$\sum_{i=1}^{J} \lambda_i p_{i0} = \sum_{i=1}^{J} \lambda_i - \sum_{i=1}^{J}\sum_{j=1}^{J} \lambda_i p_{ij} = \alpha, \tag{2.25}$$

where the second equality follows from the traffic equation [cf. (2.1) or (2.22)].

We now use Theorem 2.9 to prove that the product form $\pi(\mathbf{x})$ in Theorem 2.1 is indeed the equilibrium distribution of $\{\mathbf{X}(t)\}$. In particular, we want to verify (2.20) and (2.21). Note that here (2.21) is implied by

$$\sum_{i=1}^{J} q(\mathbf{x}, \mathbf{x} + \mathbf{e}^i) = \sum_{i=1}^{J} \tilde{q}(\mathbf{x}, \mathbf{x} + \mathbf{e}^i) \tag{2.26}$$

and

$$q(\mathbf{x}, \mathbf{x} - \mathbf{e}^i) + \sum_{j=1}^{J} q(\mathbf{x}, \mathbf{x} - \mathbf{e}^i + \mathbf{e}^j) = \tilde{q}(\mathbf{x}, \mathbf{x} - \mathbf{e}^i) + \sum_{j=1}^{J} \tilde{q}(\mathbf{x}, \mathbf{x} - \mathbf{e}^i + \mathbf{e}^j),$$

which, with the substitution of the transition rates, reduce to, respectively, (2.25) and (2.22).

To verify (2.20) amounts to verifying the following:

$$\pi(\mathbf{x})q(\mathbf{x}, \mathbf{x} + \mathbf{e}^i) = \pi(\mathbf{x} + \mathbf{e}^i)\tilde{q}(\mathbf{x} + \mathbf{e}^i, \mathbf{x}),$$
$$\pi(\mathbf{x})q(\mathbf{x}, \mathbf{x} - \mathbf{e}^i) = \pi(\mathbf{x} - \mathbf{e}^i)\tilde{q}(\mathbf{x} - \mathbf{e}^i, \mathbf{x}),$$
$$\pi(\mathbf{x})q(\mathbf{x}, \mathbf{x} - \mathbf{e}^i + \mathbf{e}^j) = \pi(\mathbf{x} - \mathbf{e}^i + \mathbf{e}^j)\tilde{q}(\mathbf{x} - \mathbf{e}^i + \mathbf{e}^j, \mathbf{x}).$$

Substituting into the above equations the transition rates $q(\cdot, \cdot)$ and $\tilde{q}(\cdot, \cdot)$ specified above, we have

$$\pi(\mathbf{x})\alpha_i = \pi(\mathbf{x} + \mathbf{e}^i)\mu_i(x_i + 1)\alpha_i/\lambda_i,$$
$$\pi(\mathbf{x})\mu_i(x_i)p_{i0} = \pi(\mathbf{x} - \mathbf{e}^i)\lambda_i p_{i0},$$
$$\pi(\mathbf{x})\mu_i(x_i)p_{ij} = \pi(\mathbf{x} - \mathbf{e}^i + \mathbf{e}^j)\mu_j(x_j + 1)\lambda_i p_{ij}/\lambda_j.$$

These are then easily verified using the product-form solution in Theorem 2.1.

Therefore, making use of the notion of time-reversal, in particular Theorem 2.9, we have not only provided an alternative proof to Theorem 2.1, but also obtained more results out of the analysis. In particular, we have shown that the time-reversal of the original network is also a Jackson network. Furthermore, since the independent Poisson arrival processes of the time-reversal are the exit processes of the original network, we get a full characterization of the latter, in the same spirit as we did before for the $M/M(n)/1$ queue.

Theorem 2.11 Consider the open Jackson network in Theorem 2.1.

(i) Its time-reversal is another Jackson network characterized as follows: The external arrivals follow a Poisson process with rate α, each arrival joins node i independently with probability $\tilde{r}_{0i} = \lambda_i p_{i0}/\alpha$ [cf. (2.24)]; internal jobs are routed following the routing probabilities in (2.23); service rates at all nodes are the same as in the original network. The time-reversal has the same equilibrium distribution as the original network, as specified in Theorem 2.1.

(ii) The exit processes from the original network are independent Poisson processes, with rate $\lambda_i p_{i0}$ for node i. Furthermore, the past history of the process of jobs leaving the network up to time t is independent of $\mathbf{X}(t)$, the state of the network at time t.

From the above analysis, we also know under what conditions the original network satisfies reversibility: It suffices to insist that the original network and its time-reversal have the same routing probabilities, $\tilde{r}_{0i} = p_{0i}$, $\tilde{r}_{i0} = p_{i0}$, and $\tilde{r}_{ij} = p_{ij}$.

Corollary 2.12 The stationary Markov chain $\{\mathbf{X}(t)\}$ associated with the open Jackson network is reversible if for all $i, j = 1, \ldots, J$, we have

$$\lambda_i p_{i0} = \alpha p_{0i}, \quad \text{and} \quad \lambda_i p_{ij} = \lambda_j p_{ji}.$$

Similarly, for closed Jackson networks, we have the following.

Corollary 2.13 The stationary Markov chain $\{\mathbf{X}(t)\}$ associated with the closed Jackson network is reversible, if for all $i, j = 1, \ldots, J$ we have

$$v_i p_{ij} = v_j p_{ji}.$$

Remark 2.14 Note that in the case of closed networks, $(v_i)_{i=1}^{J}$, normalized to unity, can be viewed as the equilibrium distribution of the routing chain: a discrete-time Markov chain. The given condition in Corollary 2.13 is nothing but the detailed balance equation for this routing chain. Hence, Corollary 2.13 indicates that the closed Jackson network is reversible if its routing chain is reversible. Corollary 2.12 can be similarly interpreted, with the state space of the routing chain properly augmented to include node 0: the external world.

2.7 Notes

Jackson networks originated from the studies of J.R. Jackson [6, 7], motivated by modeling the operation in a job shop. Gordon and Newell [5] considered the case of closed networks. The open, closed, and semiopen models presented here are based on these early works. The convolution algorithm was first proposed by Buzen in [4]. The mean value analysis belongs to Reiser and Lavenberg; see [9]. The technique of analyzing the network along with its time-reversal, based on Theorem 2.9, discussed in Section 2.6, is due originally to Kelly [8].

The main feature of Jackson networks, namely, the product-form equilibrium distribution, in fact, extends to much more general networks. For instance, jobs can be divided into several classes, the routing mechanism need not be Markovian, a variety of service disciplines can be allowed (e.g., processor sharing, last come first served). Indeed, the product form characterizes a class of networks that connect together the so-called *quasi-reversible* queues (Kelly [8]).

There is a rich body of literature that applies Jackson networks to the modeling and analysis of manufacturing systems under various configurations and operating schemes, e.g., flow lines, job shops, and flexible manufacturing systems. The book of Buzacott and Shanthikumar [1] presents the detailed modeling of many such systems. Also see Buzacott and Yao [2], and Buzacott, Shanthikumar and Yao [3].

2.8 Exercises

1. Suppose in the open Jackson network each node has an infinite number of servers, hence, $\mu_i(n) = n\mu_i$, where μ_i is the server rate of each server [cf. Remark 2.2]. Show that the product-form result in Theorem 2.1 holds, with each Y_i (the marginal) following a Poisson distribution, with parameter $\rho_i := \lambda_i/\mu_i$.

2. Repeat the above problem for the closed and semiopen models, and derive the joint distributions.

3. There are two ways to derive the throughput of a semiopen Jackson network. On the one hand, it is equal to $\alpha P[\|\mathbf{X}\| = K]$, where α is the external arrival rate and K is the buffer limit. On the other hand, the throughput can also be derived as equal to $\sum_{n=1}^{K} TH(n) \cdot P[\|\mathbf{X}\| = n]$; refer to Proposition 2.7. Verify that these two expressions are equal.

4. Consider an open Jackson network with each node i having a constant service rate: $\mu_i(n) = \mu_i$ for all $n \geq 1$ and all i (case (i) in Remark 2.2). Derive the expected total number of jobs in the network, and the expected response time: the total time a job spends in the network.

5. Repeat the above problem for the semiopen model.

6. Prove Corollaries 2.12 and 2.13.

References

[1] BUZACOTT, J.A. AND SHANTHIKUMAR, J.G. 1993. *Stochastic Models of Manufacturing Systems*, Prentice Hall, Englewood Cliffs, NJ.

[2] BUZACOTT, J.A. AND YAO, D.D. 1986. On Queueing Network Models of Flexible Manufacturing Systems, *Queueing Systems: Theory and Applications*, **1**, 5–27.

[3] BUZACOTT, J.A., SHANTHIKUMAR, J.G., AND YAO, D.D. 1994. *Stochastic Modeling and Analysis of Manufacturing Systems*, Chapter 1, D.D. Yao (ed.), Springer-Verlag, New York.

[4] BUZEN, J.B. 1973. Computational Algorithms for Closed Queueing Networks with Exponential Servers. *Comm. ACM*, **16**, 527–531.

[5] GORDON, W.J. AND NEWELL, G. F. 1967. Closed Queueing Networks with Exponential Servers. *Operations Research*, **15**, 252–267.

[6] JACKSON, J.R. 1957. Networks of Waiting Lines. *Operations Research*, **5**, 518–521.

[7] JACKSON, J.R. 1963. Jobshop-Like Queueing Systems. *Management Science*, **10**, 131–142.

[8] KELLY, F.P. 1979. *Reversibility and Stochastic Networks*. Wiley, New York.

[9] REISER, M. AND LAVENBERG, S.S. 1980. Mean Value Analysis of Closed Multichain Queueing Networks. *J. Assoc. Comp. Mach.*, **27**, 313–322.

3

Stochastic Comparisons

We continue our study of Jackson networks, but shift to focusing on their structural properties: those that describe the qualitative behavior of the network. We want to demonstrate that the Jackson network has the capability to capture the essential qualitative behavior of the system, to make it precise, and to bring out explicitly the role played by different resources and control parameters.

From Chapter 2, in particular, Proposition 2.7, and (2.10), we have seen the central role played by the throughput function, not only in the closed network but also in the open and semiopen networks. It measures the rate of job completions from the entire network as a function of the total number of jobs circulating in the network, and it also relates directly to the service rates at all the nodes. Hence, throughout the chapter, we shall focus on the throughput function, in the setting of the closed network. Most of the results derived here, however, extend readily to the semiopen model, and hence the open model as well.

In what follows, the monotonicity of the throughput function is first studied in Section 3.1, through the notion of equilibrium rate and its linkage to the likelihood ratio ordering—both topics were introduced in Chapter 1. In Section 3.2, we further exploit this connection, and use coupling techniques to establish concavity/convexity properties of the throughput function. The important case of networks with multiserver nodes is the focus of Section 3.3, where we establish properties with respect to the number of servers and the service rates. Properties that correspond to resource sharing, in terms of aggregation of servers and nodes, are studied in Section 3.4. Connections to majorization and arrangement orderings, as motivated by the allocation or assignment of workload, are studied in Section 3.5.

Throughout the chapter we continue to use the terms "increasing" and "decreasing" to mean "nondecreasing" and "nonincreasing," respectively.

3.1   Monotonicity

First recall the notion of equilibrium rate, introduced in Section 1.3, Definition 1.11: Suppose X is a nonnegative, integer-valued random variable. Let $p(n) = P(X = n) > 0$ for all $n \in \mathcal{N} = \{0, 1, \ldots, N\}$, where N can be infinite. Then, the equilibrium rate of X is defined as follows:

$$r_X(0) = 0, \qquad r_X(n) = p(n-1)/p(n), \quad n = 1, \ldots, N.$$

Next, notice that in all three models of the Jackson network in Chapter 2, the routing probabilities play only an indirect role through the solution to the traffic equations, $(\lambda_i)_i$ or $(v_i)_i$. Hence, we can assume, without loss of generality, that $\lambda_i = 1$ or $v_i = 1$ for all i; via replacing the service rates $\mu_i(n)$ by $\mu_i(n)/\lambda_i$ or by $\mu_i(n)/v_i$. (For simplicity, though, below we shall continue using $\mu_i(n)$ to denote the modified service rates.) This way, all the results, including the equilibrium distributions and the throughput functions will remain unchanged. Note that this modification is tantamount to assuming that all networks under discussion have a *serial* (or *tandem*) configuration, since the arrivals rates or the visit ratios to the stations are all equal. (In a closed network, this is more precisely a *cyclic* configuration.)

With these observations, following (2.4) we can write the service rate at each node i as the equilibrium rate of Y_i:

$$\mu_i(n) = P[Y_i = n - 1]/P[Y_i = n - 1] = r_{Y_i}(n).$$

On the other hand, following (2.10), we can express the throughput function as the equilibrium rate of $|\mathbf{Y}|$, the summation of the independent Y_i's:

$$TH(N) = \frac{P[|\mathbf{Y}| = N - 1]}{P[|\mathbf{Y}| = N]} = r_{|\mathbf{Y}|}(N). \tag{3.1}$$

That is, the production rate of the network is nothing but some sort of "convolution" of the service rates at the nodes.

3.1.1   PF₂ Property

The increasingness in the equilibrium rate, $r_X(n-1) \le r_X(n)$ for all n, expressed in terms of the pmf of X, is $p(n-1)p(n+1) \le p^2(n)$ for all n. In other words, the pmf is log-concave in n. This is known as the PF_2 property (Pólya frequency of order two). We use $X \in PF_2$ to denote that the pmf of a random variable satisfies the PF_2 property.

Definition 3.1 $X \in PF_2$ if and only if $r_X(n)$, the equilibrium rate of X, is increasing in n.

Example 3.2 Consider a birth–death queue (refer to Section 1.3, in particular, Example 1.12) with unit arrival rate and state-dependent service rate $\mu(n)$. The number of jobs in system Y satisfies the PF_2 property if and only if $\mu(n)$ is increasing in n.

It is known that the PF_2 property is preserved under convolution. That is, for a set of *independent* random variables $\{Y_1, \ldots, Y_J\}$, if $Y_i \in PF_2$ for all i, then $|\mathbf{Y}| \in PF_2$. (This will be proved later as part of the proof of a second-order property in Proposition 3.19, also see Proposition 3.23.) Hence, applying this fact to (3.1), we have the following.

Theorem 3.3 The throughput function $TH(N)$ is increasing in N, the total number of jobs in the network, if for every node i in the network, the service rate $\mu_i(n)$ is an increasing function.

Hence, if the Jackson network models a production system, then the above result says that the overall production rate is increasing in the system's WIP level if the service rate at each node is an increasing function of its local WIP level.

3.1.2 Likelihood Ratio Ordering

Recall that in the birth–death queue of Example 1.15, increasing the service rate function $\mu(n)$, in the pointwise sense, will decrease the equilibrium number of jobs in the system, in the sense of likelihood ratio ordering. Equivalently, for two random variables X, Y, the likelihood ratio ordering $X \geq_{\ell r} Y$ can be viewed as corresponding to two birth–death queues with the same arrival rate, and with state-dependent service rates $r_X(n)$ and $r_Y(n)$, and $r_Y(n) \geq r_X(n)$.

We now address the other monotonicity issue raised earlier: whether speeding up the service at each node will translate into a higher throughput of the network. The gist of this question, based on the relation in (3.1), is whether the likelihood ratio ordering is preserved under convolution. The answer to this is, in fact, known: The key condition is that the service rates at all nodes have to be increasing functions. The precise statement is the following.

Lemma 3.4 Suppose $Y_1 \geq_{\ell r} Y_2$, and $Z \in PF_2$ is independent of Y_1 and Y_2. Then $Y_1 + Z \geq_{\ell r} Y_2 + Z$.

In contrast, if the likelihood ratio ordering in the above lemma is changed to the stochastic ordering, which is weaker than the likelihood ordering (Lemma 1.14), then the PF_2 property is *not* needed. That is, $Y_1 \geq_{st} Y_2$

implies $Y_1 + Z \geq_{st} Y_2 + Z$, for any Z that is independent of Y_1 and Y_2, which is a standard result (e.g., [11]).

Below, we restate the above lemma in the context of a two-node closed (cyclic) network as shown by Figure 3.1, and supply a proof.

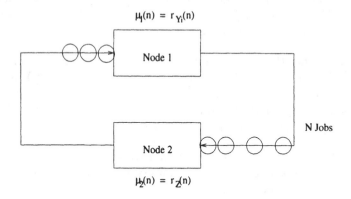

FIGURE 3.1. A closed cyclic network

Proposition 3.5 Let Y_1, Y_2, and Z be the random variables in Lemma 3.4. Construct a two-node cyclic network with N jobs. Let the service rates at the two nodes be the equilibrium rates of Y_1 (for node 1) and Z (for node 2). Next, construct a second network, with the service rate at node 1 increased to the equilibrium rate of Y_2 (since $Y_1 \geq_{\ell r} Y_2$); other things stay unchanged. Then the second network has a higher throughput.

Proof. For $i, j = 1, 2$, let X_j^i denote the number of jobs at node j in network i. Then for $i = 1, 2$,

$$P[X_1^i = n] = \frac{P[Y_i = n]P[Z = N - n]}{P[Y_i + Z = N]},$$

and hence,

$$r_{X_1^i}(n) = r_{Y_i}(n)/r_Z(N + 1 - n).$$

Therefore, $Y_1 \geq_{\ell r} Y_2$ implies $X_1^1 \geq_{\ell r} X_1^2$, and hence $N - X_1^1 \leq_{\ell r} N - X_1^2$, i.e., $X_2^1 \leq_{\ell r} X_2^2$. This, in turn, implies $X_2^1 \leq_{st} X_2^2$, from Lemma 1.14. Therefore, we have $\mathsf{E}r_Z(X_2^1) \leq \mathsf{E}r_Z(X_2^2)$, since $r_Z(\cdot)$ is an increasing function ($Z \in PF_2$). But the two expectations are exactly the throughputs of the two networks. (Note that in the two-node cyclic network under consideration, the two nodes have equal throughput, which is also the throughput of the network.) □

From the above proof, we note that the PF_2 property (or equivalently, the increasingness of the service rate at node 2) is *not* needed for the likelihood ratio ordering of the number of jobs at the two nodes. (Recall that neither is PF_2 needed in the birth–death queue of Example 1.15.) The PF_2 property is needed only to translate the likelihood ratio ordering into the monotonicity in throughput, via stochastic ordering.

Corollary 3.6 In the network of Proposition 3.5, increasing the service rate at node 1 decreases the number of jobs there (X_1), and hence increases the number of jobs at node 2 (X_2), both in the sense of likelihood ratio ordering. This holds regardless of whether the service rate at node 2 is an increasing function.

Remark 3.7 In fact, from the proof of Proposition 3.5, we can write, X_1^i and X_2^i, for $i = 1, 2$, as conditional random variables:

$$X_1^i = [Y_i | Y_i + Z = N], \quad X_2^i = [Z | Y_i + Z = N].$$

Then the result in Corollary 3.6 takes the following form:

$$Y_1 \geq_{\ell r} Y_2 \quad \Rightarrow \quad [Y_1 | Y_1 + Z = N] \geq_{\ell r} [Y_2 | Y_2 + Z = N]$$
$$\text{and} \quad [Z | Y_1 + Z = N] \leq_{\ell r} [Z | Y_2 + Z = N],$$

regardless of whether $Z \in PF_2$.

To extend the above results to a network with more than two nodes, all we need is to note that the expression in (3.1) can be rewritten as

$$TH(n) = r_{Y_B + Y_{\bar{B}}}(n), \tag{3.2}$$

where B is a subset of nodes, $B \subset \{1, \ldots, J\}$, and $\bar{B}$ is the complement set, and

$$Y_B := \sum_{i \in B} Y_i, \quad Y_{\bar{B}} := \sum_{i \in \bar{B}} Y_i.$$

This way, the original network (with J nodes) is reduced to a two-node cyclic network, with the two nodes having service rates $r_{Y_B}(n)$ and $r_{Y_{\bar{B}}}(n)$, each representing an aggregation of the nodes in B and $\bar{B}$, respectively.

Theorem 3.8 In a closed Jackson network, suppose the service rates at all nodes are increasing functions. Then, increasing the service rate functions, in a pointwise sense, at a subset (say, B) of nodes will increase the equilibrium number of jobs at each of the nodes *not* in B, in the sense of likelihood ratio ordering. Consequently, the throughput will be increased.

Remark 3.9 Notice that the above result regarding the increased number of jobs applies to *each* of the nodes that is not in B. To see this, pick one particular node that is not in B, and aggregate all the other nodes, including those in B, into a single node. Since the service rates at all nodes are increasing functions, increasing those functions (in a pointwise sense) that correspond to the nodes in B will translate into an increased service rate function of the aggregated node. (This explains why the nodes in B are required to have increasing service rate functions, which was not required in the two-node model in Proposition 3.5 and Corollary 3.6.) Now we can apply Corollary 3.6 to this aggregated two-node network.

3.1.3 Shifted Likelihood Ratio Ordering

From Theorem 3.8, we know that speeding up the service at one or several nodes will increase the overall production rate of the network, provided that all nodes in the network have increasing service rate functions. In applications, this need not always be the case. (For instance, suppose a node represents a clean room in a semiconductor fabrication facility. A higher level of WIP means longer delay; and delay beyond a certain level could degrade the quality of the waiting jobs (wafers), due to contamination, thus creating additional complications to the production process and slowing it down.) So the question we ask here is, when a network contains nodes where the service rates are not necessarily increasing functions, how do we speed up such nodes so as to increase the throughput of the whole network.

Based on the earlier analysis, we know an equivalent way to ask the same question: Can the likelihood ratio ordering be strengthened, so that it is preserved under convolution *without* the PF_2 property? From Definition 1.13, we know that strengthening the likelihood ratio ordering is tantamount to strengthening the pointwise dominance relation among equilibrium rates. This motivates the "shifted" dominance relation below:

Definition 3.10 For two functions $f, g : \mathcal{N} \mapsto \Re_+$, write $f \leq_\uparrow g$ if $f(m) \leq g(n)$ for all $m \leq n$, $m, n \in \mathcal{N}$.

The following properties are easily verified ($f \leq g$ denotes the usual pointwise ordering):

Lemma 3.11 The shifted dominance $\leq_\uparrow$ satisfies the following properties:

(i) $f \leq_\uparrow g$ implies $f \leq g$;

(ii) If either f or g is an increasing function, then $f \leq_\uparrow g$ if and only if $f \leq g$;

(iii) $f \leq_\uparrow g$ if and only if there exists an increasing function h, such that $f \leq h \leq g$.

Based on the ordering $\leq_\uparrow$ among equilibrium rates, we can define a strengthened likelihood ratio ordering.

Definition 3.12 Two random variables X and Y are said to be ordered under *shifted likelihood ratio ordering*, written $X \succeq_\uparrow^{\ell r} Y$, if their equilibrium rates are ordered as $r_X \leq_\uparrow r_Y$.

Properties of the shifted likelihood ratio ordering follow immediately from combining Lemma 3.11 with Definition 3.12.

Proposition 3.13 The shifted likelihood ratio ordering satisfies the following properties:

(i) $X \geq_{\uparrow}^{\ell r} Y$ implies $X \geq_{\ell r} Y$;

(ii) If either $X \in PF_2$ or $Y \in PF_2$, then $X \geq_{\uparrow}^{\ell r} Y$ if and only if $X \geq_{\ell r} Y$;

(iii) $X \geq_{\uparrow}^{\ell r} Y$ if and only if there exists a random variable $Z \in PF_2$ such that $X \geq_{\ell r} Z \geq_{\ell r} Y$.

Now we have the analogue of Lemma 3.4 without requiring the PF_2 property.

Lemma 3.14 If for $i = 1, 2$, Y_i and Z_i are independent, then $Y_1 \geq_{\uparrow}^{\ell r} Y_2$, $Z_1 \geq_{\uparrow}^{\ell r} Z_2$ implies $Y_1 + Z_1 \geq_{\uparrow}^{\ell r} Y_2 + Z_2$.

Although the above can also be rephrased in the context of a two-node cyclic network, similar to Proposition 3.5, it cannot be proved by simply adapting the proof there. Nevertheless, there is another way to prove Proposition 3.5, which we illustrate below; and this new approach *will* extend, quite readily, to the proof of Lemma 3.14.

For $i, j = 1, 2$, let $X_j^i(t)$ denote the number of jobs at node j in network i at time t. Note that $X_1^i(t) + X_2^i(t) = N$ for $i = 1, 2$ and for all t. Let $D_j^i(t)$ denote the cumulative number of departures (i.e., service completions), up to time t, from node j in network i. Recall in Proposition 3.5 that nodes 1 of the two networks have service rates $r_{Y_1}(\cdot)$ and $r_{Y_2}(\cdot)$, respectively, and $r_{Y_1} \leq r_{Y_2}$, while nodes 2 in both networks have the same service rate $r_Z(\cdot)$, which is an increasing function. Below we establish a pathwise stochastic ordering among the state (X) and the counting processes (D) of the two networks. This result then implies the throughput result in Proposition 3.5, since clearly, for $i = 1, 2$, the throughput of network i is

$$TH_i = \lim_{t \to \infty} D_1^i(t)/t = \lim_{t \to \infty} D_2^i(t)/t.$$

Proposition 3.15 Suppose the two networks in Proposition 3.5 start from the same initial state, i.e., $X_1^1(0) = X_1^2(0)$ [hence, $X_2^1(0) = X_2^2(0)$], and $D_j^1(0) = D_j^2(0) = 0$ for $j = 1, 2$. Then

$$\{X_1^1(t)\} \geq_{st} \{X_1^2(t)\}, \quad \{(D_1^1(t), D_2^1(t))\} \leq_{st} \{(D_1^2(t), D_2^2(t))\}.$$

Proof. With an approach similar to that in Example 1.17, we use uniformization to "couple" the two networks as well as to discretize time, and inductively construct the paths of the stochastic processes in question such that the desired order relations are maintained at all times.

For $j = 1, 2$, write r_{Y_j} as r_j for simplicity. Let

$$\eta := \max_k \{r_2(k)\} + r_Z(N) := \eta_1 + \eta_2$$

be the uniformization constant. Let $0 = \tau_0 < \tau_1 < \cdots$ be a sequence of Poisson event epochs with occurrence rate η. Let $\{U_k; k = 0, 1, 2, \ldots\}$ be a sequence of i.i.d. random variables uniformly distributed on the interval $[-\eta_2, \eta_1]$, and independent of the Poisson event epochs $\{\tau_k\}$. Below, using the two sequences $\{\tau_k\}$ and $\{U_k\}$, we generate the paths of the processes in the two networks.

Let $x_j^i(k)$ and $d_j^i(k)$ denote the sample values of $X_j^i(\tau_k)$ and $D_j^i(\tau_k)$ through the construction. Initially, set $x_1^i(0) = X_1^i(0)$ and $d_j^i(0) = 0$ for $i, j = 1, 2$. Inductively, suppose the construction has been carried out to τ_k, and the following relations hold:

$$x_1^1(k) \geq x_1^2(k); \qquad d_j^1(k) \leq d_j^2(k), \quad j = 1, 2. \tag{3.3}$$

[The first inequality above implies $x_2^1(k) = N - x_1^1(k) \leq N - x_1^2(k) = x_2^2(k)$.] We specify the construction at τ_{k+1}, and show that the relations in (3.3) also hold at τ_{k+1}. Note that for $t \in (\tau_k, \tau_{k+1})$, $x_j^i(t) = x_j^i(k)$, $d_j^i(t) = d_j^i(k)$, $i, j = 1, 2$. That is, in between two consecutive Poisson event epochs there are no changes in either network, following the construction.

At τ_{k+1}, for $i = 1, 2$, set

$$x_1^i(k+1) = x_1^i(k) - 1_1^i + 1_2^i,$$
$$x_2^i(k+1) = N - x_1^i(k+1),$$
$$d_j^i(k+1) = d_j^i(k) + 1_j^i, \quad j = 1, 2,$$

where 1_j^i are indicator functions defined as

$$1_1^i = 1\{U_k \in (0, r_i(x_1^i(k))]\}, \quad 1_2^i = 1\{U_k \in [-r_Z(x_2^i(k)), 0]\}.$$

(Hence, the two indicator functions correspond to whether or not there is a service completion at either node 1 or node 2 in the two networks.) This construction is valid, i.e., the constructed processes are indeed the correct probabilistic replicas of the original processes, since the service completions occur with the right rates: $r_i(x_1^i(k))$ for node 1 and $r_Z(x_2^i(k))$ for node 2 in network $i = 1, 2$.

We next show that the induction hypotheses in (3.3) also hold at τ_{k+1}. If $x_1^1(k) > x_1^2(k)$, then, since either quantity can change only by at most one unit, regardless of what happens, $x_1^1(k+1) \geq x_1^2(k+1)$ always holds. On the other hand, if $x_1^1(k) = x_1^2(k)$, then

$$r_1(x_1^1(k)) \leq r_2(x_1^2(k)),$$

implying $1_1^1 \leq 1_1^2$, and hence $x_1^1(k+1) \geq x_1^2(k+1)$.

For the counting processes, $d_2^1(k+1) \leq d_2^2(k+1)$ obviously holds, since $r_Z(x_2^1(k)) \leq r_Z(x_2^2(k))$ follows from the increasingness of $r_Z(\cdot)$ and $x_2^1(k) \leq x_2^2(k)$. That $d_1^1(k+1) \leq d_1^2(k+1)$ holds is also obvious when $x_1^1(k) = x_1^2(k)$, since $r_1 \leq r_2$. On the other hand, when $x_1^1(k) > x_1^2(k)$, we have

$$d_1^2(k) - d_1^1(k) = [x_1^1(k) - x_1^2(k)] + [d_2^2(k) - d_2^1(k)] > 0,$$

making use of the induction hypothesis, $d_2^2(k) \geq d_2^1(k)$. Hence, in both cases, $d_1^1(k+1) \leq d_1^2(k+1)$ still holds. $\qquad\square$

Now we can adapt the above proof to prove a network analogue of Lemma 3.14.

Proposition 3.16 Consider the two cyclic networks in Proposition 3.5. Suppose the service rates at node 1 and node 2 are, respectively, equal to the equilibrium rates of Y_i and Z_i, in network i, for $i = 1, 2$. When Y_i and Z_i satisfy the conditions in Lemma 3.14, the counting processes in the two networks satisfy

$$\{(D_1^1(t), D_2^1(t))\} \leq_{st} \{(D_1^2(t), D_2^2(t))\}.$$

Proof. Let r_j^i denote the service rate at node j in network i, for $i, j = 1, 2$. Then, $r_j^1 \leq_\uparrow r_j^2$ for $j = 1, 2$, following the conditions on Y_i and Z_i in Lemma 3.14. Proceed by adapting the proof of Proposition 3.15 as follows:

(i) Let the uniformization constant be $\eta := \max_k\{r_1^2(k)\} + \max_k\{r_1^2(k)\}$.

(ii) Apply induction only to the d_j^i's, i.e., remove the part on x_j^i in (3.3).

(iii) Replace r_i by r_1^i, and r_Z by r_2^i.

As the two nodes are now symmetrical, arguing for one of them will suffice. For instance, for node 1, the key is $x_1^1(k) \leq x_1^2(k) \Rightarrow r_1^1(x_1^1(k)) \leq r_1^2(x_1^2(k))$ (since $r_1^1 \leq_\uparrow r_1^2$); and $x_1^1(k) > x_1^2(k) \Rightarrow d_1^1(k) < d_1^2(k)$ (just as in the last part of the proof of Proposition 3.15). $\qquad\square$

Extension of the above result to more than two nodes is immediate, making use of the expression in (3.2), as in Theorem 3.8.

Theorem 3.17 In a closed Jackson network, where the service rates are not necessarily increasing functions, increasing the service rates at a subset of nodes, in the sense of the shifted ordering in Definition 3.10, will increase the throughput of the network.

Remark 3.18 Note that the stronger result of stochastic ordering among the (node-based) departure processes in Proposition 3.16 does not apply to the general network in Theorem 3.17: It will apply only to any aggregated two-node version of the network.

3.2 Concavity and Convexity

In Theorem 3.3 we have established that the throughput function is an increasing function of the network WIP level, provided that all nodes have increasing service rate functions. We also know that this result is equivalent to the preservation of the PF_2 property (or the increasingness of the equilibrium rates) under convolution of the pmfs of independent random variables.

Here we study the concavity/convexity of the throughput function, with respect to the network WIP level. We will show that not only the increasingness of the equilibrium rates but also the increasing concavity and increasing convexity are preserved under convolution.

Below we start with a two-node cyclic network, and present a coupling proof of the increasing concavity of the departure processes, and hence the throughput, with respect to the job population. In contrast to the monotonicity proof, such as the one in Proposition 3.5, where the coupling is applied to two networks, here we need to construct *four* networks simultaneously.

Proposition 3.19 Consider four two-node cyclic networks, indexed by the superscript i, with job population $N^1 = N$, $N^2 = N^3 = N + 1$, and $N^4 = N + 2$. The service rates at the two nodes are $\mu_1(\cdot)$ and $\mu_2(\cdot)$, common for all four networks. Both μ_1 and μ_2 are increasing and concave functions. For $j = 1, 2$ and $i = 1, 2, 3, 4$, let $D_j^i(t)$ be the cumulative number of service completions up to time t, and $X_j^i(t)$, the number of jobs at time t, in network i at node j. Assume initially $X_1^i(0) = N^i$, $X_2^i(0) = 0$, and $D_j^i(0) = 0$ for all i, j. Then,

$$\{(D_1^1(t) + D_1^4(t), D_2^1(t) + D_2^4(t))\} \leq_{st} \{(D_1^2(t) + D_1^3(t), D_2^2(t) + D_2^3(t))\} \tag{3.4}$$

and

$$\{(D_1^1(t), D_2^1(t))\} \leq_{st} \{(D_1^i(t), D_2^i(t))\}, \quad i = 2, 3. \tag{3.5}$$

Proof. The uniformization procedure is similar to the proof of Proposition 3.15. Let the uniformization constant be

$$\eta := 2\mu_1(N + 2) + 2\mu_2(N + 2) := \eta_1 + \eta_2.$$

Let $\{\tau_k\}$ be the sequence of Poisson event epochs with rate η; let $\{U_k\}$ be an i.i.d. sequence of uniform random variables on $[-\eta_2, \eta_1]$, independent of $\{\tau_k\}$.

Again, let $x_j^i(k)$ and $d_j^i(k)$ denote the sample values of the state and the counting processes at τ_k generated by the construction. Initially, as specified in the theorem, for $i = 1, 2, 3, 4$, set $x_1^i(0) = N^i$, and $x_2^i(0) = 0$; set $d_j^i(0) = 0$ for $j = 1, 2$.

Inductively, suppose the construction has been carried out up to τ_k. As induction hypotheses, suppose

$$d_j^1(k) + d_j^4(k) \le d_j^2(k) + d_j^3(k), \quad j = 1, 2; \tag{3.6}$$

and, in addition,

$$x_j^1(k) \le x_j^2(k) \le x_j^4(k), \quad x_j^1(k) \le x_j^3(k); \quad j = 1, 2. \tag{3.7}$$

Note that the relations in (3.7) will guarantee, via the construction below, the increasingness property in (3.5).

The construction at τ_{k+1} takes the same form as in the coupling proof of Proposition 3.15. For simplicity, write $x_j^i := x_j^i(k)$ and $d_j^i := d_j^i(k)$. For $i = 1, 2, 3, 4$, set

$$x_1^i(k+1) = x_1^i - 1_1^i + 1_2^i,$$
$$x_2^i(k+1) = N^i - x_1^i(k+1),$$
$$d_j^i(k+1) = d_j^i + 1_j^i, \quad j = 1, 2.$$

Here, however, the indicator functions are specified differently:

$$1_1^i = 1\{U_k \in (0, \mu_1(x_1^i)]\}, \quad 1_2^i := 1\{U_k \in (-\mu_2(x_2^i), 0]\}; \quad i = 1, 2, 4;$$

and

$$1_1^3 = 1\{U_k \in (0, \mu_1(x_1^1)]\}$$
$$+ 1\{U_k \in (\mu_1(x_1^2), \mu_1(x_1^2) + \mu_1(x_1^3) - \mu_1(x_1^1)]\},$$
$$1_2^3 = 1\{U_k \in (-\mu_2(x_2^1), 0]\}$$
$$+ 1\{U_k \in (-\mu_2(x_2^2), -\mu_2(x_2^2) - \mu_2(x_2^3) + \mu_2(x_2^1)]\}.$$

That is, 1_j^i signifies a service completion at node j in network i. Clearly, the above construction yields the correct service completion rates in all four networks. Note, for instance, that in 1_1^3, the two nonoverlapping intervals add up to a total length of $\mu_1(x_1^3)$, the right service rate for node 1 in network 3. (In particular, based on the induction hypothesis and the increasingness of the service rates, we have $\mu_1(x_1^1) \le \mu_1(x_1^2)$ and $\mu_1(x_1^3) \ge \mu_1(x_1^1)$.) The construction for network 3 might appear somewhat unusual; but it is crucial, as will become evident below.

That (3.7) holds at τ_{k+1} is obvious, following the construction. So we focus on (3.6). Due to symmetry, we need to argue only for $j = 1$. Consider two cases:

(i) Suppose $x_1^1 + x_1^4 \le x_1^2 + x_1^3$. This, along with (3.7) and the increasing concavity of μ_1, implies

$$\mu_1(x_1^2) + \mu_1(x_1^3) - \mu_1(x_1^1) \ge \mu_1(x_1^4).$$

Therefore, we have:

(ia) $1_1^4 = 1 \Rightarrow$ either $1_1^2 = 1$ or $1_1^3 = 1$ (or both),

in addition to the more obvious implication

(ib) $1_1^1 = 1 \Rightarrow 1_1^2 = 1_1^3 = 1$.

In words, the construction guarantees, in this case, that a service completion (at node 1) in network 1 implies a service completion in both networks 2 and 3 [(ib)], while a service completion in network 4 implies a service completion in either network 2 or 3, or both [(ia)]. Hence, (3.6) also holds at τ_{k+1}.

(ii) Suppose $x_1^1 + x_1^4 > x_1^2 + x_1^3$. Then, together with the induction hypothesis in (3.6) for $j = 2$, this implies

$$
\begin{aligned}
d_1^2 + d_1^3 &= d_2^2 + d_2^3 - (x_1^2 + x_1^3) + (2n + 2) \\
&> d_2^1 + d_2^4 - (x_1^1 + x_1^4) + (2n + 2) = d_1^1 + d_1^4. \quad (3.8)
\end{aligned}
$$

So, although we do not have (ia) in this case, we still have (ib), which is sufficient, due to the strict inequality in (3.8). In other words, the construction rules out the possibility of having service completions in both networks 1 and 4, without any service completion in either network 2 or network 3. The worst thing that can happen is a service completion in network 4 but no service completion in the other networks. But this poses no harm, due to the strict inequality in (3.8).

The induction is now completed, and hence the proof.   □

It is not difficult to observe that the above proof applies equally well to the convex case, i.e., the service rates are increasing and convex functions. The only change is the conclusion: increasing convexity for the counting processes, and hence the throughput, with respect to the job population.

Corollary 3.20 In the setting of Proposition 3.19, if the service rates at all nodes are increasing and convex functions, then the conclusion there changes to increasing convexity, i.e., the inequality in (3.4) is reversed, while (3.5) stays the same.

Remark 3.21 The linear case corresponds to an infinite-server node (i.e., the number of servers at each node is at least N), with each server having a constant service rate. Since linearity is both convex and concave, Proposition 3.19 and Corollary 3.20 both apply. That is, (3.4) should be satisfied as an equality. This implies that the throughput is also linear in the job population.

To summarize, the main result in this section can be stated as follows:

Theorem 3.22 In a closed Jackson network, if the service rates at all nodes are increasing and concave (respectively increasing and convex) functions, then the throughput is increasing and concave (respectively increasing and convex) in the job population.

From the relation of the throughput function and the equilibrium rate of a convolution of independent random variables [refer to (3.1)], we know that the conclusion of Proposition 3.19 and more generally that of Theorem 3.22 can be interpreted as the preservation of concavity and convexity, in addition to monotonicity, of the equilibrium rates under convolution.

Proposition 3.23 Let $\{Y_1, \ldots, Y_J\}$ be a set of independent random variables, each with an increasing and concave (increasing and convex) equilibrium rate. Then, the equilibrium rate of $Y_1 + \cdots + Y_J$ is also increasing and concave (increasing and convex).

3.3   Multiple Servers

An important special case of an increasing and concave service rate function is that in which a node has multiple, say c, parallel servers. Suppose each server has a constant service rate ν; then $\mu(n) = \min\{n, c\}\nu$ is increasing and concave (in n) (since min is an increasing and concave function). Adapting the proof in the last section, Proposition 3.19 in particular, we can establish a concavity result with respect to the number of parallel servers.

Proposition 3.24 Suppose the four networks in Proposition 3.19 all have the same job population N, and the same service rate at node 2, $\mu_2(\cdot)$, which is an increasing and concave function. They differ at node 1, which has c^i parallel servers, each with a constant service rate $\nu > 0$, for network i, $i = 1, 2, 3, 4$. Suppose $c^1 = c$, $c^2 = c^3 = c + 1$, $c^4 = c + 2$. Then the conclusions in Proposition 3.19, in particular (3.4) and (3.5), still hold. Consequently, the throughput is increasing and concave in the number of servers c (at node 1).

Proof. Follow essentially the proof of Proposition 3.19, with necessary modifications. Keep the induction hypothesis in (3.6), but change the one in (3.7) to

$$x_1^1(k) \geq x_1^2(k) \geq x_1^4(k), \quad x_1^3(k) \geq x_1^4(k), \tag{3.9}$$

which then implies

$$x_2^1(k) \leq x_2^2(k) \leq x_2^4(k), \quad x_2^3(k) \leq x_2^4(k),$$

since all four networks have the same job population. Hence, the construction for node 2 stays the same as in the proof of Proposition 3.19, while the construction for node 1 needs to be modified.

Omit the argument k from $x_j^i(k)$. The service rate at node 1 is now $\mu_1(x_1^i) = \min\{x_1^i, c^i\}\nu$, for $i = 1, 2, 3, 4$. Write these as $r^i := (x_1^i \wedge c^i)\nu$.

Without loss of generality, set $\nu = 1$. First observe that we have

$$r^1 \leq r^2 \quad \text{or} \quad r^4 \leq r^3 \quad \text{or both.} \tag{3.10}$$

To see this, consider two cases: (a) $x_1^4 \geq c + 2$, and (b) $x_1^4 \leq c + 1$. In (a), we have $r^2 = c + 1 > c = r^1$, taking into account (3.9). In (b), we have $r^4 = x_1^4 \leq r^3$, since $x_1^4 \leq x_1^3$ and $r^3 = x_1^3 \wedge (c + 1)$.

Therefore, if both $r^1 \leq r^2$ and $r^4 \leq r^3$, follow the construction (for node 1) in Proposition 3.19, with

$$1_1^i = 1\{U_k \in (0, r^i]\}, \quad i = 1, 2, 3, 4. \tag{3.11}$$

If $r^1 \leq r^2$ but $r^4 > r^3$, there are two subcases: If $r^3 \geq r^1$, then follow 1_1^i above for $i = 1, 3, 4$, while for $i = 2$, set

$$1_1^2 = 1\{U_k \in (0, r^1]\} + 1\{U_k \in (r^3, r^2 + r^3 - r^1]\}.$$

Otherwise, i.e., if $r^3 < r^1$, then follow 1_1^i in (3.11) for $i = 1, 2, 3$ instead, while for $i = 4$, set

$$1_1^4 = 1\{U_k \in (0, r^3]\} + 1\{U_k \in (r^1, r^1 + r^4 - r^3]\}.$$

If $r^4 \leq r^3$ but $r^1 > r^2$, the construction is similar: Replace 1_1^2 and 1_1^4 above by 1_1^3 and 1_1^1, respectively, according to whether $r^2 \geq r^4$ or $r^2 < r^4$.

Regardless, however, we still have, as in the proof of Proposition 3.19,

$$x_1^1 + x_1^4 \leq x_1^2 + x_1^3 \quad \Rightarrow \quad r^1 + r^4 \leq r^2 + r^3. \tag{3.12}$$

The above follows trivially if either $x_1^4 \geq c + 2$ or $x_1^1 \leq c$. So suppose $x_1^4 \leq c + 1 \leq x_1^1$. Then, $r^1 + r^4 = c + x_1^4$, which, under the induction hypothesis in (3.9), is clearly dominated by

$$r^2 + r^3 = [x_1^2 \wedge (c + 1)] + [x_1^3 \wedge (c + 1)]$$
$$= (x_1^2 + x_1^3) \wedge [(x_1^2 \wedge x_1^3) + c + 1] \wedge (2c + 2).$$

Therefore, (3.12) ensures the validity of the argument for case (i) in the proof of Proposition 3.19, while (3.10) supports the argument in case (ii). □

Consider again the two-node cyclic network in Proposition 3.19. Suppose node 1 has c parallel servers, each with a service rate r; and the other node has an increasing and concave service rate function. Following essentially the proof there, we can show that the throughput function is increasing and concave in r (for any given job population N). Let the four networks differ in the service rate (of each server) at node 1, which equals $r^1 = r$ for network 1, $r^2 = r^3 = r + \Delta$ for networks 2 and 3 (where $\Delta > 0$ is a real value), and $r^4 = r + 2\Delta$ for network 4. Keep the induction hypothesis in (3.6), while limit the one in (3.7) to $j = 2$ only (which then implies a relationship, with the inequalities reversed, for $j = 1$, since the four networks now have the same job population). The service rates at node 1 is $\mu_1^i(x_1^i) = (x_1^i \wedge c)r^i$ in network i. The rest of the construction and the proof simply follow those in the proof of Proposition 3.19.

To extend this to a more general closed Jackson network, suppose a node i has c_i servers, each with service rate μ_i, and the visit frequency is v_i. Then, letting $r_i = \mu_i/v_i$ and aggregating the other nodes into a single node, we get back to the two-node network above, and reach the conclusion that the throughput is increasing and concave in r_i. Furthermore, the *reciprocal* of the throughput is decreasing and *convex* in r_i (the reciprocal of a concave function is convex). Note that the reciprocal throughput, when multiplied by the total job population N, yields the average response time (i.e., sojourn time in the network). Since following Proposition 3.24, the throughput function is concave in the number of servers at a multiple-server node, say c_i, the average job delay is also decreasing and convex in c_i. To summarize, we have the following theorem.

Theorem 3.25 In a closed Jackson network, suppose a node i has c_i servers, each with service rate μ_i, and the visit frequency is v_i. Set $r_i = \mu_i/v_i$. Suppose the service rates at the other nodes are increasing and concave functions. Then, for each N,

(i) the throughput function $TH(N, c_i, r_i)$ is increasing and concave in c_i and in r_i;

(ii) the reciprocal throughput $TH^{-1}(N, c_i, r_i)$, and hence the average response time $N/TH(N, c_i, r_i)$, is decreasing and convex in c_i, and in r_i.

Write $\rho_i = v_i/\mu_i = 1/r_i$, and $\rho := (\rho_i)_{i=1}^J$. When all nodes have single servers, the above can be strengthened to a *joint* convexity with respect to ρ. To see this, first recall that the network can be equivalently viewed as having a cyclic configuration, with each node i having a service rate $r_i = 1/\rho_i$. Let $T_i(n)$ be the time epoch of the nth service completion at node i. Let $\{U_i(n) \in [0,1]; i = 1, \ldots, J; n = 1, 2, \ldots\}$ be a sequence of i.i.d. uniform random variables. Then the service time of job n at node i can be expressed as $\rho_i[-\ln U_i(n)]$. Assume initially that all N jobs are in front of node 1. We have the following recursion:

$$T_i(n) = \rho_i[-\ln U_i(n)] + \max\{T_{i-1}(n), T_i(n-1)\}, \qquad (3.13)$$

for all n and all i, with the understanding that $T_i(n) \equiv 0$ if $n \leq 0$, and $T_0(n) = T_J(n - N)$ (i.e., both node index i and job index n are cyclic, with modulus J and N, respectively). Since max is increasing and convex, the above recursion preserves increasingness and convexity, with respect to $\rho = (\rho_i)_{i=1}^J$. Let $D_i(t)$ denote the cumulative number of service completions from node i up to t. Then,

$$\lim_{n\to\infty} \frac{T_i(n)}{n} = \lim_{t\to\infty} \frac{T_i(D_i(t))}{D_i(t)} = \lim_{t\to\infty} \frac{t}{D_i(t)} = TH^{-1}(N,\rho).$$

(The existence of the limits above can be rigorously justified; see the Notes section.)

Theorem 3.26 In a closed Jackson network where each node i has a single server with fixed service rate μ_i and visit frequency v_i, the reciprocal throughput function $TH^{-1}(N,\rho)$, and hence the average job delay, are increasing and convex in $\rho = (\rho_i)_{i=1}^{J}$, where $\rho_i = v_i/\mu_i$.

3.4   Resource Sharing

3.4.1   Aggregation of Servers

Recall Example 1.16, where we showed that reducing the number of servers in an $M/M/c/N$ queue while maintaining the overall service capacity (i.e., maximum output rate) will reduce the number of jobs in the system in the sense of likelihood ratio ordering, as well as increase the throughput (Exercise 11 of Chapter 1). Here we show that similar results also hold in a network setting. Although the server aggregation in Example 1.16 is a special case of increasing the service rate function (in the pointwise sense), and hence belongs to the context of Theorem 3.8, more results can be obtained in this special case.

Proposition 3.27 Consider a two-node cyclic network with N jobs: Node 1 has c parallel servers, each with a constant service rate $1/c$; node 2 has an increasing service rate function $\mu_2(\cdot)$. Index this network by superscript 0. Now, replace the c servers at node 1 by c' ($c' < c \le N$) servers, each with a constant service rate $1/c'$. Index this network by superscript 1. Further reduce the job population of network 1 to $N - (c - c')$, and index the network by superscript 2. Let X_j^i be the number of jobs at node j in network i; $j = 1, 2$; $i = 0, 1, 2$. Then,

$$X_1^0 \ge_{\ell r} X_1^1, \quad X_2^0 \le_{\ell r} X_2^1; \quad X_j^0 \ge_{\ell r} X_j^2, \quad j = 1, 2.$$

Furthermore, let TH^i, $i = 0, 1, 2$, denote the throughput of the three networks. Then,

$$TH^2 \le TH^0 \le TH^1.$$

Proof. Without loss of generality, assume $c' = c - 1$. Let Y_1 and Y_2 denote two random variables with equilibrium rates

$$r_{Y_1}(n) = \min\{n,c\}/c \quad \text{and} \quad r_{Y_2}(n) = \min\{n, c-1\}/(c-1),$$

respectively. Then, clearly,

$$r_{Y_2}(n-1) \le r_{Y_1}(n) \le r_{Y_2}(n).$$

Let Z denote a third random variable, independent of Y_1 and Y_2, with equilibrium rate $r_Z(n) = \mu_2(n)$. Then, similar to the proof of Proposition 3.5, we have

$$r_{X_1^0}(n) = r_{Y_1}(n)/r_Z(N+1-n)$$
$$\le r_{Y_2}(n)/r_Z(N+1-n) = r_{X_1^1}(n)$$

for all n. Hence, $X_1^0 \ge_{\ell r} X_1^1$, which also implies $X_2^0 \le_{\ell r} X_2^1$. On the other hand,

$$r_{X_1^0}(n) \le r_{Y_1}(n)/r_Z(N-n) = r_{X_1^2}(n),$$

since r_Z is an increasing function; and

$$r_{X_2^0}(n) = r_Z(n)/r_{Y_1}(N+1-n)$$
$$\le r_Z(n)/r_{Y_2}(N-n) = r_{X_2^2}(n).$$

(When $n = N$, define $r_{X_j^2}(N) = \infty$, for $j = 1, 2$.) Therefore, $X_j^0 \ge_{\ell r} X_j^2$ for $j = 1, 2$.

The throughput relations now follow immediately, since the throughput for network i is equal to $\mathrm{E} r_Z(X_2^i)$ for $i = 0, 1, 2$, and r_Z is an increasing function. □

Corollary 3.28 For any multiserver node in the B part of the network considered in Theorem 3.8, reducing the number of servers while maintaining the maximal service capacity will lead to the same conclusions. Furthermore, if the the total number of jobs in the network is also reduced by the same number (as the number of reduced servers), then the throughput of the network will decrease, and so will the number of jobs in the node, in the sense of likelihood ratio ordering.

3.4.2 Aggregation of Nodes

Consider again a closed, cyclic network with two nodes and N jobs. There are $c_j \ge 1$ parallel servers at node j, each with a service rate μ_j, $j = 1, 2$. Suppose we now formulate a second network by aggregating the two nodes into a single node, with $c := c_1 + c_2$ parallel servers. The total number of jobs is still N. Each job still requires two stages of service as before; specifically, the service times are exponentially distributed with mean $1/\mu_j$ at stage j, for $j = 1, 2$. But now the service of both stages of any job can be handled by any one of the c servers.

Note that in the second network, although there is only one node, there are still two stages of service for each of the N jobs. In particular, each job goes through the two stages of service in sequence, stage 1 followed by stage 2 followed by stage 1, and so on; and at any time each job is either in stage 1 (waiting or being processed for stage 1 service) or in stage 2. Indeed, we can think of the N jobs as formulating two queues, one for each stage, while each of the c servers can take jobs from both queues.

Use the superscript i ($= 1, 2$) to index the two networks. Let $D_j^i(t)$ be the cumulative number of service completions up to time t from stage j in network i. We want to show that

$$\{(D_1^1(t), D_2^1(t)\} \leq_{st} \{(D_1^2(t), D_2^2(t)\}. \tag{3.14}$$

Again, we use coupling. Let the uniformization constant be $\eta := c\mu := (c_1+c_2)(\mu_1 \vee \mu_2)$. Let $\{\tau_k\}$ be the sequence of Poisson event epochs with rate η; let $\{U_k\}$ be a sequence of i.i.d. uniform random variables on $[-c\mu, c\mu]$, independent of $\{\tau_k\}$.

For $i, j = 1, 2$, let $d_j^i(k)$ denote the sample values of the counting processes at τ_k generated by the construction; let $x_j^i(k)$ denote the number of stage j jobs in network i at τ_k. Initially, for $i = 1, 2$, set $x_1^i(0) = N$, and $x_2^i(0) = 0$; set $d_j^i(0) = 0$ for $j = 1, 2$.

As induction hypothesis, suppose

$$(d_1^1(k), d_2^1(k)) \leq (d_1^2(k), d_2^2(k)). \tag{3.15}$$

We want to show that the above also holds at τ_{k+1}.

To specify the construction in network 2, we assume the following mode of operation: Among the c servers, at any time there will be no more than c_1 servers working on stage 1 jobs unless the number of stage 2 jobs is less than c_2; and similarly, at any time there will be no more than c_2 servers working on stage 2 jobs unless the number of stage 1 jobs is less than c_1. Hence, for instance, initially all c servers start with stage 1 jobs (suppose $N > c$). As some of the servers complete their service, they may continue, on the same job, into stage 2. But once the number of servers working on stage 2 jobs reaches c_2, any server that completes a stage 1 job will have to switch to another stage 1 job.

Due to work conservation, the operation of network 2 assumed above will deplete work at the same rate as any other work-conserving schemes (i.e., no server will be idle as long as there is a job to be processed), and hence the same throughput (i.e., the same rate of depleting $jobs$) as well. Consider, for instance, another mode of operation: Every server works on the two stages of service of each job in sequence; after completing both stages consecutively at a server, a job will join the end of the queue and the server will pick up another job from the front of the queue. Let $D^2(t)$ be the number of jobs completed up to time t, under this mode of operation (in network 2). For $j = 1, 2$, let S_j denote the generic exponential service time at stage j; let $S = S_1 + S_2$. Then, the long-run average rate of work depletion is

$$\lim_{t \to \infty} \left[S(1) + \cdots + S(D^2(t)) \right] / t$$
$$= \lim_{t \to \infty} \left[S(1) + \cdots + S(D^2(t)) \right] / D^2(t) \cdot \lim_{t \to \infty} D^2(t)/t$$
$$= \left(\frac{1}{\mu_1} + \frac{1}{\mu_2} \right) \cdot TH,$$

where $S(i)$ are i.i.d. replicas of S, and TH denotes the throughput.

On the other hand, the long-run average rate of work depletion under the operation mode specified earlier can be similarly derived as follows:

$$\lim_{t \to \infty} [S_1(1) + \cdots + S_1(D_1^2(t)) + S_2(1) + \cdots + S_2(D_2^2(t))]/t$$
$$= \frac{1}{\mu_1} \lim_{t \to \infty} D_1^2(t)/t + \frac{1}{\mu_2} \lim_{t \to \infty} D_1^2(t)/t$$
$$= \left(\frac{1}{\mu_1} + \frac{1}{\mu_2} \right) \cdot TH',$$

where

$$TH' = \lim_{t \to \infty} D_1^2(t)/t = \lim_{t \to \infty} D_2^2(t)/t.$$

Since the two work-depletion rates are the same, we have $TH' = TH$.

Therefore, we specify the construction in network 2, assuming the mode of operation specified earlier, and verify the induction hypothesis (3.15) at τ_{k+1}, and thereby establish the desired relation in (3.14).

Again, dropping the argument k from $x_j^i(k)$ and $d_j^i(k)$, we set, for $i = 1, 2$,

$$x_1^i(k+1) = x_1^i - 1_1^i + 1_2^i, \quad x_2^i(k+1) = x_2^i - 1_2^i + 1_1^i;$$

and

$$d_j^i(k+1) = d_j^i + 1_j^i, \quad j = 1, 2;$$

where

$$1_1^1 = 1\{U_k \in (-(x_1^1 \wedge c_1)\mu_1, 0]\},$$
$$1_2^1 = 1\{U_k \in (0, (x_2^1 \wedge c_2)\mu_2]\};$$
$$1_1^2 = 1\{U_k \in (-(x_1^2 \wedge [c_1 + (c_2 - x_2^2)^+])\mu_1, 0]\},$$
$$1_2^2 = 1\{U_k \in (0, (x_2^2 \wedge [c_2 + (c_1 - x_1^2)^+])\mu_2]\}.$$

Note that in network 2, the numbers of jobs in the two stages of service are, respectively, $x_1^2 \wedge [c_1 + (c_2 - x_2^2)^+]$ and $x_2^2 \wedge [c_2 + (c_1 - x_1^2)^+]$.

Consider two cases: If $x_1^2 \geq x_1^1$, then $1_1^2 \geq 1_1^1$, and hence $d_1^2(k+1) \geq d_1^1(k+1)$. On the other hand, if $x_1^2 < x_1^1$, then

$$d_1^2 = d_2^2 + N - x_1^2 > d_2^1 + N - x_1^1 = d_1^1,$$

which also guarantees $d_1^2(k+1) \geq d_1^1(k+1)$, no matter what transition takes place at τ_{k+1}. The proof of $d_2^2(k+1) \geq d_2^1(k+1)$ follows similarly.

It is easy to see that if we add a third common node into the two networks, the above proof can be slightly modified (adding $x_3^2(k) \geq x_3^1(k)$ to the induction hypothesis) to reach the same conclusion. The third node can have any kind of service rate function, not necessarily monotone. This third node can represent the aggregation of the other parts of a more general network.

Finally, suppose for $j = 1, 2$, node j has a visit frequency v_j and c_j servers, each with service rate μ_j. We can equivalently view these two nodes as having the same, unit, visit frequency, but with modified service rates μ_1/v_1 and μ_2/v_2. Aggregating the two nodes then yields one with $c_1 + c_2$ servers, and jobs with mean service times v_1/μ_1 and v_2/μ_2 for the two stages of service. The cyclic network result above then applies.

To summarize, we have the following result.

Theorem 3.29 In a closed Jackson network, aggregating a set, say B, of multiple-server nodes into a single node will increase the overall network throughput. Furthermore, it will reduce the total number of jobs in B, in the sense of likelihood ratio ordering. In an open Jackson network, the aggregation will reduce the total number of jobs in B, and hence in the network, in the sense of likelihood ratio ordering.

3.5 Arrangement and Majorization

Definition 3.30 Let $x = (x_i)_{i=1}^n$ and $y = (y_i)_{i=1}^n$ be two n-vectors of real-valued components. The arrangement ordering, denoted by $x \geq_A y$, is defined as follows: x can be obtained from y through successive pairwise interchanges of its components, with each interchange resulting in a decreasing order of the two interchanged components. A function $\psi : \Re^n \mapsto \Re$ that preserves the arrangement ordering is termed an "arrangement increasing function"; that is, ψ is arrangement increasing if $\psi(x) \geq \psi(y)$ for any $x \geq_A y$.

For instance, $(4, 5, 3, 1) \geq_A (4, 3, 5, 1) \geq_A (4, 1, 5, 3)$. Note that $x \geq_A y$ implies that x is necessarily a permutation of the components of y.

Definition 3.31 Two n-vectors x and y are ordered under majorization, denoted by $x \leq_m y$, if

$$\sum_{j=1}^i x_{[j]} \leq \sum_{j=1}^i y_{[j]}, \ i = 1, \ldots, n-1; \quad \sum_{j=1}^n x_{[j]} = \sum_{j=1}^n y_{[j]};$$

where $x_{[j]}$ and $y_{[j]}$ denote the jth-largest components of x and y, respectively. Replacing the last equality above by $\leq$ results in "weak majorization," denoted by $x \leq_{\text{wm}} y$. A function ψ is termed Schur convex if for any x and y, $x \leq_m y \Rightarrow \psi(x) \leq \psi(y)$. In addition, $x \leq_{\text{wm}} y$ if and only if $\psi(x) \leq \psi(y)$ for any ψ that is increasing and Schur convex. Finally, a function ψ is Schur concave if $-\psi$ is Schur convex.

Remark 3.32 (i) From the definition of majorization, we have $x \leq_m x^\pi$ as well as $x \geq_m x^\pi$ for any permutation π. Hence, if a Schur convex/concave function ψ has a symmetric domain (i.e., x belongs to the domain if and only if x^π belongs to the domain), then ψ is necessarily symmetric, i.e., $\psi(x) = \psi(x^\pi)$ for any permutation π. Below, we shall assume that this is indeed the case.

(ii) Suppose a pair of components of x is ordered: $x_i > x_j$. Then obviously, $x \geq_m x - \epsilon e^i + \epsilon e^j$, where e^i and e^j denote the ith and the jth unit vectors, and $0 < \epsilon < x_i - x_j$. In other words, x can be made smaller, in the sense of majorization, by shifting a small positive amount (ϵ) from one of its larger component (x_i) to a smaller component (x_j) (ϵ should be small enough that after the shift the larger component is still larger). Call such an operation on the vector x "transposition." Then clearly, $x \leq_m y$ if and only if x can be reached from y through a sequence of such transpositions, each resulting in a vector that is majorized by the vector from the previous transposition.

(iii) There are two implications from the above. One is that to establish the Schur convexity of a function, we need only to establish the dominance relation between the function values at x and its transposition. In other words, Schur convexity is essentially a property with respect to each *pair* of components. The other implication is that for a differentiable function ψ, Schur convexity is equivalent to

$$x_i > x_j \quad \Rightarrow \quad \psi_i(x) \geq \psi_j(x),$$

for any $i \neq j$ and any x, where ψ_i and ψ_j denote the partial derivatives with respect to x_i and x_j.

To motivate the results below, let $(v_i)_{i=1}^J$ be the visit ratios in a closed Jackson network, with $\sum_{i=1}^J v_i = 1$. Suppose each job carries one unit of work, which can be allocated (shared) among the J servers (nodes). Hence, we can regard $(v_i)_{i=1}^J$ as one such allocation, under which node i receives an exponentially distributed amount of work, with mean v_i (which is processed at rate $\mu_i(\cdot)$ as specified before). Below we address the issue of allocating the workload among the nodes, in terms of the visit ratios. Note that since the visit ratios are determined by the routing matrix, via the traffic equation, changing the visit ratios amounts to changing the routing matrix. However, since the product-form equilibrium distribution depends on the routing matrix only through the visit ratios, here we choose to focus on the visit ratios as the starting point.

Lemma 3.33 Consider a two-node closed Jackson network in equilibrium, with N jobs. For $j = 1, 2$, let $\mu_j(\cdot)$ and v_j be, respectively, the service rate function and the visit ratio at node j. Suppose $v_1 \geq v_2$. Let X_1 and $N - X_1$ be the number of jobs at node 1 and node 2, respectively. In addition, consider a second two-node network (indexed by a superscript "$\prime$"), which differs from the first one only in visit ratios: $v_1' = v_1 - \Delta$ and $v_2' = v_2 + \Delta$, where $0 < \Delta \leq v_1 - v_2$.

(i) Suppose $\mu_1(n) \geq \mu_2(n)$ for all n, and let $\Delta = v_1 - v_2$ (i.e., the second network is obtained through interchanging the visit ratios at the two nodes in the first network). Then,

$$\min(X_1, N - X_1) \geq_{st} \min(X_1', N - X_1').$$

(ii) Suppose $\mu_1(n) = \mu_2(n)$ for all n. Then,

$$\min(X_1, N - X_1) \leq_{st} \min(X_1', N - X_1').$$

Proof. For (i), we want to show that

$$P(X_1 \leq n) + P(X_1 \geq N - n) \leq P(X_1' \leq n) + P(X_1' \geq N - n) \quad (3.16)$$

for all $n \leq \lfloor \frac{N}{2} \rfloor$, where $\lfloor \frac{N}{2} \rfloor$ denotes the integer part of $N/2$. Write $\rho = v_1/v_2 \geq 1$, and for $j = 1, 2$, $M_j(0) = 1$, $M_j(n) = \mu_j(1) \cdots \mu_j(n)$ for $n \geq 1$. The above can be written explicitly as

$$\frac{\sum_{k=0}^{n} \frac{\rho^k}{M_1(k)M_2(N-k)} + \frac{\rho^{N-k}}{M_1(N-k)M_2(k)}}{\sum_{k=0}^{\lfloor \frac{N}{2} \rfloor} \frac{\rho^k}{M_1(k)M_2(N-k)} + \frac{\rho^{N-k}}{M_1(N-k)M_2(k)}}$$

$$\leq \frac{\sum_{k=0}^{n} \frac{\rho^{N-k}}{M_1(k)M_2(N-k)} + \frac{\rho^k}{M_1(N-k)M_2(k)}}{\sum_{k=0}^{\lfloor \frac{N}{2} \rfloor} \frac{\rho^{N-k}}{M_1(k)M_2(N-k)} + \frac{\rho^k}{M_1(N-k)M_2(k)}},$$

which can be directly verified, taking into account $\mu_1(n) \geq \mu_2(n)$.

To prove (ii), write $M(n) = M_1(n) = M_2(n)$, since $\mu_1(n) = \mu_2(n)$ for all n. We want to establish (3.16) with the inequality reversed. In place of (3.5), it suffices to show that

$$\frac{\sum_{k=0}^{n} \frac{\rho^k + \rho^{N-k}}{M(k)M(N-k)}}{\sum_{k=0}^{\lfloor \frac{N}{2} \rfloor} \frac{\rho^k + \rho^{N-k}}{M(k)M(N-k)}} \quad (3.17)$$

is decreasing in ρ, since $\rho = v_1/v_2 \geq (v_1 - \Delta)/(v_2 + \Delta)$. Again, this can be directly verified. $\square$

Theorem 3.34 Consider the two networks in Lemma 3.33. Suppose $v_1 \geq v_2$.

(i) If in addition to the conditions in Lemma 3.33(i), $\mu_2(n)$ is concave in n, and $\mu_1(n) - \mu_2(n)$ is increasing in n, then $TH(N) \geq TH'(N)$ for any given N.

(ii) If in addition to the conditions in Lemma 3.33(ii), $\mu_1(n) = \mu_2(n)$ is concave in n, then $TH(N) \leq TH'(N)$ for any given N.

Proof. To show (i), following (2.11), we have

$$TH(N) = \mathsf{E}\mu_1(X_1)/v_1 = \mathsf{E}\mu_2(N - X_1)/v_2,$$

and hence, assuming $v_1 + v_2 = 1$ without loss of generality,

$$
\begin{aligned}
TH(N) &= \mathsf{E}[\mu_1(X_1) + \mu_2(N - X_1)] \\
&= \mathsf{E}[\mu_1(X_1) - \mu_2(X_1) + \mu_2(X_1) + \mu_2(N - X_1)]. \quad (3.18)
\end{aligned}
$$

Replacing X_1 by X_1' yields a similar expression for $TH'(N)$.

Now, a larger visit ratio is equivalent to a smaller service rate, hence $v_1 \geq v_2$ implies $X_1 \geq_{\ell r} X_1'$, and hence $X_1 \geq_{st} X_1'$. Since $\mu_1(n) - \mu_2(n)$ is increasing in n as assumed, we have

$$\mu_1(X_1) - \mu_2(X_1) \geq_{st} \mu_1(X_1') - \mu_2(X_1'). \quad (3.19)$$

On the other hand, since $\mu_2(n)$ is concave in n, and

$$\min(X_1, N - X_1) \geq_{st} \min(X_1', N - X_1')$$

(from Lemma 3.33(i)), we have

$$\mu_2(X_1) + \mu_2(N - X_1) \geq_{st} \mu_2(X_1') + \mu_2(N - X_1'), \quad (3.20)$$

following a simple coupling argument.

The desired result $TH(N) \geq TH'(N)$ then follows from summing up (3.19) and (3.20), taking expectations on both sides, and comparing against (3.18).

To prove (ii), (3.19) becomes trivial (both sides are zero), and (3.20), with the inequality reversed, follows from Lemma 3.33(ii). $\square$

Note that the required conditions in Theorem 3.34 are easily satisfied by the case of multiple parallel servers. Let c_1 and c_2 be the number of servers at the two nodes. Suppose $c_1 \geq c_2$, and suppose all servers have the same service rate ν. Then, in addition to the increasing concavity of both $\mu_1(n)$ and $\mu_2(n)$, we also have

$$\mu_1(n) - \mu_2(n) = \nu[\min\{n, c_1\} - c_2]^+$$

increasing in n.

In Theorem 3.34(i) (as well as Lemma 3.33), the second network is obtained through interchanging the workload assignment of the first network (i.e., v_2 to node 1 and v_1 to node 2). Since $\mu_1(n) \geq \mu_2(n)$ for all n, the result simply states that the throughput is increased by assigning a higher workload to a faster node (or, in the multiserver case, a node with more servers, provided that all servers have the same service rate). Following Definition 3.30, when $v_1 \geq v_2$, we can write

$$v = (v_1, v_2) \geq_A v' = (v_2, v_1).$$

Then the result in Theorem 3.34(i) says that if the two nodes in the closed network are numbered in decreasing order of their service rates, i.e., $\mu_1(n) \geq \mu_2(n)$ for all n, then $v = (v_1, v_2) \geq_A v' = (v_2, v_1)$ implies $TH(v) \geq TH(v')$. In other words, the throughput function is an arrangement increasing function.

Furthermore, when $v_1 > v_2$ and $\Delta < v_1 - v_2$, we can write

$$v = (v_1, v_2) \geq_m v' = (v_1 - \Delta, v_2 + \Delta).$$

Since $v \geq_m v'$ implies $TH(v) \leq TH(v')$ following Theorem 3.34(ii), the throughput is a Schur concave function of the visit ratio v.

Note that both arrangement and majorization orderings are pairwise orderings; that is, $x \geq_A y$ if and only if there exists a sequence of vectors $z_1, \ldots, z_k$ such that

$$x \geq_A z_1 \geq_A \cdots \geq_A z_k \geq_A y,$$

and each pair of neighboring vectors above differ only in two components; and the same holds for $x \geq_m y$. Hence, the results in Theorem 3.34 extend immediately to a J-node network as follows. Repeatedly apply the theorem, each time to one pair of nodes in the network (while holding the other nodes fixed): Switch the visit ratios so that the faster node is visited more often, or make the visit ratios closer to each other through the transposition operation. Either way will result in a higher overall throughput from the two nodes (since the theorem holds for any N), and hence a higher throughput of the entire network (provided that the service rates are increasing functions).

Theorem 3.35 In a closed Jackson network with J nodes and N jobs, where the service rates at all nodes are increasing and concave functions,

(i) if the nodes are ordered such that $\mu_i(n) - \mu_{i+1}(n)$ is increasing in n, for all $i = 1, \ldots, J-1$, then the throughput is arrangement increasing in the visit ratios $(v_i)_{i=1}^J$;

(ii) if all nodes have the same service rate function, i.e., $\mu_i(n) = \mu(n)$ for all i and all n, then the throughput is Schur concave in $(v_i)_{i=1}^J$.

In the special case of multiple parallel servers, we can allow the servers at different nodes to have different service rates, say ν_i for node i, by treating

$\rho_i = v_i/\nu_i$ as v_i. The arrangement and majorization results then apply to ρ_i.

Corollary 3.36 In the network of Theorem 3.35, suppose node i has c_i parallel servers, each with rate ν_i. Let $\rho_i = v_i/\nu_i$ for all $i = 1, \ldots, J$.

(i) If the nodes are ordered such that $c_1 \geq \cdots \geq c_J$, then the throughput is arrangement increasing in $(\rho_i)_{i=1}^J$.

(ii) If all nodes have the same number of servers (i.e., $c_i = c$ for all i), then the throughput is Schur concave in $(\rho_i)_{i=1}^J$. In particular, given $\sum_{j=1}^J \rho_j = w$, a constant, the balanced loading, i.e., $\rho_i = w/J$ for all i, yields the highest throughput.

3.6 Notes

In Section 3.1 the notion of equilibrium rate and its role in linking together the throughput function of a closed Jackson network and the likelihood ratio ordering were first developed in Shanthikumar and Yao [16, 17]. The motivation there was to address the issue of whether speeding up the servers at one or more nodes in a closed Jackson network will yield an increased throughput. Based on the equilibrium rate, the issue can be equivalently stated as the preservation of the likelihood ratio ordering under convolution, which, for PF_2 distributions, is a standard result; see, e.g., Keilson and Sumita [9]. The extension of this to *shifted* likelihood ratio ordering (Section 3.1.3), without the PF_2 property, is due to Shanthikumar and Yao [16]. In both cases, the results have made it precise how to speed up servers in a closed network so as to increase the overall throughput.

For a long period of time it was widely believed that the throughput function of a closed Jackson network is concave in the job population; but proofs, based on rather tedious algebra, were available only for special cases, e.g., single-server nodes with constant service rates. With the machinery of equilibrium rate, Shanthikumar and Yao transformed this into comparisons of the service completion processes in two-node cyclic networks. The proof was then based on coupling arguments. The general result was first established in [20], upon which much of Section 3.2 is based.

That the PF_2 property is preserved under convolution is a classical result; see Karlin and Proschan [8]. As pointed out at the end of Section 3.2, the increasing concavity/convexity of the throughput function, via equilibrium rates, in fact has extended this classical result to the preservation of second-order properties (i.e., concavity and convexity), in addition to the first-order (monotonicity) property PF_2.

The coupling arguments in the proof of Proposition 3.19, as well as in similar proofs of other results in Section 3.2 and Section 3.3, are based on the basic idea of constructing sample paths of stochastic processes under comparison on a common probability space, and demonstrating that they possess certain desired relations. Refer to Kamae, Krengel, and O'Brien [7]. Note, however, that the concavity result in Proposition 3.19 requires the construction and comparison of *four* processes (as opposed to the usual three points that define concavity of a function: two end points and their convex combination). This type of concavity falls into the framework of a notion of *stochastic* convexity/concavity developed in Shaked and Shanthikumar [12, 13]. In contrast, the convexity through the recursion in (3.13) relates more directly to the usual functional (deterministic) convexity. It corresponds to a stronger notion of stochastic convexity developed in Shanthikumar and Yao [22]. This type of stochastic convexity also holds for other quequeing networks, e.g., those with kanban control; refer to Glasserman and Yao [3], and [4] (Chapter 5).

The result regarding the aggregation of servers in Section 3.4.1 first appeared in [17] (where it was called "server reduction"). One application of the bounds in Proposition 3.27 is to bound the throughput of a closed network with multiserver nodes by the throughput of one with single-server nodes (see Shanthikumar and Yao [21]), since the computation of the latter is much simpler; refer to the algorithms in Section 2.5 The result in Section 3.4.2 concerning aggregation of nodes is new.

On at least two occasions (in Section 3.3 and Section 3.4) we expressed the throughput as the limit of the service-completion (counting) processes over time. The existence of the limit was implicitly assumed. This can be rigorously justified based on the subadditive ergodic theory; refer to Glasserman and Yao [5]; also see [4] (Chapter 7).

The majorization and arrangement results in Section 3.5 are motivated by the allocation or assignment of workload in manufacturing systems. Although it was known that balanced loading in a closed Jackson network with single-server nodes (constant service rate) maximizes the throughput, Yao [24] was the first to establish the Schur concavity of the throughput function and related the workload allocation and assignment problems to the majorization and arrangement orderings. Similar results in related or more general settings appeared in Shanthikumar [14], Shanthikumar and Stecke [15], Yao [25], and Yao and Kim [26]. Other related results in the optimal allocation of servers in manufacturing systems appeared in Shanthikumar and Yao [18, 19].

Marshall and Olkin [10] is a standard reference for majorization and related topics. Refer to [1, 2] for a treatment that is more formal and extensive than in this chapter, on stochastic majorization and arrangement orders, along with stochastic convexity.

3.7 Exercises

1. Show that the stochastic ordering is preserved under convolution, i.e., if $X_i \geq_{st} Y_i$ for $i = 1, 2$, with X_1 (Y_1) independent of X_2 (Y_2), then $X_1 + X_2 \geq_{st} Y_1 + Y_2$.

2. Construct a counterexample to illustrate that if the stochastic ordering in the above problem is changed to the likelihood ratio ordering, the result no longer holds.

3. Verify the properties in Lemma 3.11 and Proposition 3.13.

4. Let $\{X(t), t \geq 0\}$ be a pure death process, with death rate $\mu(n) > 0$ when $X(t) = n > 0$ and $\mu(0) = 0$. Suppose $X(0) = x > 0$, and we want to show that the process is increasing in the initial state x. Do we need to assume any monotonicity condition on the rates $\mu(n)$?

5. Suppose we want to show that the pure death process in the above problem is increasing and concave in the initial state x. Spell out any condition that is required of the rates $\mu(n)$, and give the full proof.

6. Directly verify Remark 3.21, i.e., derive the throughput function of a closed Jackson network with each node having an infinite number of parallel servers, and verify that the throughput is linear in N, the total number of jobs in the network.

7. Consider the multiple-sever (c), finite-buffer (N) queue $M/M/c/N$, with arrival and service rates denoted by λ and μ, and $\rho := \lambda/\mu$. Adapt Theorem 3.25 to derive the properties of the throughput from this queue with respect to N, c, and ρ.

8. For the $M/M/c/N$ queue above, make use of Proposition 3.27 to derive an upper bound and a lower bound on the throughput, with either bound expressed in terms of the throughput of a single-server queue.

9. Verify the inequality in (3.5). Verify that the expression in (3.17) is decreasing in ρ.

10. Prove that a symmetric and convex function $f(x_1, \ldots, x_n)$ is Schur convex, and hence in particular that max is a Schur convex function. Illustrate by a counterexample that if f is symmetric but only convex in each variable, then f need not be Schur convex.

11. Prove that if f is symmetric, submodular, and convex in each variable, then f is Schur convex. f being submodular means that

$$f(\ldots, x_i + \delta_i, \ldots, x_j + \delta_j, \ldots) + f(\ldots, x_i, \ldots, x_j, \ldots)$$
$$\leq f(\ldots, x_i + \delta_i, \ldots, x_j, \ldots) + f(\ldots, x_i, \ldots, x_j + \delta_j, \ldots),$$

for any $\delta_i, \delta_j \geq 0$.

12. Prove that increasing convexity is preserved under composition, i.e., $h(f(x_1, \ldots, x_n))$ is an increasing and convex function if both $h(x)$ and $f(x_1, \ldots, x_n)$ are increasing and convex functions.

13. Prove that increasing convexity is preserved under composition, i.e., $h(f(x_1, \ldots, x_n))$ is an increasing and convex function if both $h(x)$ and $f(x_1, \ldots, x_n)$ are increasing and convex functions.

References

[1] Chang, C.S. and Yao, D.D. 1993. Rearrangement, Majorization, and Stochastic Scheduling. *Mathematics of Operations Research*, **18**, 658–684.

[2] Chang, C.S., Shanthikumar, J.G., and Yao, D.D. 1994. Stochastic Convexity and Stochastic Majorization. In: *Stochastic Modeling and Analysis of Manufacturing Systems*, D.D. Yao (ed.), Springer-Verlag, New York, 189–231.

[3] Glasserman, P. and Yao, D.D. 1994. A GSMP Framework for the Analysis of Production Lines. In: *Stochastic Modeling and Analysis of Manufacturing Systems*, D.D. Yao (ed.), Springer-Verlag, New York, 1994, 131–186.

[4] Glasserman, P. and Yao, D.D. 1994. *Monotone Structure in Discrete-Event Systems*. Wiley, New York.

[5] Glasserman, P. and Yao, D.D. 1995. Subadditivity and Stability of a Class of Discrete-Event Systems. *IEEE Transactions on Automatic Control*, **40**, 1514–1527.

[6] Gordon, W.J. and Newell, G. F. 1967. Closed Queueing Networks with Exponential Servers. *Operations Research*, **15**, 252–267.

[7] Kamae, T. Krengel, U., and O'Brien, G.L. 1977. Stochastic Inequalities on Partially Ordered Spaces. *Annals of Probability*, **5**, 899–912.

[8] Karlin, S. and Proschan, F. 1960. Polya-Type Distributions of Convolutions. *Ann. Math. Stat.*, **31**, 721–736.

[9] Keilson, J. and Sumita, U. 1982. Uniform Stochastic Ordering and Related Inequalities. *Canadian Journal of Statistics*, **10**, 181–198.

[10] Marshall, A.W. and Olkin, I. 1979. *Inequalities: Theory of Majorization and Its Applications*. Academic Press, New York.

[11] Ross, S.M. 1983. *Stochastic Processes*. Wiley, New York.

[12] Shaked, M. and Shanthikumar, J.G. 1988. Stochastic Convexity and Its Applications. *Advances in Applied Probability*, **20**, 427–446.

[13] Shaked, M. and Shanthikumar, J.G. 1990. Convexity of a Set of Stochastically Ordered Random Variables. *Advances in Applied Probability*, **22**, 160–167.

[14] Shanthikumar, J.G. 1987. Stochastic Majorization of Random Variables with Proportional Equilibrium Rates. *Advances in Applied Probability*, **19**, 854–872.

[15] Shanthikumar, J.G. and Stecke, K.E. 1986. Reducing the Work-in-Process Inventory in Certain Class of Flexible Manufacturing Systems. *European J. of Operational Res.*, **26**, 266–271.

[16] Shanthikumar, J.G. and Yao, D.D. 1986. The Preservation of Likelihood Ratio Ordering under Convolution. *Stochastic Processes and Their Applications*, **23**, 259–267.

[17] Shanthikumar, J.G. and Yao, D.D. 1986. The Effect of Increasing Service Rates in Closed Queueing Networks. *Journal of Applied Probability*, **23**, 474–483.

[18] Shanthikumar, J.G. and Yao, D.D. 1987. Optimal Server Allocation in a System of Multiserver Stations. *Management Science*, **34**, 1173–1180.

[19] Shanthikumar, J.G. and Yao, D.D. 1988. On Server Allocation in Multiple-Center Manufacturing Systems. *Operations Research*, **36**, 333–342.

[20] Shanthikumar, J.G. and Yao, D.D. 1988. Second-Order Properties of the Throughput in a Closed Queueing Network. *Mathematics of Operations Research*, **13**, 524–534.

[21] Shanthikumar, J.G. and Yao, D.D. 1988. Throughput Bounds for Closed Queueing Networks with Queue-Dependent Service Rates. *Performance Evaluation*, **9**, 69–78.

[22] Shanthikumar, J.G. and Yao, D.D. 1991. Strong Stochastic Convexity: Closure Properties and Applications. *Journal of Applied Probability*, **28** (1991), 131–145.

[23] Stoyan, D. 1983. *Comparison Methods for Queues and Other Stochastic Models*. Wiley, New York.

[24] Yao, D.D. 1984. Some Properties of the Throughput Function of Closed Networks of Queues. *Operations Research Letters*, **3**, 313–317.

[25] Yao, D.D. 1987. Majorization and Arrangement Orderings in Open Networks of Queues. *Annals of Operations Research*, **9**, 531–543.

[26] Yao, D.D. and Kim, S.C. 1987. Reducing the Congestion in a Class of Job Shops, *Management Science*, **34**, 1165–1172.

4
Kelly Networks

The central approach used in this chapter, just as in Section 2.6, is to examine a stationary Markov chain together with its time-reversal: another Markov chain defined through reversing the time index of the original chain. Applying this technique to the Jackson network in Section 2.6, we not only recovered the product-form distributions derived earlier in Chapter 2 but also established the Poisson property of the exit processes from the network. In particular, we showed that in equilibrium the Jackson network preserves the Poisson property of the input processes to the network, such that the exit processes are also Poisson. This "Poisson-in-Poisson-out" property is the main motivation here for defining and studying a class of so-called *quasi-reversible* queues. This class is an extension of the basic $M/M(n)/1$ queues, which constitute the nodes in a Jackson network, to allow multiclass jobs and more general arrival and service disciplines, while still preserving the Poisson-in-Poisson-out property. It turns out that a general multiclass network, known as a Kelly network, which connects a set of quasi-reversible queues through some very general routing schemes, still enjoys the basic properties of the Jackson network, namely, the product-form equilibrium distribution and the Poisson-in-Poisson-out property.

Below, we start with presenting the definition and properties of a quasi-reversible queue in Section 4.1, and then focusing on a special class of quasi-reversible queues, known as symmetric queues, in Section 4.2. The general multiclass Kelly network is discussed in Section 4.3, where the treatment is essentially an extension of the approach developed in Section 2.6 for the Jackson network. In the last two sections we discuss two related topics: (a) an approach to identifying Poisson flows in the Kelly network (Section

4.4), and (b) what is known as "arrival theorems" (Section 4.5): results stating that arrivals to any node in a network of quasi-reversible queues observe time averages (although these arrivals in general are not Poisson processes).

4.1   Quasi-Reversible Queues

Let us first recall the two key relations, (2.20) and (2.21), discussed in Section 2.6, between a stationary Markov chain and its time-reversal :

$$\pi(i)\tilde{q}(i,j) = \pi(j)q(j,i), \quad \forall i,j \in \mathbf{S},\ i \neq j; \tag{4.1}$$

and

$$q(i) := \sum_{j \neq i} q(i,j) = \sum_{j \neq i} \tilde{q}(i,j) := \tilde{q}(i), \quad \forall i \in \mathbf{S}. \tag{4.2}$$

Specifically, we have (reproducing Theorem 2.9) the following theorem.

Theorem 4.1 Suppose π is a probability distribution with support set $\mathbf{S}$, and Q and $\tilde{Q}$ are two rate matrices that satisfy the relation in (4.1) and (4.2). Then, Q and $\tilde{Q}$ are the rate matrices of a stationary Markov chain and its time-reversal, and π is the equilibrium distribution of both.

The key steps in the approach outlined in Section 2.6 are as follows. First, based on the original Markov chain, usually it is not difficult to identify the time-reversal in terms of specifying its transition rates $\tilde{q}$: As the original Markov chain moves from state i to state j, its time-reversal moves back from state j to state i. (This phenomenon is perhaps best observed from a videotape that is played backwards. For instance, a character that opens a door and then steps into a room will be observed as first backing out from a room and then closing the door.) Second, the identified transition rates of the time-reversal have to satisfy the equation in (4.2). When this verification is done, we can then make use of the equation in (4.1) to derive the equilibrium distribution π. Hence, we can start with a "conjecture" about the time-reversal and the associated product-form equilibrium distribution, and use (4.1) and (4.2) as guidance to "fine tune" our conjecture until it satisfies both (4.1) and (4.2). Below we illustrate these steps through two scenarios.

In the arguments that led to Theorem 2.11, we have seen that it is possible to satisfy (4.2) by breaking the equation into several subequations. In particular, we can express (2.26) and (2.6) as follows:

$$\sum_{y \in A(x)} q(\mathbf{x}, \mathbf{y}) = \sum_{y \in A(x)} \tilde{q}(\mathbf{x}, \mathbf{y}) \tag{4.3}$$

and

$$\sum_{y \in D_i(\mathbf{x})} q(\mathbf{x}, \mathbf{y}) = \sum_{y \in D_i(\mathbf{x})} \tilde{q}(\mathbf{x}, \mathbf{y}), \tag{4.4}$$

where $A(\mathbf{x})$ and $D_i(\mathbf{x})$ denote two subsets of states that can be reached from state $\mathbf{x}$ due, respectively, to a job's entering the network (i.e., from outside to one of the nodes in the network) and to a job's departure from node i (including both leaving the network and going to the other nodes). These two sets of equations can then be put together to yield

$$\sum_y q(\mathbf{x}, \mathbf{y})$$
$$= \left[\sum_{y \in A(\mathbf{x})} + \sum_i \sum_{y \in D_i(\mathbf{x})} \right] q(\mathbf{x}, \mathbf{y})$$
$$= \left[\sum_{y \in A(\mathbf{x})} + \sum_i \sum_{y \in D_i(\mathbf{x})} \right] \tilde{q}(\mathbf{x}, \mathbf{y})$$
$$= \sum_y \tilde{q}(\mathbf{x}, \mathbf{y}),$$

which is what is required in (4.2).

Now, (4.3) and (4.4) imply more. Along with [cf. (4.1)]

$$\pi(\mathbf{x}) q(\mathbf{x}, \mathbf{y}) = \pi(\mathbf{y}) \tilde{q}(\mathbf{y}, \mathbf{x}),$$

they lead to

$$\pi(\mathbf{x}) \sum_{y \in A(\mathbf{x})} q(\mathbf{x}, \mathbf{y}) = \sum_{y \in A(\mathbf{x})} \pi(\mathbf{y}) q(\mathbf{y}, \mathbf{x}),$$

and

$$\pi(\mathbf{x}) \sum_{y \in D_i(\mathbf{x})} q(\mathbf{x}, \mathbf{y}) = \sum_{y \in D_i(\mathbf{x})} \pi(\mathbf{y}) q(\mathbf{y}, \mathbf{x}).$$

The two equations above are sometimes referred to as "partial balance" equations (cf. Remark 1.5), since they fall in between the full balance equations (which define the invariant distribution) and the detailed balance equations (which define reversibility): stronger than the former but weaker than the latter.

On the other hand, there are cases in which it is easy to satisfy (4.1), but it will not be possible to satisfy (4.2). (This usually implies that there is no product-form distribution.) As an example, consider a single-server queue that serves several different classes of jobs under a first-come-first-served (FCFS) discipline. Arrivals follow independent Poisson processes, with rate λ_c for class c jobs; service times are independent and exponentially distributed with mean $1/\mu_c$ for class c jobs. Let $\mathbf{x} = (c_1, \ldots, c_n)$ denote a typical state, where c_i denotes the class of the ith job in the system, where the first job (c_1) is in service and the last job (c_n) the most recently arrived. From $\mathbf{x}$, the possible transitions are to $(c_1, \ldots, c_n, c)$ with rate λ_c, and to $(c_2, \ldots, c_n)$ with rate μ_{c_1}.

Under time-reversal, the same state $\mathbf{x} = (c_1, \ldots, c_n)$ has a different interpretation: c_n is the job in service, while c_1 is the job at the end of the queue. Hence, the time-reversal will transit from $\mathbf{x}$ to $(c, c_1, \ldots, c_n)$ with rate λ_c, and to $(c_1, \ldots, c_{n-1})$ with rate μ_{c_n}.

Based on these transition rates, it seems that we can postulate the following product-form distribution:

$$\pi(c_1, \ldots, c_n) = A \cdot \frac{\lambda_{c_1} \cdots \lambda_{c_n}}{\mu_{c_1} \cdots \mu_{c_n}}, \tag{4.5}$$

where A is a normalizing constant (assuming $A^{-1} < \infty$). Indeed, this will satisfy (4.1), along with the transition rates specified above.

However, (4.2) cannot be satisfied: We have

$$\sum_{\mathbf{y}} q(\mathbf{x}, \mathbf{y}) = \mu_{c_1} + \sum_c \lambda_c,$$

while

$$\sum_{\mathbf{y}} \tilde{q}(\mathbf{x}, \mathbf{y}) = \mu_{c_n} + \sum_c \lambda_c.$$

So, unless the service rates are *homogeneous* among job classes, i.e., $\mu_c = \mu$ for all c, (4.2) is violated. Indeed, it is easy to verify that the product-form solution postulated above will *not* satisfy the full balance equation

$$\left(\mu_{c_1} + \sum_c \lambda_c\right)\pi(c_1, \ldots, c_n)$$
$$= \lambda_{c_n}\pi(c_1, \ldots, c_{n-1}) + \sum_c \mu_c \pi(c, c_1, \ldots, c_n),$$

unless $\mu_c = \mu$ for all c.

When $\mu_c = \mu$ for all c, the product-form distribution in (4.5) is indeed the invariant distribution. Furthermore, the model can be generalized as follows. Suppose that each arriving job joins the system at the ℓth position, with probability $\delta(n+1, \ell)$, when the system is in state $\mathbf{x} = (c_1, \ldots, c_n)$. (The jobs originally in positions ℓ through n, move to $\ell+1$ through $n+1$.) Each job requires an i.i.d. exponential time (with unit mean) for service. In state $\mathbf{x}$, the server serves at rate $\phi(n)$, of which a fraction $\gamma(n, \ell)$ is applied to the job in the ℓth position. Here, naturally, we require that for any n,

$$\sum_{\ell=1}^{n} \delta(n, \ell) = \sum_{\ell=1}^{n} \gamma(n, \ell) = 1.$$

The δ and γ functions together can model a variety of service disciplines. For instance, letting

$$\delta(n, \ell) = 1[\ell = n] \quad \text{and} \quad \gamma(n, \ell) = 1[\ell = 1]$$

represents the FCFS discipline, while

$$\delta(n, \ell) = 1[\ell = 1] \quad \text{and} \quad \gamma(n, \ell) = 1[\ell = 1]$$

specifies the last-come-first-served (LCFS) discipline.

With this generalization, the time-reversal is another multiclass queue that reverses the roles of the two functions: γ specifies the arrival mechanism, and δ specifies the service mechanism. (Hence, when FCFS is the discipline in the original system, as considered earlier, it becomes an LCFS discipline when time is reversed.) Specifically, let $A_\ell^c(\mathbf{x})$ denote the state that can be reached from $\mathbf{x} = (c_1, \ldots, c_n)$ due to a class c job's arrival to the ℓth position and $D_\ell(\mathbf{x})$ the state reached due to a job's leaving the system from the ℓth position. We have

$$q(\mathbf{x}, A_\ell^c(\mathbf{x})) = \lambda_c \delta(n + 1, \ell),$$
$$q(\mathbf{x}, D_\ell(\mathbf{x})) = \phi(n)\gamma(n, \ell),$$

for the original system; and

$$\tilde{q}(\mathbf{x}, A_\ell^c(\mathbf{x})) = \lambda_c \gamma(n + 1, \ell),$$
$$\tilde{q}(\mathbf{x}, D_\ell(\mathbf{x})) = \phi(n)\delta(n, \ell),$$

for the time-reversal . Then, the invariant distribution

$$\pi(c_1, \ldots, c_n) = A \cdot \frac{\lambda_{c_1} \cdots \lambda_{c_n}}{\phi(1) \cdots \phi(n)} \tag{4.6}$$

satisfies (4.1), while (4.2) is satisfied due to

$$\sum_{\ell=1}^{n+1} q(\mathbf{x}, A_\ell^c(\mathbf{x})) = \lambda_c = \sum_{\ell=1}^{n+1} \tilde{q}(\mathbf{x}, A_\ell^c(\mathbf{x})) \tag{4.7}$$

and

$$\sum_{\ell=1}^{n} q(\mathbf{x}, D_\ell(\mathbf{x})) = \phi(n) = \sum_{\ell=1}^{n} \tilde{q}(\mathbf{x}, D_\ell(\mathbf{x})).$$

Interestingly, in the special case of LCFS, we can allow class-dependent service rates: μ_c for class c jobs. What plays a key role here is the *symmetry* of LCFS: The γ and δ functions are the same. That is, both the original system and the time-reversal follows LCFS. It is easily verified that in this case, the invariant distribution is

$$\pi(c_1, \ldots, c_n) = A \cdot \frac{\rho_{c_1} \cdots \rho_{c_n}}{\phi(1) \cdots \phi(n)}, \tag{4.8}$$

where $\rho_c = \lambda_c/\mu_c$; and

$$\sum_{\ell=1}^{n} q(\mathbf{x}, D_\ell(\mathbf{x})) = \mu_{c_n} \phi(n) = \sum_{\ell=1}^{n} \tilde{q}(\mathbf{x}, D_\ell(\mathbf{x})),$$

that is, in state $\mathbf{x} = (c_1, \ldots, c_n)$, both the original system and its time-reversal serve job c_n dedicate the entire capacity to it.

When both (4.1) and (4.2) are satisfied, we not only get a product-form distribution, but also a complete characterization of the system under time-reversal, and through the latter, a Poissonian characterization of the departure processes from the original system, provided that the arrivals to the original system are Poisson processes. This last property is what is often referred to as Poisson-in-Poisson-out. We have seen this kind of result in Section 2.6 for an open Jackson network. In the above example of a single-server queue with multiclass jobs, when the service rates are homogeneous among classes, we can draw similar conclusions: In addition to the product form of the stationary distribution, the departure processes are independent Poisson processes, with rate λ_c for class c jobs, exactly the same as the arrival processes, and past departures up to time t are independent of the system state at time t.

Let us examine the Poisson-in-Poisson-out property more closely. First, this is a property strictly in the context of *queueing applications*, as opposed to the more generic Theorem 4.1, which applies to all stationary Markov chains. Second, Poisson-in-Poisson-out requires more than the conditions in (4.1) and (4.2): The Poisson input processes, in the case of the open Jackson network, correspond to (4.3), with both sides equal to α, the overall arrival rate to the network; and in the case of the single-server multiclass queue (homogeneous services), they correspond to (4.7).

We can generalize this to a class of *quasi-reversible* queues.

Definition 4.2 A stationary Markovian queue—more precisely, a queue with a state process that is a stationary Markov chain—is termed *quasi-reversible*, if the queue and its time-reversal satisfy the condition

$$\lambda_c = \sum_{\mathbf{y} \in A^c(\mathbf{x})} q(\mathbf{x}, \mathbf{y}) = \sum_{\mathbf{y} \in A^c(\mathbf{x})} \tilde{q}(\mathbf{x}, \mathbf{y}), \tag{4.9}$$

for each class c and each state $\mathbf{x}$, where $A^c(\mathbf{x})$ denotes the set of states that can be reached from $\mathbf{x}$ due to a class c job's arrival, and λ_c denotes the arrival rate of class c jobs.

Remark 4.3 (i) Note that the condition in (4.9) insists that the rate λ_c is a constant, *independent* of the state $\mathbf{x}$. This, along with the Markov property of the state process, implies that the arrival processes, in both the original system and the time-reversal, are Poisson processes.

(ii) In the applications below we often postulate that the time-reversal of a queue is of a certain form, and that both are quasi-reversible queues, and then verify our postulation. In doing so, we must verify all three conditions in (4.1), (4.2), and (4.9). Note that (4.9) is a strengthening of (4.3): It requires that each side of the equation in (4.9) be a constant independent of the state $\mathbf{x}$. Furthermore, either side of the equation in (4.9) is part of the corresponding side of the equation in (4.2). Hence, once (4.9) is verified, it suffices to verify that the remaining parts of the two sides in (4.2) are equal, such as the equation in (4.4).

Theorem 4.4 A quasi-reversible queue has the following properties:

(i) The arrival processes, in both the original system and the time-reversal, are Poisson processes, with rate λ_c for class c jobs, and independent among the classes.

(ii) The departure processes, in both the original system and the time-reversal, are Poisson processes, with rate λ_c for class c jobs, and independent among the classes, i.e., probabilistically the same as the arrival processes.

(iii) Past departures up to time t, in both the original system and the time-reversal, are independent of the state of the system at time t.

Applying Theorem 4.4 to the open Jackson network and the single-server, multiclass queue, we have the following corollary.

Corollary 4.5 The open Jackson network (in equilibrium) is itself a quasi-reversible queue.

(Note that in the Jackson network, the Poisson-in-Poisson-out property means that the the exit job stream as a whole, i.e., with rate α, is Poisson, and as a set of individual processes—breaking down over the nodes—are also *independent* Poisson processes.)

Corollary 4.6 The single-server, multiclass queue specified earlier—with the general arrival and service mechanisms characterized by the δ and γ functions and the state-dependent service rate $\phi(n)$—is quasi-reversible, with the invariant distribution following (4.5).

Later we will demonstrate that quasi-reversible queues, such as the single-server, multiclass queues considered above, can be connected together, through a quite general job routing mechanism, to form a network, and still maintain the quasi-reversible nature of the network, as well as the product-form invariant distribution. In fact, we have already seen a preview of this type of result (Theorem 2.11 and Corollary 4.5) through the Jackson network, which connects together the simplest, single-class $M/M(n)/1$ queues, each of which is clearly a quasi-reversible queue.

But before we do that, we will continue in the next section to exploit the symmetry (in terms of $\delta \equiv \gamma$) associated with the special case of LCFS, as noted earlier. We will show that when a quasi-reversible queue possesses this kind of symmetry, the service times not only can be state-dependent, they can also follow distributions that are much more general than exponentials.

4.2 Symmetric Queues

A queue is termed *symmetric* if $\delta \equiv \gamma$; that is, the same function specifies both the arrival and the service mechanisms. The purpose of this section is to show that for symmetric queues the result in Corollary 4.6 can be extended to a general class of service-time distributions, called *PH-distributions*.

4.2.1 Phase-Type Distributions and Processes

The so-called *phase-type* distribution, or simply PH-distribution, is defined through a Markov chain with state space $\{0, 1, \ldots, m\}$, and a rate matrix

$$\begin{pmatrix} 0 & \mathbf{0} \\ s & S \end{pmatrix}, \tag{4.10}$$

where $s = (s_i)_{i=1}^m$ is a column vector, and $S = [S_{ij}]_{i,j=1}^m$ is a nonsingular matrix. Hence, with probability 1, the chain will be absorbed into state 0 from any initial state. Let $a = (a_i)_{i=1}^m$ be the initial distribution, $\sum_{i=1}^m a_i = 1$. The time to absorption is then said to follow a PH-distribution, denoted by $\mathrm{PH}(a, S)$.

Alternatively, we can view the PH-distribution as the service requirement of a job. It consists of a set of m phases (or stages). Each phase i ($i = 1, \ldots, m$) corresponds to an exponential distribution with rate

$$\sigma_i := \sum_{j \neq i} S_{ij} + s_i.$$

At service initiation, one of the m phases is selected, following the probability distribution $a = (a_i)_i$. When a phase i service is completed, the job either continues into another phase $j \neq i$ with probability S_{ij}/σ_i, or quits service with probability s_i/σ_i. In the latter case, the job has completed its service requirement.

Clearly the PH-distribution models a wide class of distributions that are composites of exponentials. In particular, it can model the serial and parallel connections of exponential phases: If phase i and phase j are connected in series, then $S_{ij}/\sigma_i = 1$; if phases j and k are connected in parallel following phase i, then $S_{ij} > 0$ and $S_{ik} > 0$. Hence, PH-distribution includes as special cases Coxian and generalized Erlang distributions.

We now modify the Markov chain that defines the PH-distribution: whenever the chain reaches state 0, it starts all over again, from the same initial distribution a. The transition rates of this modified chain are

$$q(i, j) = S_{ij} + s_i a_j, \quad i, j = 1, \ldots, m, \ i \neq j;$$

and the corresponding rate matrix is $Q = S + sa'$. Let $v = (v_i)_{i=1}^m$ denote the invariant distribution: $vQ = 0$ and $\sum_{i=1}^m v_i = 1$. We shall refer to this stationary Markov chain as a PH-*process*.

Note that alternatively, the PH-process can be viewed as a regenerative process, with the time between regenerations following the $PH(a, S)$ distribution. Let $\mu := v's$; μ is then the rate of regeneration, or the reciprocal of the mean regeneration time.

Next consider the time-reversal of the above PH-process.

Lemma 4.7 The time-reversal of the $PH(s, S)$ process is also a PH-process $PH(\tilde{a}, \tilde{S})$, characterized by

$$\tilde{a}_i = s_i v_i / \mu, \quad \text{and} \quad \tilde{S}_{ij} = v_j S_{ji} / v_i.$$

From the above characterization, we have

$$\tilde{s}_i = -\sum_j \tilde{S}_{ij} = -\sum_j v_j S_{ji} / v_i = \mu a_i / v_i,$$

where the last equality follows from $v'Q = 0$. The proof of the above lemma is a simple verification of the conditions in Lemma 1.3, in particular, that

$$\tilde{q}(i, j) := \tilde{S}_{ij} + \tilde{s}_i \tilde{a}_j$$

indeed form a rate matrix $\tilde{Q} = \tilde{S} + \tilde{s}\tilde{a}'$, and satisfy (4.1), i.e.,

$$v_i q(i, j) = v_j \tilde{q}(j, i), \forall i \neq j.$$

Based on Lemma 1.3 (also refer to Remark 2.10), we have

$$\sum_{j \neq i} q(i, j) = \sum_{j \neq i} \tilde{q}(i, j),$$

that is,

$$\sum_{j \neq i} S_{ij} + s_i \sum_{j \neq i} a_j = \sum_{j \neq i} \tilde{S}_{ij} + \tilde{s}_i \sum_{j \neq i} \tilde{a}_j,$$

which, along with $s_i a_i = \tilde{s}_i \tilde{a}_i$, yields

$$\sum_{j \neq i} S_{ij} + s_i = \sum_{j \neq i} \tilde{S}_{ij} + \tilde{s}_i, \tag{4.11}$$

an equation that will become useful below.

4.2.2   Multiclass M/PH/1 Queue

Consider a single-server queue that serves a set of different classes of jobs. Arrivals follow independent Poisson processes, with rate λ_c for class c jobs. Service times are independent and follow PH-distributions, $PH(a^c, S^c)$ for class c jobs. (A tip on notation: When class is the only index, it is denoted by a subscript, as in λ_c; otherwise, it is denoted by a superscript, as in a^c

and S^c, allowing the addition of subscripts, such as a_i^c and S_{ij}^c.) Hence, with the service times following PH-distributions, the model is a generalization of the one considered in the last section, where the service times follow exponential distributions.

Let $\mathbf{x} = (c_\ell, i_\ell)_{\ell=1}^n$ denote a typical state, where c_ℓ and i_ℓ denote, respectively, the class identity of the job in the ℓth position and the phase index of its service. As in the model of the last section, in state $\mathbf{x}$, the server works at rate $\phi(n)$, of which a fraction $\gamma(n, \ell)$ is applied to the job at the ℓth position. However, here the same γ function also specifies the arrival mechanism (instead of the δ function). That is, an arriving job that finds the system in state $\mathbf{x}$ joins the ℓth position with probability $\gamma(n+1, \ell)$. (As noted in the last section, the symmetric mechanism must exclude FCFS, which requires $\delta(n, \ell) = \mathbf{1}[\ell = n]$ for arrivals, while $\gamma(n, \ell) = \mathbf{1}[\ell = 1]$ for services.)

To analyze the above queue, we relate it to its time-reversal, applying Theorem 4.1. We postulate that the time-reversal is a similar multiclass queue, with the same independent Poisson arrival processes, with rate λ_c for class c jobs, and the same arrival and service mechanisms governed by the function $\gamma(\cdot, \cdot)$. (The service rate function $\phi(n)$ also remains the same in the time-reversal.) The service times, however, follow the *reversed* PH-distributions, PH($\tilde{a}^c, \tilde{S}^c$) for class c jobs, with the parameters $\tilde{a}^c$ and $\tilde{S}^c$ relating to a^c and S^c of the original queue in exactly the same manner as specified in the last subsection.

Let $A_{\ell,i}^c(\mathbf{x})$ denote the state reached from state $\mathbf{x}$ due to a class c job's arriving at the queue, joining the ℓ th position and starting service in phase i. Let $D_\ell(\mathbf{x})$ denote the state reached from state $\mathbf{x}$ due to a job's leaving the queue from the ℓth position. To lighten up notation, here and below, we will assume, without loss of generality, that the job at the ℓth position has class index c (instead of c_ℓ) and phase index i (instead of i_ℓ). Let $I_{\ell,j}(\mathbf{x})$ denote the state reached from state $\mathbf{x}$ due to a (class c) job's completing its current service phase (i) and moving into another phase j; the class (c) and position (ℓ) indices remain unchanged.

Then, the transition rates of the stationary Markov chain associated with the original queue can be specified as follows:

$$q(\mathbf{x}, A_{\ell,i}^c(\mathbf{x})) = \lambda_c \gamma(n+1, \ell) a_i^c,$$
$$q(\mathbf{x}, D_\ell(\mathbf{x})) = \phi(n) \gamma(n, \ell) s_i^c,$$
$$q(\mathbf{x}, I_{\ell,j}(\mathbf{x})) = \phi(n) \gamma(n, \ell) S_{ij}^c.$$

The transition rates for the time-reversal are similar: We replace a_i^c, s_i^c, and S_{ij}^c by $\tilde{a}_i^c$, $\tilde{s}_i^c$, and $\tilde{S}_{ij}^c$, and the two sets of parameters are related through

Lemma 4.7. Hence,

$$\tilde{q}(\mathbf{x}, A_{\ell,i}^c(\mathbf{x})) = \lambda_c \gamma(n+1, \ell) s_i^c v_i^c / \mu_c,$$
$$\tilde{q}(\mathbf{x}, D_\ell(\mathbf{x})) = \phi(n) \gamma(n, \ell) \mu_c a_i^c / v_i^c,$$
$$\tilde{q}(\mathbf{x}, I_{\ell,j}(\mathbf{x})) = \phi(n) \gamma(n, \ell) v_j^c S_{ji}^c / v_i^c.$$

Theorem 4.8 Consider the symmetric, multiclass M/PH/1 queue specified above. Let $v^c = (v_i^c)_{i=1}^m$ and μ_c be the invariant distribution and the rate of regeneration associated with the PH(a^c, S^c) process. Let $\rho_c = \lambda_c / \mu_c$.

(i) Its stationary distribution has the following product form:

$$\pi(c_1, i_1, \ldots, c_n, i_n) = A \prod_{\ell=1}^n [\rho_{c_\ell} v_{i_\ell}^{c_\ell} / \phi(\ell)],$$

where

$$A := \left(\sum_{(c_1, i_1, \ldots, c_n, i_n)} \prod_{\ell=1}^n [\rho_{c_\ell} v_{i_\ell}^{c_\ell} / \phi(\ell)] \right)^{-1}$$

is the normalizing constant (assuming that the summation is finite).

(ii) It is a quasi-reversible queue. In particular, its time-reversal is another symmetric, multiclass M/PH/1 queue, with the same (Poisson) arrival processes, but with the service times following the (reversed) PH($\tilde{a}^c, \tilde{S}^c$) processes. And, the departure processes from the original and the time-reversal queues are independent Poisson processes, with rate λ^c for class c jobs, exactly the same as the arrival processes.

Proof. We need to verify (4.1) and (4.2) in order to apply Theorem 4.1, and to verify (4.9) to claim the Poissonian departure processes in (ii).

To verify (4.1), based on the given product-form distribution and the transition rates specified earlier, the desired relation

$$\pi(\mathbf{x}) q(\mathbf{x}, \mathbf{y}) = \pi(\mathbf{y}) \tilde{q}(\mathbf{y}, \mathbf{x})$$

reduces to the following, respectively for $\mathbf{y} = A_{\ell,i}^c(\mathbf{x})$, $D_\ell(\mathbf{x})$, and $I_{\ell,j}(\mathbf{x})$:

$$\lambda_c a_i^c = \rho_c v_i^c \tilde{s}_i^c,$$
$$\rho_c v_i^c s_i^c = \lambda_c \tilde{a}_i^c,$$
$$v_i^c S_{ij}^c = v_j^c \tilde{S}_{ji}^c.$$

All three follow immediately from the relations between the PH(a^c, S^c) and the PH($\tilde{a}^c, \tilde{S}^c$) distributions (Lemma 4.7).

To verify (4.2), note it takes the following form here:

$$\sum_{c,\ell,i} q(\mathbf{x}, A_{\ell,i}^c(\mathbf{x})) + \sum_{\ell,j\neq i} q(\mathbf{x}, I_{\ell,j}(\mathbf{x})) + \sum_{\ell} q(\mathbf{x}, D_\ell(\mathbf{x}))$$

$$= \sum_{c,\ell,i} \tilde{q}(\mathbf{x}, A_{\ell,i}^c(\mathbf{x})) + \sum_{\ell,j\neq i} \tilde{q}(\mathbf{x}, I_{\ell,j}(\mathbf{x})) + \sum_{\ell} \tilde{q}(\mathbf{x}, D_\ell(\mathbf{x})). \qquad (4.12)$$

First we have

$$\sum_i q(\mathbf{x}, A_{\ell,i}^c(\mathbf{x})) = \sum_i \tilde{q}(\mathbf{x}, A_{\ell,i}^c(\mathbf{x})) = \lambda_c \gamma(n+1, \ell), \qquad (4.13)$$

since $\sum_i a_i^c = \sum_i \tilde{a}_i^c = 1$. Summing over ℓ on both sides verifies (4.9). Next, making use of (4.11), we have

$$\sum_{j\neq i} q(\mathbf{x}, I_{\ell,j}(\mathbf{x})) + q(\mathbf{x}, D_\ell(\mathbf{x}))$$

$$= \phi(n)\gamma(n, \ell) \Big[\sum_{j\neq i} S_{ij}^c + s_i^c \Big]$$

$$= \phi(n)\gamma(n, \ell) \Big[\sum_{j\neq i} \tilde{S}_{ij}^c + \tilde{s}_i^c \Big]$$

$$= \sum_{j\neq i} \tilde{q}(\mathbf{x}, I_{\ell,j}(\mathbf{x})) + \tilde{q}(\mathbf{x}, D_\ell(\mathbf{x})).$$

Hence, the desired relation in (4.12) follows.   □

Suppose we are interested only in the distribution of the number of jobs of the different classes in the system. We can sum over all the phase indices and redefine the state as $\mathbf{z} = (z_c)$, with z_c denoting the number of class c jobs in the system.

Corollary 4.9 The stationary distribution of the number of jobs in the M/PH/1 queue in Theorem 4.8 is

$$\pi((z_c)) = A \frac{\prod_c (\rho_c)^{z_c}}{\phi(1)\cdots\phi(|\mathbf{z}|)},$$

where A is a normalizing constant (not necessarily the same as the one in Theorem 4.8), and $|\mathbf{z}| = \sum_c z_c$.

Since the above distribution has a multinomial form, we know that (in stationarity) the class identities of the jobs are independent. Specifically, each job in system belongs to class c with probability $\rho_c / \sum_c \rho_c$.

The symmetric queue includes several special cases of particular interest, which we list below, including the LCFS case noted in the last section.

(i) Last-come-first-served (LCFS): This can be modeled by

$$\gamma(n, \ell) = 1[\ell = 1], \quad \text{and} \quad \phi(n) = 1.$$

That is, the server has a constant, unit service rate, which is always applied to the job at the head of the line, and any new arrival also joins the head of the line. Hence, the most recent arrival will always preempt the job currently in service, which will join the queue and become the next (to be served) in the line. When its turn comes again, the service is *resumed*, that is, starting from the point when it was interrupted. For instance, suppose it was in phase i when interrupted by another arriving job, it will start in phase i when the service is resumed.

(ii) Processor-sharing (PS): The service capacity is equally shared among all jobs that are present in the system. That is,

$$\gamma(n, \ell) = \frac{1}{n}, \quad \text{and} \quad \phi(n) = 1.$$

(iii) Infinite-server (IS): That is,

$$\gamma(n, \ell) = \frac{1}{n}, \quad \text{and} \quad \phi(n) = n.$$

(Here, the term "infinite-server" is a misnomer; more accurately, this is a single server with infinite capacity: one unit for each job present in the system.)

Note that in all three cases, each job is served immediately on arrival, and thus no job has to wait in queue for service.

Remark 4.10 The interesting fact from Corollary 4.9 is that for symmetric queues the invariant distribution for the number of jobs in the system remains the same whether the service times follow exponential distributions or the more general PH-distributions: Only the *means* of the service times $1/\mu_c$ appear in the product form. This is what is often referred to as an *insensitivity* result: It is insensitive to the distributions, and only the means matter.

4.3   A Multiclass Network

The network under consideration here is an extension of the Jackson network to include jobs with different classes. The network still has a set of J nodes, indexed by the subscripts i and j. The jobs are now identified by classes, indexed by superscripts c and d. External jobs arrive at the network following a Poisson process with rate α. Each arriving job joins node j as a class c job with probability p_{0j}^c. After completing service at node i, a class c job goes to node j and becomes a class d job, with probability p_{ij}^{cd}, and leaves ("exits") the network with probability $p_{i0}^c = 1 - \sum_{j,d} p_{ij}^{cd}$.

Let λ_i^c be the overall traffic rate of class c jobs that visit node i. We have

$$\lambda_i^c = \alpha p_{0i}^c + \sum_{j,d} \lambda_j^d p_{ji}^{dc}, \tag{4.14}$$

for each class c and each node i. This is the set of traffic equations for the multiclass network.

We are not particularly concerned with the service mechanism internal to the queue at each node, except that we assume that each queue *in isolation* will behave like a quasi-reversible queue (including symmetric queues) similar to the ones studied in the last two sections. (We will return to this point later.) Hence, the network is essentially a system that connects together m such queues with the routing matrix specified above.

Let $\{\mathbf{X}(t) = (\mathbf{X}_i(t))_{i=1}^J\}$ denote the continuous-time Markov chain associated with the network. Set $\mathbf{x} := (\mathbf{x}_i)_{i=1}^J$ and $\mathbf{y} := (\mathbf{y}_i)_{i=1}^J$, with $\mathbf{x}_i := (x_i^c)_c$ and $\mathbf{y}_i := (y_i^c)_c$ denoting the typical states of the queue at node i (similar to those in the last two sections). Let $A_i^c(\mathbf{x})$ denote the set of states that can be reached from $\mathbf{x}$ by a class c job's *arrival* to node i, from external sources; let $E_i^c(\mathbf{x})$ denote a similar set of states due to a class c job's *exiting* the network from node i; let $T_{ij}^{cd}(\mathbf{x})$ denote the set of states reached from $\mathbf{x}$ due to a class c job's *transition* from node i to node j and becoming a classed d job.

Now, consider node i in isolation. The input is a set of independent Poisson arrival processes, with rate λ_i^c for class c jobs. Denote the states of the associated (stationary) Markov chain by $\mathbf{x}_i$ and $\mathbf{y}_i$. Let $\pi_i(\mathbf{x}_i) := (\pi_i(x_i^c))_c$ denote the stationary distribution. Denote the transition rates of the Markov chain by $q_i(\mathbf{x}_i, \mathbf{y}_i)$. In particular,

$$\sum_{\mathbf{y}_i \in A_i^c(\mathbf{x}_i)} q_i(\mathbf{x}_i, \mathbf{y}_i) = \lambda_i^c,$$

and the sum is independent of the state $\mathbf{x}_i$. Hence, given a transition from state $\mathbf{x}_i$ to state $\mathbf{y}_i \in A_i^c(\mathbf{x}_i)$, it is due to a class c job's arrival with probability $q_i(\mathbf{x}_i, \mathbf{y}_i)/\lambda_i^c$.

Let $\tilde{q}_i(\mathbf{x}_i, \mathbf{y}_i)$ denote the transition rates of the time-reversal of the above Markov chain. We have

$$\pi_i(\mathbf{x}_i)\tilde{q}_i(\mathbf{x}_i, \mathbf{y}_i) = \pi_i(\mathbf{y}_i)q_i(\mathbf{y}_i, \mathbf{x}_i).$$

We now specify the stationary Markov chain associated with the original network, in terms of its transition rates, expressed in relation to the transition rates of the Markov chains associated with each node described above. We have

$$q(\mathbf{x}, \mathbf{y}) = q_i(\mathbf{x}_i, \mathbf{y}_i)p_{i0}^c, \qquad \mathbf{y} \in E_i^c(\mathbf{x}), \ \mathbf{y}_i \in E_i^c(\mathbf{x}_i), \qquad (4.15)$$

corresponding to a class c job leaving the network from node i;

$$q(\mathbf{x}, \mathbf{y}) = q_i(\mathbf{x}_i, \mathbf{y}_i)(\alpha p_{0i}^c / \lambda_i^c), \qquad \mathbf{y} \in A_i^c(\mathbf{x}), \ \mathbf{y}_i \in A_i^c(\mathbf{x}_i), \qquad (4.16)$$

corresponding to an external arrival (of class c) at node i; and

$$q(\mathbf{x}, \mathbf{y}) = q_i(\mathbf{x}_i, \mathbf{y}_i)q_j(\mathbf{x}_j, \mathbf{y}_j)(p_{ij}^{cd} / \lambda_j^d), \qquad (4.17)$$

for

$$\mathbf{y} \in T_{ij}^{cd}(\mathbf{x}), \quad \mathbf{y}_i \in E_i^c(\mathbf{x}_i), \quad \mathbf{y}_j \in A_j^d(\mathbf{x}_j),$$

which corresponds to an internal transition of a class c job from node i to node j, where it switches to class d.

Next, consider the time-reversal of the above Markov chain. We postulate that it corresponds to a similar network, with a Poisson external arrival process at rate α. Each arrival independently joins node i as a class c job with probability $\tilde{p}_{0i}^c = \lambda_i^c p_{i0}^c / \alpha$. Internal jobs follow the following routing probabilities:

$$\tilde{p}_{i0}^c = \alpha p_{0i}^c / \lambda_i^c, \quad \tilde{p}_{ij}^{cd} = \lambda_j^d p_{ji}^{dc} / \lambda_i^c. \qquad (4.18)$$

Note that

$$\tilde{p}_{i0}^c + \sum_{j,d} \tilde{p}_{ij}^{cd} = \left(\alpha p_{0i}^c + \sum_{j,d} \lambda_j^d p_{ji}^{dc} \right) \Big/ \lambda_i^c = 1$$

follows from (4.14). On the other hand, to verify $\sum_{i,c} \tilde{p}_{0i}^c = 1$, from $p_{i0}^c = 1 - \sum_{j,d} p_{ij}^{cd}$, we have

$$\sum_{i,c} \lambda_i^c p_{i0}^c = \sum_{i,c} \lambda_i^c - \sum_{i,c} \lambda_i^c \sum_{j,d} p_{ij}^{cd} = \sum_{i,c} \lambda_i^c - \sum_{i,c} \sum_{j,d} \lambda_j^d p_{ji}^{dc} = \alpha,$$

where the last equality follows from taking summation $\sum_{i,c}$ on both sides of (4.14).

In the time-reversed network, the queue at each node i is simply the time-reversal of the original queue (i.e., the queue at node i of the original network), and the corresponding Markov chain follows the transition rates $\tilde{q}_i(\cdot, \cdot)$. Hence, we can write down the expressions for the transition rates of the Markov chain corresponding to the time-reversed network $\tilde{q}(\cdot, \cdot)$, similar to the expressions for $q(\cdot, \cdot)$ in (4.15), (4.16), and (4.17), by replacing q_i with $\tilde{q}_i$ and p with $\tilde{p}$.

It turns out that the key condition that leads to the results in Theorem 4.11 below is that the queue at each node i, in isolation, is a quasi-reversible queue. In particular, this implies that the following must be satisfied (for each i):

$$\lambda_i^c := \sum_{y_i \in A_i^c(x_i)} q_i(x_i, y_i) = \sum_{y_i \in A_i^c(x_i)} \tilde{q}_i(x_i, y_i), \qquad (4.19)$$

in addition to

$$\sum_{y_i} q_i(x_i, y_i) = \sum_{y_i} \tilde{q}_i(x_i, y_i), \qquad (4.20)$$

as required by (4.2). Now, (4.20) can be written as

$$\left[\sum_c \sum_{y_i \in A_i^c(x_i)} + \sum_c \sum_{y_i \in E_i^c(x_i)} \right] q_i(x_i, y_i)$$

$$= \left[\sum_c \sum_{y_i \in A_i^c(x_i)} + \sum_c \sum_{y_i \in E_i^c(x_i)} \right] \tilde{q}_i(x_i, y_i),$$

which, in view of (4.19), is reduced to

$$\sum_c \sum_{y_i \in E_i^c(x_i)} q_i(x_i, y_i) = \sum_c \sum_{y_i \in E_i^c(x_i)} \tilde{q}_i(x_i, y_i). \qquad (4.21)$$

Theorem 4.11 Suppose each node i in the multiclass network specified earlier is a quasi-reversible queue while operating in isolation; in particular, (4.19) and (4.20) are satisfied.

(i) The stationary distribution of the Markov chain associated with the network has the following product form:

$$\pi(\mathbf{x}) = \prod_{i=1}^{J} \pi_i(\mathbf{x}_i),$$

where $\pi_i(\mathbf{x}_i)$ is the stationary distribution of the queue at node i while operating in isolation and fed with a set of independent Poisson streams with rates $(\lambda_i^c)_c$.

(ii) The time-reversal of the Markov chain corresponds to another multiclass network: External arrivals follow a Poisson process with rate α; each arrival independently joins node i as a class c job with probability $\tilde{p}_{0i}^c = \lambda_i^c p_{i0}^c / \alpha$; and internal jobs follow the routing probabilities in (4.18).

(iii) The original network is itself a quasi-reversible queue. In particular, the exit processes from the original network are independent Poisson processes, just like the arrival processes. Furthermore, the past history of jobs leaving the network up to time t is independent of the network state at time t.

Proof. Again, we are required to verify (4.1) and (4.2), and also (4.9). For (4.1), it suffices to verify the relation

$$\pi(\mathbf{x})q(\mathbf{x},\mathbf{y}) = \pi(\mathbf{y})\tilde{q}(\mathbf{y},\mathbf{x})$$

for each $\mathbf{y}$ that belongs to the three subsets of the state space: $E_i^c(\mathbf{x})$, $A_i^c(\mathbf{x})$, and $T_{ij}^{cd}(\mathbf{x})$. For instance, for $\mathbf{y} \in T_{ij}^{cd}(\mathbf{x})$, this amounts to verifying

$$\pi(\mathbf{x})q_i(\mathbf{x}_i,\mathbf{y}_i)q_j(\mathbf{x}_j,\mathbf{y}_j)p_{ij}^{cd}/\lambda_j^d = \pi(\mathbf{y})\tilde{q}_i(\mathbf{y}_i,\mathbf{x}_i)\tilde{q}_j(\mathbf{y}_j,\mathbf{x}_j)\tilde{p}_{ji}^{dc}/\lambda_i^c.$$

Substituting the given product form for $\pi(\mathbf{x})$ and $\pi(\mathbf{y})$ above, we have

$$\pi_i(\mathbf{x}_i)\pi_j(\mathbf{x}_j)q_i(\mathbf{x}_i,\mathbf{y}_i)q_j(\mathbf{x}_j,\mathbf{y}_j)p_{ij}^{cd}/\lambda_j^d$$
$$= \pi_i(\mathbf{y}_i)\pi_j(\mathbf{y}_j)\tilde{q}_i(\mathbf{y}_i,\mathbf{x}_i)\tilde{q}_j(\mathbf{y}_j,\mathbf{x}_j)\tilde{p}_{ji}^{dc}/\lambda_i^c.$$

But the above follows from applying (4.1) to the queues at node i and node j and their time-reversals, as well as the relation between p and $\tilde{p}$. When $\mathbf{y}$ belongs to the other two subsets, the verification is similar.

We next verify (4.2), that is,

$$\sum_{\mathbf{y}} q(\mathbf{x},\mathbf{y}) = \sum_{\mathbf{y}} \tilde{q}(\mathbf{x},\mathbf{y}),$$

which takes the following form:

$$\left[\sum_{i,c}\sum_{\mathbf{y}\in A_i^c(\mathbf{x})} + \sum_{i,c}\sum_{\mathbf{y}\in E_i^c(\mathbf{x})} + \sum_{i,c}\sum_{j,d}\sum_{\mathbf{y}\in T_{ij}^{cd}(\mathbf{x})} \right] q(\mathbf{x},\mathbf{y})$$

$$= \left[\sum_{i,c}\sum_{\mathbf{y}\in A_i^c(\mathbf{x})} + \sum_{i,c}\sum_{\mathbf{y}\in E_i^c(\mathbf{x})} + \sum_{i,c}\sum_{j,d}\sum_{\mathbf{y}\in T_{ij}^{cd}(\mathbf{x})} \right] \tilde{q}(\mathbf{x},\mathbf{y}).$$

$$(4.22)$$

Below we show that (4.22) is satisfied in two pieces: One corresponds to external jobs arriving at the *network*, which equates the first term on both sides of the above equation, and the other corresponds to service completions at *each node i*.

First, making use of (4.19), we have

$$\sum_{\mathbf{y}\in A_i^c(\mathbf{x})} q(\mathbf{x},\mathbf{y}) = \sum_{\mathbf{y}\in A_i^c(\mathbf{x})} q_i(\mathbf{x}_i,\mathbf{y}_i)\alpha p_{0i}^c \Big/ \lambda_i^c = \alpha p_{0i}^c$$

and

$$\sum_{\mathbf{y}\in A_i^c(\mathbf{x})} \tilde{q}(\mathbf{x},\mathbf{y}) = \sum_{\mathbf{y}\in A_i^c(\mathbf{x})} \tilde{q}_i(\mathbf{x}_i,\mathbf{y}_i)\alpha\tilde{p}_{0i}^c \Big/ \lambda_i^c = \alpha\tilde{p}_{0i}^c = \lambda_i^c p_{i0}^c.$$

Since

$$\sum_{i,c} \alpha p_{0i}^c = \sum_{i,c} \lambda_i^c p_{i0}^c = \alpha,$$

we have

$$\sum_{i,c} \sum_{\mathbf{y} \in A_i^c(\mathbf{x})} q(\mathbf{x}, \mathbf{y}) = \sum_{i,c} \sum_{\mathbf{y} \in A_i^c(\mathbf{x})} \tilde{q}(\mathbf{x}, \mathbf{y}) = \alpha, \qquad (4.23)$$

which is the first equation we want in order to establish (4.22). Furthermore, we have also verified (4.9).

To derive the second equation, we have,

$$\sum_{\mathbf{y} \in E_i^c(\mathbf{x})} q(\mathbf{x}, \mathbf{y}) + \sum_{j,d} \sum_{\mathbf{y} \in T_{ij}^{cd}(\mathbf{x})} q(\mathbf{x}, \mathbf{y})$$

$$= \left[p_{i0}^c + \sum_{j,d} p_{ij}^{cd} \right] \sum_{\mathbf{y}_i \in E_i^c(\mathbf{x}_i)} q_i(\mathbf{x}_i, \mathbf{y}_i)$$

$$= \sum_{\mathbf{y}_i \in E_i^c(\mathbf{x}_i)} q_i(\mathbf{x}_i, \mathbf{y}_i). \qquad (4.24)$$

Note that the last equation above follows from $p_{i0}^c = 1 - \sum_{j,d} p_{ij}^{cd}$; whereas in the first equation, we have made use of the (internal) transition rate in (4.17), along with the quasi-reversibility condition in (4.19):

$$\sum_{\mathbf{y} \in T_{ij}^{cd}(\mathbf{x})} q(\mathbf{x}, \mathbf{y})$$

$$= \sum_{j,d} p_{ij}^{cd} \sum_{\mathbf{y}_i \in E_i^c(\mathbf{x}_i)} q_i(\mathbf{x}_i, \mathbf{y}_i) \sum_{\mathbf{y}_j \in A_j^d(\mathbf{x}_j)} q_j(\mathbf{x}_j, \mathbf{y}_j) / \lambda_j^d$$

$$= \sum_{j,d} p_{ij}^{cd} \sum_{\mathbf{y}_i \in E_i^c(\mathbf{x}_i)} q_i(\mathbf{x}_i, \mathbf{y}_i).$$

Similarly, for the time-reversal, we have

$$\sum_{\mathbf{y} \in E_i^c(\mathbf{x})} \tilde{q}(\mathbf{x}, \mathbf{y}) + \sum_{j,d} \sum_{\mathbf{y} \in T_{ij}^{cd}(\mathbf{x})} \tilde{q}(\mathbf{x}, \mathbf{y})$$

$$= \sum_{\mathbf{y}_i \in E_i^c(\mathbf{x})} \tilde{q}_i(\mathbf{x}_i, \mathbf{y}_i) \left[\alpha p_{0i}^c + \sum_{j,d} \lambda_j^d p_{ji}^{dc} \right] / \lambda_i^c$$

$$= \sum_{\mathbf{y}_i \in E_i^c(\mathbf{x})} \tilde{q}_i(\mathbf{x}_i, \mathbf{y}_i), \qquad (4.25)$$

where the last equation follows from the traffic equation in (4.14). In view of (4.21), we know that the right-hand sides of (4.24) and (4.25) are equal, which is exactly the desired second equation for establishing (4.22). □

In summary, to make use of the above theorem, the key is to verify that (4.19) and (4.20) are satisfied for each node i, i.e., the queue when operating in isolation is quasi-reversible. This verification was illustrated earlier for a multiclass $M/PH/1$ queue under various service disciplines; refer to (4.12) and (4.13). Exercise 5 is another application of Theorem 4.11, to what is known as the Kumar–Seidman network, which will be studied in detail in Chapter 8.

To conclude this section, we note that in many applications, jobs will follow a non-Markovian routing scheme: The next node to visit depends on the nodes visited in the past. The simplest example is deterministic routing. Suppose each class c job visits the following nodes in *sequence*:

$$i_1^c \to i_2^c \to \cdots \to i_k^c \to \cdots \to i_{t_c}^c, \tag{4.26}$$

that is, entering the network through node i_1^c, followed by visiting nodes $i_2^c, \ldots, i_k^c, \ldots$ in sequence, and leaving the network after visiting $i_{t_c}^c$. Note that we allow some of the stations to be visited more than once. Also note that no job ever changes class. Suppose external arrivals follow Poisson processes, with rate α_c for class c jobs, and $\alpha = \sum_c \alpha_c$.

In this case, the routing matrix is a 0-1 "incidence" matrix, completely specified by the deterministic routing in (4.26). Also, the arrival rate of class c jobs to each node i is simply determined as

$$\lambda_i^c = \alpha_c \sum_{k=1}^{t_c} \mathbf{1}[i_k^c = i],$$

replacing the traffic equations in (4.14). The time-reversal is a similar network, with the route of each class c in (4.26) reversed, i.e., starting from node $i_{t_c}^c$ and ending with node i_1^c.

The product-form result in Theorem 4.11 remains unchanged, and the network is still a quasi-reversible queue itself, with the exiting jobs forming a set of independent Poisson processes, with rate α_c for class c jobs.

Note that this mechanism also applies to routings in which the next node to visit depends *probabilistically* on nodes visited in the past: Simply index each different route by a different class. Hence, in general, the routing of each class c can be represented by a graph, and the class will then be split into as many (sub)classes as there are routes between the source-destination pair(s).

4.4 Poisson Flows

From Theorem 4.11 (in particular, part (iii)), we know that the *exit* processes from the multiclass network of the last section are Poisson processes. It turn out that this result can also be used to identify internal flows within the network that are also Poisson processes. The idea is quite simple: Suppose we can identify an internal flow, say jobs moving from node i to node j, as an exit process from some subnetwork (that includes node i). Then, applying Theorem 4.11, we can claim that the flow is a Poisson process. This clearly requires that the subnetwork never be visited again by the flow of jobs in question. Equivalently, we can also require the flow in question to be an external arrival process to the subnetwork that includes node j. In other words, the flow of jobs in question will have never visited that subnetwork before entering node j.

Another way to look at this is as follows. Suppose after visiting node j the jobs could either directly or indirectly go back to node i. Then, past arrivals to node j (from node i) will enhance the feedback to node i, and in turn enhance future arrivals to node j: a positive feedback effect that violates the Poisson property.

To formalize the argument, we need to aggregate the routing probabilities over job classes. For all nodes i and j, let

$$p_{ij} = \sum_{c,d} p_{ij}^{cd}.$$

Next, for each node i, let $I(i)$ denote the set of nodes that are the *direct* input to node i, i.e.,

$$I(i) = \{j = 1, \ldots, J : r_{ji} > 0\}.$$

Similarly,

$$O(j) = \{i = 1, \ldots, J : r_{ji} > 0\}$$

denotes the set of nodes that are the *direct* output of node j. Define $\mathcal{I}(i)$ recursively as follows:

- $I(i) \subset \mathcal{I}(i)$;

- If $k \in \mathcal{I}(i)$, then $I(k) \subset \mathcal{I}(i)$.

That is, $\mathcal{I}(i)$ denotes the set of nodes that will directly or indirectly ever send any job to node i. Define $\mathcal{O}(j)$ in a similar way, as the set of nodes that will directly or indirectly ever receive any job from node j. Clearly, $i \in \mathcal{I}(j)$ if and only if $j \in \mathcal{O}(i)$.

If $i \in \mathcal{I}(j)$, we say there is a *path* from node i to node j, denoted by $i \rightarrow j$. Clearly, $\rightarrow$ is transitive: $i \rightarrow j$ and $j \rightarrow k$ implies $i \rightarrow k$. When a path starts and ends with the same node, say $i \rightarrow i$, call it a *loop*. Clearly, i is on a loop if and only if $i \in \mathcal{I}(i)$, or equivalently $i \in \mathcal{O}(i)$.

Lemma 4.12 $i \rightarrow j$ is part of a loop if and only if $j \in \mathcal{I}(i) \cap \mathcal{O}(i)$.

Proof. $i \rightarrow j$ is part of a loop if and only if we have both $i \rightarrow j$ and $j \rightarrow i$, i.e., $j \in \mathcal{O}(i)$ and $j \in \mathcal{I}(i)$. □

Theorem 4.13 In a network of quasi-reversible queues, the flow of jobs from node i to node j (allowing class changes) is a Poisson process, provided that $j \notin \mathcal{I}(i)$ (i.e., $i \rightarrow j$ is *not* part of a loop), in which case $\bar{\mathcal{I}} \equiv \mathcal{I}(i) \cup \{i\}$ constitutes a quasi-reversible subnetwork, and so does $\bar{\mathcal{O}} \equiv \mathcal{O}(j) \cup \{j\}$.

Proof. When $j \notin \mathcal{I}(i)$, the flow of jobs from node i to node j becomes the *exit* process, via node i, from a subnetwork that consists of the nodes in $\bar{\mathcal{I}}$. In particular, none of those jobs after visiting node j will ever return to $\bar{\mathcal{I}}$. This is because those jobs can visit the nodes in $\mathcal{O}(j)$ only after visiting node j, and for any $k \in \mathcal{O}(j)$, we must have $k \notin \mathcal{I}(i)$; for otherwise we would have reached a contradiction: $j \in \mathcal{I}(i)$. Since $\bar{\mathcal{I}}$ is a (sub-)network of quasi-reversible queues, it is itself quasi-reversible as well. Since the flow of jobs in question is one of its exit processes, the flow must be a Poisson process. The quasi-reversibility of $\bar{\mathcal{O}}$ follows from considering the time-reversal. $\square$

4.5 Arrival Theorems

Consider a single-node queue in isolation. From Theorem 4.4, we know that if the queue is quasi-reversible, then it necessarily has Poisson arrivals and hence, Poisson departures, since the time-reversal is also a quasi-reversible queue. As a matter of fact, quasi-reversibility also implies that arrivals will observe the *time-average* behavior of the queue. More precisely, the probability that the queue is in state $\mathbf{x}$ at the arrival epochs (of jobs to the queue) is exactly the same as $\pi(\mathbf{x})$, the ergodic, invariant distribution, which is the long-run average proportion of time that the system is in state $\mathbf{x}$. (In general, an "arrival theorem" refers to the result that the state observed at arrival epochs follows the ergodic distribution.)

To prove the above claim, consider the departure process from the queue. Let $\pi^D(\mathbf{x})$ denote the probability that the system is in state $\mathbf{x}$ at the departure epoch of a job. (Note that the state does *not* include the departing job. Hence, "at the departure epoch" should read as "immediately *after* the departure.") Suppose the departing job belongs to class c. Then, $\pi^D(\mathbf{x})$ is equal to the proportion of class c jobs departing from the queue leaving it in state $\mathbf{x}$. The overall rate of class c jobs leaving the queue is equal to the input rate λ_c [cf. (4.9)]. Hence, conditioning upon the state that transits into $\mathbf{x}$ due to a class c job's departure, relating to the time-reversal, and recognizing that $\mathbf{y} \in A^c(\mathbf{x})$ if and only if $\mathbf{x} \in D^c(\mathbf{y})$, we have

$$
\pi^D(\mathbf{x}) = \frac{1}{\lambda_c} \sum_{\mathbf{y} \in A^c(\mathbf{x})} \pi(\mathbf{y}) q(\mathbf{y}, \mathbf{x})
$$

$$
= \frac{1}{\lambda_c} \sum_{\mathbf{y} \in A^c(\mathbf{x})} \pi(\mathbf{x}) \tilde{q}(\mathbf{x}, \mathbf{y})
$$

$$
= \frac{1}{\lambda_c} \pi(\mathbf{x}) \sum_{\mathbf{y} \in A^c(\mathbf{x})} \tilde{q}(\mathbf{x}, \mathbf{y})
$$

$$
= \pi(\mathbf{x}), \tag{4.27}
$$

where the last equality makes use of (4.9). Hence, departing jobs observe the ergodic (or time-average) distribution $\pi(\mathbf{x})$. Since this argument also applies to the time-reversal, in which the departing jobs are just the arrivals to the original queue, we have established the claim that arrivals to a quasi-reversible queue observe time averages. That is, the state observed immediately *before* an arrival (hence, the arriving job is also excluded from the state description) follows the ergodic distribution as well.

To the extent that the arrival process to a quasi-reversible queue is necessarily Poisson, the above argument in fact reestablishes a well-known result in queueing theory, called *Poisson arrivals see time averages* (PASTA); refer to [24].

When quasi-reversible queues are connected into a network, as in Section 4.3 (in particular, Theorem 4.11), the arrival process to a *node* (or the departure process from it), in general, is no longer Poisson (in particular, if the flow of jobs is part of a loop, following the discussion in the last section). Nevertheless, in a network of quasi-reversible queues, arrivals to each node still *see time averages*. Indeed, this follows essentially from the same argument as above for an individual quasi-reversible queue [cf. (4.27)], making use of the product-form distribution in Theorem 4.11.

Specifically, let $\mathbf{x} = (\mathbf{x}_i)_{i=1}^{J}$ be the state left behind by a class c job's leaving node i and on its way to joining node j as a class d job. Note that here the state $\mathbf{x}$ does *not* include the departing job. That is, $\mathbf{x}$ is the state immediately *after* the job's leaving node i and immediately *before* its joining node j. The overall rate for this process of jobs in transition is $\lambda_i^c p_{ij}^{cd}$. Then, the only states $\mathbf{y}$ that can transit to $\mathbf{x}$ are $\mathbf{y} \in A_i^c(\mathbf{x})$. Specifically, $\mathbf{y}_i \in A_i^c(\mathbf{x}_i)$, and $\mathbf{y}_k = \mathbf{x}_k$ for $k \neq i$. Making use of the transition rates in (4.17), we can derive the probability of state $\mathbf{x}$ as observed by the jobs in transition, denoted by $\pi^T(\mathbf{x})$, as follows:

$$
\begin{aligned}
\pi^T(\mathbf{x}) \\
&= \frac{1}{\lambda_i^c p_{ij}^{cd}} \sum_{\mathbf{y} \in A_i^c(\mathbf{x})} \pi(\mathbf{y}) q(\mathbf{y}, \mathbf{x}) \\
&= \frac{1}{\lambda_i^c p_{ij}^{cd}} \prod_{k \neq i} \pi_k(\mathbf{x}_k) \sum_{\mathbf{y}_i \in A_i^c(\mathbf{x}_i)} \pi_i(\mathbf{y}_i) q_i(\mathbf{y}_i, \mathbf{x}_i) \sum_{\mathbf{y}_j \in A_j^d(\mathbf{x}_j)} q_j(\mathbf{y}_j, \mathbf{x}_j) \frac{p_{ij}^{cd}}{\lambda_j^d} \\
&= \frac{1}{\lambda_i^c} \prod_{k \neq i} \pi_k(\mathbf{x}_k) \cdot \pi_i(\mathbf{x}_i) \sum_{\mathbf{y}_i \in A_i^c(\mathbf{x}_i)} \tilde{q}_i(\mathbf{x}_i, \mathbf{y}_i) \\
&= \prod_{k \neq i} \pi_k(\mathbf{x}_k) \pi_i(\mathbf{x}_i) = \pi(\mathbf{x}).
\end{aligned}
\tag{4.28}
$$

Theorem 4.14 In the network of quasi-reversible queues in Theorem 4.11, the arrival to (or departure from) any node observes time averages, with the arrived or departed job itself excluded. In particular, the probability that the network is in state $\mathbf{x}$, as observed by any job in transition between nodes (with the job excluded from the state description), is equal to $\pi(\mathbf{x})$, the ergodic distribution.

The situation in a closed Jackson network is similar. Consider the closed network of Theorem 2.3. Pick node 1, for instance. Suppose that immediately *before* the departure of a job from node 1, the state is $\mathbf{x}$. Then, the departing job leaves behind a state $\mathbf{x} - \mathbf{e}_1$, which corresponds to a network with one fewer job. Using an argument similar to the one that led to (4.27), we have,

$$\pi^D(\mathbf{x} - \mathbf{e}_1) = \mathsf{E}\{\mathbf{1}[\mathbf{X} = \mathbf{x}]\mu_1(X_1)\}/\mathsf{E}[\mu_1(X_1)].$$

Making use of the relations in (2.9) and (2.10), we have

$$\mathsf{E}\{\mathbf{1}[\mathbf{X} = \mathbf{x}]\mu_1(X_1)\}$$

$$= \mu_1(x_1)\mathsf{P}[Y_1 = x_1]\frac{\mathsf{P}[\mathbf{Y}_{-1} = \mathbf{x}_{-1}]}{\mathsf{P}[|\mathbf{Y}| = N]}$$

$$= v_1\mathsf{P}[Y_1 = x_1 - 1]\frac{\mathsf{P}[\mathbf{Y}_{-1} = \mathbf{x}_{-1}]}{\mathsf{P}[|\mathbf{Y}| = N]}$$

$$= \frac{\mathsf{P}[Y_1 = x_1 - 1]\mathsf{P}[\mathbf{Y}_{-1} = \mathbf{x}_{-1}]}{\mathsf{P}[|\mathbf{Y}| = N - 1]} \cdot \frac{v_1\mathsf{P}[|\mathbf{Y}| = N - 1]}{\mathsf{P}[|\mathbf{Y}| = N]}$$

$$= \mathsf{P}[X_1 = x_1 - 1, \mathbf{X}_{-1} = \mathbf{x}_{-1}] \cdot v_1 TH(N).$$

Since [cf. (2.11)]

$$\mathsf{E}[\mu_1(X_1)] = v_1 TH(N),$$

we have

$$\pi^D(\mathbf{x} - \mathbf{e}_1) = \mathsf{P}[X_1 = x_1 - 1, \mathbf{X}_{-1} = \mathbf{x}_{-1}] = \pi_{N-1}(\mathbf{x} - \mathbf{e}_1), \qquad (4.29)$$

where π_{N-1} denotes the ergodic distribution of a closed network with $N-1$ jobs. That is, the probability of the state left behind by the departing job equals the ergodic probability of the same state in a network with one fewer job.

Theorem 4.15 In a closed Jackson network, the arrival to (or departure from) any node observes time averages, with the job itself excluded. In particular, the probability that the network is in state $\mathbf{x}$ immediately before an arrival (or immediately after a departure) epoch at a node i is equal to the ergodic distribution, of a closed network with one fewer job, in state $\mathbf{x} - \mathbf{e}_i$.

4.6   Notes

Kelly [8, 9] pioneered the technique of examining simultaneously a stationary Markov chain and its time-reversal (in the context of Theorem 4.1), the notion of a quasi-reversible queue, and the application of both to multiclass networks. Also refer to [10, 11]. Kelly's work builds upon the earlier work of Whittle [22], where the notion of partial balance was first developed. (Also see Whittle [21, 23].) The Poisson-in-Poisson-out property in certain queueing systems was first noted by Burke [5, 6].

Prior to Kelly's work, the most general model in the product-form framework was due to Baskett, Chandy, Muntz, and Palacios [2], known as the BCMP network, where they identified product-form results with multiclass networks under four types of service disciplines: LCFS, PS, IS and FCFS, with the first three allowing class-dependent and nonexponential service times. From Kelly's work, we now know that these three disciplines are special cases of symmetric queues, and that product-form results are widely present in networks of quasi-reversible queues.

The materials in Section 4.1 are drawn mostly from Chapter 1 of Kelly [10]. The symmetric queues in Section 4.2 differ slightly from those of Kelly's (Chapter 3 of [10]) in that we replace his gamma distributions by the more general PH-distributions. (This is also the treatment in Walrand [20], Chapter 3.) Neuts [16] is a standard reference for the PH-distributions and related queueing applications and computational techniques. The insensitivity result mentioned in Remark 4.10 was due originally to Barbour [1]. The multiclass network in Section 4.3 is similar in generality to the model discussed in Walrand [20], Chapter 4. Our proof is more detailed. The non-Markovian routing discussed at the end of that section is the routing scheme in Kelly's ([10]) original multiclass network.

The *nonloop criterion* in Theorem 4.13 is well known; e.g., Walrand [20] pp. 166–167; also see [17]. The proof of the arrival theorems in Section 4.5 essentially follows the original approach of Kelly [10]. Prior to Kelly's work, arrival theorems were best known in the setting of closed networks, where the phenomenon of having to exclude the arriving job itself from the observed state is more striking; see, e.g., Sevcik and Mitrani [19], Lavenberg and Reiser [12].

There is an extensive body of literature on PASTA (Wolff [24]), and more generally, ASTA (arrivals see time averages); refer to Brémaud [3], Brémaud, Kannurpatti, and Mazumdar [4], Melamed and Whitt [14, 13], Melamed and Yao [15], among others. The tutorial article [15] includes an extensive coverage of the literature and surveys both the elementary approach (based on Riemann–Stieltjes integral) and the point-process martingale approach, and also covers applications to networks of quasi-reversible queues.

For a more recent and comprehensive treatment of product-form networks, with a level of generality that is beyond the scope of this chapter, we refer the reader to Chao et al. [7] and Serfozo [18].

4.7 Exercises

1. Explain whether or not the following queues are quasi-reversible:

   (i) $M(n)/M(n)/1$,

   (ii) $M/M(n)/1$,

   (iii) $M/M/c$.

2. Consider the single-server, finite-buffer queue $M/M/1/N$.

   (i) Show that the state process (number of jobs systems) satisfies reversibility.

   (ii) Show that the departure process (service completions) is *not* Poisson; hence, the queue is *not* quasi-reversible.

   (iii) Does (ii) contradict (i)? Explain.

3. Consider the $M/M/1$ queue with feedback, i.e., every job, upon service completion, will return to the end of the queue with probability $1 - p$, and will leave the system with probability p. Let the arrival and service rates be λ and μ. Let $X(t)$ denote the number of jobs in the system at t; let $F(t)$ and $D(t)$ denote, respectively, the cumulative number of feedback jobs and departures (i.e., jobs leaving the system) up to t.

   (i) Derive the stationary (ergodic) distribution of X. What is the required traffic condition (in terms of λ, μ, and p)? Show that X is reversible.

   (ii) Explain whether or not the process F is Poisson. How about D?

   (iii) Do the feedback jobs see the time average (i.e., the ergodic distribution of X)?

4. Verify the product-form result in (4.8) for the LCFS discipline; in particular, modify one of the relations that lead to (4.6) as follows:

   $$\tilde{q}(\mathbf{x}, D_\ell(\mathbf{x})) = \mu_{c_\ell} \phi(n) \delta(n, \ell),$$

   i.e., allowing the class-dependent service rate μ_{c_ℓ}.

5. Consider a two-station network with four job classes, known as the Kumar–Seidman network, as shown in Figure 4.1. Class 1 jobs arrive at station 1, and after service completion continue on to station 2 as class 2. Class 3 jobs arrive at station 2, and after service completion continue on to station 1 as class 4. External arrivals (classes 1 and 3) follow Poisson processes with rates α_1 and α_3. Each station has a single server.

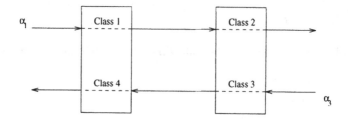

FIGURE 4.1. Kumar–Seidman network

(i) Suppose both stations follow a first-come-first-served discipline. Suppose the service times are i.i.d. exponential, dependent on the stations but independent of the job classes, with means $1/\mu_1$ and $1/\mu_2$ at the two stations. Verify that each station in isolation is a quasi-reversible queue, and derive the product-form solution for the system in equilibrium.

(ii) Suppose both stations follow a last-come-first-served discipline. Show that the results in (i) hold even when the i.i.d. exponential service times at each are class dependent (as well as station dependent).

(iii) Explain why quasi-reversibility would fail in (i) if the service times are class dependent.

References

[1] BARBOUR, A.D. 1976. Networks of Queues and the Methods of Stages. *Adv. Appl. Prob.*, **8**, 584–591.

[2] BASKETT, F., CHANDY, M., MUNTZ, R., AND PALACIOS, J. 1975. Open, Closed, and Mixed Networks of Queues with Different Classes of Customers. *J. Assoc. Comput. Mach.*, **22**, 248–260.

[3] BRÉMAUD, P. 1989. Characteristics of Queueing Systems Observed at Events and the Connection between Stochastic Intensity and Palm Probability. *Queueing Systems: Theory and Applications*, **5**, 99–112.

[4] BRÉMAUD, P., KANNURPATTI, R., AND MAZUMDAR, R. 1992. Event and Time Averages: A Review and Some Generalizations. *Adv. Appl. Prob.*, **24**, 377–411.

[5] BURKE, P.J. 1956. The Output of a Queueing System. *Operations Res.*, **4**, 699–704.

[6] BURKE, P.J. 1957. The Output of a Stationary M/M/s Queueing System. *Ann. Math. Stat.*, **39**, 1144–1152.

[7] CHAO, X., MIYAZAWA, M., AND PINEDO, M., *Queueing Networks: Customers, Signals and Product Form Solutions*, Wiley, New York, 1999.

[8] KELLY, F.P. 1975. Networks of Queues with Customers of Different Types. *J. Appl. Prob.*, **12**, 542–554.

[9] KELLY, F.P. 1976. Networks of Queues. *Adv. Appl. Prob.*, **8**, 416–432.

[10] KELLY, F.P. 1979. *Reversibility and Stochastic Networks.* Wiley, New York.

[11] KELLY, F.P. 1982. Networks of Quasi-Reversible Nodes. In *Applied Probability—Computer Science, the Interface*, Proc. of the ORSA/TIMS Boca Raton Symposium, R. Disney and T. Ott (eds.), Birkhäuser, Boston, MA.

[12] LAVENBERG, S.S. AND REISER, M. 1980. Stationary State Probabilities at Arrival Instants for Closed Queueing Networks with Multiple Types of Customers. *J. Appl. Prob.*, **17**, 1048–1061.

[13] MELAMED, B. AND WHITT, W. 1990. On Arrivals That See Time Averages: A Martingale Approach. *J. Appl. Prob.*, **27**, 376–384.

[14] MELAMED, B. AND WHITT, W. 1990. On Arrivals That See Time Averages. *Operations Res.*, **38**, 156–172.

[15] MELAMED, B. AND YAO, D.D. 1995. The ASTA Property. In: *Advances in Queueing*, J.H. Dshalalow (ed.), CRC Press, Boca Raton, FL, 195–224.

[16] NEUTS, M.F. 1981. *Matrix-Geometric Solutions in Stochastic Models.* Johns Hopkins University Press, Baltimore, MD.

[17] SERFOZO, R.F. 1989. Poisson Functionals of Markov Processes and Queueing Networks. *Adv. Appl. Prob.*, **21**, 595–611.

[18] SERFOZO, R., *Introduction to Stochastic Networks,* Springer-Verlag, New York, 1999.

[19] SEVCIK, K.C. AND MITRANI, I. 1981. The Distribution of Queueing Network States at Input and Output Instants. *J. Assoc. Comput. Mach.*, **28**, 358–371.

[20] WALRAND, J. 1988. *An Introduction to Queueing Networks.* Prentice Hall, Englewood Cliffs, NJ.

[21] WHITTLE, P. 1968. Equilibrium Distributions for an Open Migration Process. *J. Appl. Prob.*, **5**, 567–571.

[22] WHITTLE, P. 1985. Partial Balance and Insensitivity. *J. Appl. Prob.*, **22**, 168–176.

[23] WHITTLE, P. 1986. *Systems in Stochastic Equilibrium.* Wiley, New York.

[24] WOLFF, R.W. 1982. Poisson Arrivals See Time Averages. *Operations Res.*, **30**, 223–231.

5
Technical Desiderata

This chapter collects background materials for the many limit theorems in queues and queueing networks that will appear in later chapters. We start with presenting some preliminaries in basic probability theory, such as almost sure convergence and weak convergence, Donsker's theorem and Brownian motion, in Sections 5.1–5.3. Then in Sections 5.4–5.6, we focus on a pair of fundamental processes: the partial sum of i.i.d. random variables and the associated renewal counting process. The pair serves as a building block for modeling many queueing systems. We show that under different time–space scaling the pair converges differently, leading to functional versions of the strong law of large numbers and the central limit theorem. Furthermore, with additional moment conditions (on the i.i.d. random variables), we can refine these limits via functional versions of the law of iterated logarithms and strong approximations. When the generating function of the i.i.d. random variables exist, we can further characterize the convergence rate via exponential bounds.

We strive to strike a balance between mathematical rigor and accessibility to a broad audience. For example, we shall avoid the mathematical subtleties of certain concepts, proofs, and results if understanding them is not essential to understanding the main results of the later chapters. In addition, we either leave out or are sketchy about certain results and proofs that are less frequently used in the later chapters. Detailed references are given in Section 5.7 to assist those who would like to seek more comprehensive treatments of the topics covered in this chapter.

5.1   Convergence and Limits

Let $\Re^J$ denote J-dimensional Euclidean space, and $\mathcal{B}(\Re^J)$ the Borel field of $\Re^J$ (i.e., the smallest σ-field that contains all open sets in $\Re^J$). For $x = (x_j) \in \Re^J$, define

$$\|x\| := \max_{1 \le j \le J} |x_j|. \tag{5.1}$$

Let X be a mapping from a probability space $(\Omega, \mathcal{F}, \mathsf{P})$ to $(\Re^J, \mathcal{B}(\Re^J))$. If X is measurable (i.e., if $\{\omega \in \Omega : X(\omega) \in A\} \in \mathcal{F}$ for all $A \in \mathcal{B}(\Re^J)$), then we call it a (J-dimensional) random vector, defined on the probability space $(\Omega, \mathcal{F}, \mathsf{P})$ and with its range in $\Re^J$. When $J = 1$, we call X a random variable. The *distribution* of the random vector X is the probability measure P on $(\Re^J, \mathcal{B}(\Re^J))$ defined as

$$P(A) := \mathsf{P}\big(\omega \in \Omega, X(\omega) \in A\big), \qquad A \in \mathcal{B}(\Re^J),$$

and it is completely determined by the *distribution function* of X:

$$F(x) := \mathsf{P}\big(\omega \in \Omega, X(\omega) \le x\big), \qquad x \in \Re^J.$$

Below, we shall omit all references to $\omega \in \Omega$ whenever this is clear from the context. Hence, we shall write the above simply as $\mathsf{P}(X \in A)$ and $\mathsf{P}(X \le x)$.

Let X be a random vector, and let $\{X_n, n \ge 1\}$ be a sequence of J-dimensional random vectors, all defined on the probability space $(\Omega, \mathcal{F}, \mathsf{P})$. We say that X_n converges to X *with probability one* or *almost surely* as $n \to \infty$, if

$$\mathsf{P}\big(\lim_{n \to \infty} \|X_n - X\| = 0\big) = 1, \tag{5.2}$$

and that $X_n \to X$ *in probability* as $n \to \infty$ if for any $\epsilon > 0$,

$$\lim_{n \to \infty} \mathsf{P}\big(\|X_n - X\| \ge \epsilon\big) = 0. \tag{5.3}$$

Let P and P_n be the distributions of X and X_n, respectively. We say that P_n converges *weakly* to P if for every bounded and continuous function f on $\Re^J$,

$$\int_{\Re^J} f dP_n \to \int_{\Re^J} f dP \qquad \text{as } n \to \infty. \tag{5.4}$$

In this case, we say that X_n converges to X *in distribution*, or X_n converges *weakly* to X, written $X_n \overset{\mathrm{d}}{\to} X$. Note that an alternative but equivalent definition for convergence in distribution is that

$$\mathsf{P}(X_n \le x) \to \mathsf{P}(X \le x) \qquad \text{as } n \to \infty, \tag{5.5}$$

for all continuous points of $P(X \leq x)$.

We next discuss stochastic processes and their convergence. It would be helpful to relate the definitions and results below to those of random vectors. Fix $T > 0$; let $\mathcal{D}^J[0,T]$ be the space of J-dimensional real-valued functions on $[0,T]$ that are right-continuous and with left limits (RCLL). Here and throughout the later chapters the space $\mathcal{D}^J[0,T]$ is endowed with the uniform topology, with respect to the norm

$$\|x\|_T := \sup_{0 \leq t \leq T} \|x(t)\| = \sup_{0 \leq t \leq T} \max_{1 \leq j \leq J} \|x_j(t)\| \qquad (5.6)$$

for all $x \in \mathcal{D}^J$, where $x := \{x(t), t \geq 0\}$ and $x(t) := (x_j(t))_{j=1}^J$. (We note that the space $\mathcal{D}^J[0,T]$ under the uniform topology is not separable. Consequently, the so-called Skorohod topology is usually required for studying the weak convergence in $\mathcal{D}^J[0,T]$. However, the uniform topology is sufficient for our purpose, since all the limiting processes that arise from weak convergence considered in the later chapters are in $\mathcal{C}^J[0,T]$, the subspace of *continuous* functions. Refer to Section 5.7 for more details.) Let let $\mathcal{B}(\mathcal{D}^J[0,T])$ be the Borel field of $\mathcal{D}^J[0,T]$, and for simplicity, we shall write $\mathcal{B}(\mathcal{D}^J)$.

Similar to the definition of random vectors, let X be a mapping from a probability space $(\Omega, \mathcal{F}, P)$ to $(\mathcal{D}^J[0,T], \mathcal{B}(\mathcal{D}^J))$. X is said to be measurable if $\{\omega \in \Omega : X(\omega) \in A\} \in \mathcal{F}$ for all $A \in \mathcal{B}(\mathcal{D}^J)$. When X is measurable, we call it a (J-dimensional) stochastic process, with its range in $\mathcal{D}^J[0,T]$. The *distribution* of X is the probability measure P on $(\mathcal{D}^J[0,T], \mathcal{B}(\mathcal{D}^J))$, denoted as

$$P(A) := \mathsf{P}(\omega \in \Omega, X(\omega) \in A) := \mathsf{P}(X \in A), \qquad A \in \mathcal{B}(\mathcal{D}^J).$$

For any set of time points, $0 \leq t_1, \ldots, t_n \leq T$, we can define a finite dimensional distribution of X, by

$$\mathsf{P}(X(t_i) \leq x_i, i = 1, \ldots, n), \qquad \text{for all } x_1, \ldots, x_n \in \Re^J.$$

Note, however, that the probability law of X cannot be fully characterized by the set of all of its finite dimensional distributions. (Refer to Exercise 2.)

Let X be a stochastic process, and let $\{X_n, n \geq 1\}$ be a sequence of stochastic processes, all defined on the probability space $(\Omega, \mathcal{F}, \mathsf{P})$. We say that $X_n \to X$ *with probability one* or *almost surely* as $n \to \infty$ if (5.2) holds; and we say that $X_n \to X$ in probability as $n \to \infty$ if (5.3) holds for any $\epsilon > 0$. In both cases, the norm in (5.6) replaces the norm in (5.1).

Let P and P_n be the distributions of X and X_n, respectively. We say that P_n converges *weakly* to P if the convergence in (5.4) holds for every bounded and continuous real function f defined on $\mathcal{D}^J[0,T]$. In this case, we say that X_n converges to X *in distribution*, or X_n converges weakly to X, denoted by $X_n \overset{d}{\to} X$. However, an obvious extension of the equivalent definition for the weak convergence of random vectors based on (5.5) will not lead to an equivalent definition here. Specifically, the convergence of all finite-dimensional distributions,

$$\mathsf{P}(X_n(t_i) \leq x_i, i = 1, \ldots, m) \to \mathsf{P}(X(t_i) \leq x_i, i = 1, \ldots, m), \quad \text{as } n \to \infty,$$

for all $m \geq 1$, is *not* sufficient to guarantee the weak convergence of X_n to X. This should not be surprising, given, as we noted above, that the set of all finite dimensional distributions of a stochastic process does not uniquely determine its distribution. Also see Exercises 4 and 5.

Now we extend $\mathcal{D}^J[0, T]$ to $\mathcal{D}^J := \mathcal{D}^J[0, \infty)$ in a natural way, with the latter denoting the space of J-dimensional real-valued functions on $[0, \infty)$ that are right-continuous and with left limits. Hence, we can define stochastic processes in $\mathcal{D}^J$. (Necessarily, the restriction of a stochastic process in $\mathcal{D}^J$ to the interval $[0, T]$ is also a stochastic process in $\mathcal{D}^J[0, T]$ for any $T > 0$.) We say that a sequence of stochastic processes X_n in $\mathcal{D}^J$ converges to X in $\mathcal{D}^J$ almost surely (respectively, in probability, weakly) if the restrictions of X_n to $[0, T]$ converge to the restriction of X to $[0, T]$ in $\mathcal{D}^J[0, T]$ almost surely (respectively in probability, weakly) for any fixed $T > 0$.

Some of the most often used subspaces of $\Re^J$ and $\mathcal{D}^J$ in this book are $\Re^J_+ = \{x \in \Re^J, x \geq 0\}$, the nonnegative orthant of $\Re^J$; $\mathcal{C}^J[0, T]$, the space of J-dimensional real-valued continuous functions on $[0, T]$; $\mathcal{D}^J_0[0, T] = \{x \in \mathcal{D}^J[0, T], x(0) \geq 0\}$; and $\mathcal{C}^J_0[0, T] = \{x \in \mathcal{C}^J[0, T], x(0) \geq 0\}$. Similarly, set $\mathcal{C}^J = \mathcal{C}^J[0, \infty)$, $\mathcal{D}^J_0 = \mathcal{D}^J_0[0, \infty)$ and $\mathcal{C}^J_0 = \mathcal{C}^J_0[0, \infty)$.

We conclude this section by revisiting the almost sure convergence in the space $\mathcal{D}^J$, which is the mode of convergence of almost all of the limit theorems in this book. Note that from the above definition, for x and x_n ($n \geq 1$) in $\mathcal{D}^J[0, T]$, $\|x_n - x\|_T \to 0$ as $n \to \infty$ is equivalent to x_n converging to x *uniformly* in $[0, T]$. Hence, that a sequence of stochastic processes $X_n \in \mathcal{D}^J$ converges to a stochastic process X almost surely can be more explicitly stated as, for each $\omega \in \Omega$, $X_n(\omega)$ converges to $X(\omega)$ uniformly in $[0, T]$ as $n \to \infty$, or *uniformly on all compact sets* (u.o.c.) as $n \to \infty$.

5.2   Some Useful Theorems

For simplicity, we shall use a *random element* to refer to either a random vector or a stochastic process, and a *metric space* $\mathcal{S}$ will be used to refer to either $\Re^J$, $\mathcal{D}^J[0, T]$, or $\mathcal{D}^J$. Two random elements X and Y, which need not be defined on the same probability space, are said to be *equal in distribution*, or one is said to be a version of another, if their ranges are in the same space $(\mathcal{S}, \mathcal{B}(\mathcal{S}))$ and $\mathsf{P}_1(X \in A) = \mathsf{P}_2(Y \in A)$ for all $A \in \mathcal{B}(\mathcal{S})$, where $\mathcal{B}(\mathcal{S})$ is the Borel σ-field of $\mathcal{S}$, and P_1 and P_2 are the probability measures for the probability spaces on which X and Y are defined, respectively. A sequence of random elements $\{Y_n, n \geq 1\}$ is said to be a version of a sequence $\{X_n, n \geq 1\}$ if Y_n is a version of X_n for every $n \geq 1$.

We now state the Skorohod representation theorem, which converts weak convergence into almost sure convergence.

Theorem 5.1 (Skorohod Representation Theorem) Let X be a random element, and $\{X_n, n \geq 1\}$ a sequence of random elements with range in a separable metric space S. Suppose that $X_n \overset{d}{\to} X$ as $n \to \infty$. Then there exists a common probability space in which versions of X and $\{X_n, n \geq 1\}$, denoted by X' and $\{X'_n, n \geq 1\}$, are defined such that X'_n converges to X' almost surely as $n \to \infty$.

Note that when $S = \mathcal{D}^J[0, T]$ or $\mathcal{D}^J$, the almost sure convergence in the above theorem is with respect to the Skorohod topology. When $X \in \mathcal{C}^J[0, T]$, however, the almost sure convergence in Skorohod topology is equivalent to the almost sure convergence in the uniform topology. And this is the case throughout this book. Therefore, we shall not introduce the Skorohod topology. Also note that although the above theorem holds in more general metric spaces, for our purpose, it is sufficient to assume that the metric space S is either $\Re^J$, $\mathcal{D}^J[0, T]$, $\mathcal{D}^J$, or one of their subspaces. (This applies to the next three theorems as well. Those who are interested in the more general cases are referred to Whitt [13].) The proof of Theorem 5.1 in its full generality is beyond the scope of this book. However, a special case of this theorem, when X and $\{X_n, n \geq 1\}$ are random variables, is simple enough to be designated as an exercise problem (Exercise 6).

As a consequence of the Skorohod representation theorem, we next conveniently formulate deterministic versions of the continuous mapping theorem, random time-change theorem and convergence-together theorem. The proofs of these theorems are elementary and are left as exercises.

Theorem 5.2 (Continuous Mapping Theorem) Suppose X and $\{X_n, n \geq 1\}$ belong to $\mathcal{D}^J$, and f is a measurable mapping from $(\mathcal{D}^J, \mathcal{B}(\mathcal{D}^J))$ into $(\mathcal{D}^J, \mathcal{B}(\mathcal{D}^J))$. If f is continuous, then as $n \to \infty$, $X_n \to X$ u.o.c. implies $f(X_n) \to f(X)$ u.o.c.

Theorem 5.3 (Random Time-Change Theorem) Let $\{X_n, n \geq 1\}$ and $\{Y_n, n \geq 1\}$ be two sequences in $\mathcal{D}^J$. Assume that Y_n is nondecreasing with $Y_n(0) = 0$. If as $n \to \infty$, (X_n, Y_n) converges uniformly on compact sets to (X, Y) with X and Y in $\mathcal{C}^J$, then $X_n(Y_n)$ converges uniformly on compact sets to $X(Y)$, where $X_n(Y_n) = X_n \circ Y_n = \{X_n(Y_n(t)), t \geq 0\}$ and $X(Y) = X \circ Y = \{X(Y(t)), t \geq 0\}$.

Theorem 5.4 (Convergence-Together Theorem) Let $\{X_n, n \geq 1\}$ and $\{Y_n, n \geq 1\}$ be two sequences in $\mathcal{D}^J$. If as $n \to \infty$, $X_n - Y_n$ converges uniformly on compact sets to zero and X_n converges uniformly on compact sets to X in $\mathcal{C}^J$, then Y_n converges uniformly on compact sets to X.

5.3   Brownian Motion

A process $W = \{W(t), t \geq 0\} \in \mathcal{C}$ is said to be a standard Brownian motion (or a Wiener process) if $W(0) = 0$ and for every $t, s \geq 0$, the increment $W(t + s) - W(t)$ is independent of $\{W(u), 0 \leq u \leq t\}$, and is normally distributed with mean 0 and variance s. A process $X = \{X(t), t \geq 0\}$, defined by $X(t) = X(0) + \theta t + \sigma W(t)$, is called a Brownian motion (starting at $X(0)$) with drift θ and variance σ^2 if W is a standard Brownian motion, $X(0)$ is a random variable independent of W, and θ and σ are real constants. Clearly, letting $X(0) = 0$, $\theta = 0$, and $\sigma = 1$ recovers the standard Brownian motion.

To study the limiting distribution for the maximum of a Brownian motion, we introduce martingales, a very useful concept in itself. Let $\mathcal{T}$ be an index set representing either $\Re_+$ or $\mathcal{Z}_+$ (the set of nonnegative integers). Let $X = \{X_t, t \in \mathcal{T}\}$ be a real-valued process, adapted to a filtration $\mathcal{F} = \{\mathcal{F}_t, t \in \mathcal{T}\}$ (that is, $\mathcal{F}_s \subseteq \mathcal{F}_t$ for all $s, t \in \mathcal{T}$ with $s \leq t$ and $X_t \in \mathcal{F}_t$ for all $t \in \mathcal{T}$). X is called a *martingale* (w.r.t. $\mathcal{F}$) if $\mathsf{E}|X_t| < \infty$, and

$$\mathsf{E}[X_t | \mathcal{F}_s] = X_s, \tag{5.7}$$

for all $s, t \in \mathcal{T}$ with $s \leq t$. It is called a *submartingale* (respectively *supermartingale*) if the last equality is replaced by $\geq$ (respectively $\leq$). Therefore, if $\{X_t\}$ is a submartingale, then $\{-X_t\}$ is a supermartingale; if $\{X_t\}$ is a martingale, then it is both a submartingale and a supermartingale. For simplicity, below we shall omit the reference to the filtration $\mathcal{F}$ if it is the "natural" filtration generated by X, or if the filtration in question is obvious from the context.

It is easy to verify that the standard Brownian motion (W) is a martingale (the independent increment property is the key to the martingale property).

Lemma 5.5 Let X be a Brownian motion with drift μ and standard deviation σ starting at the origin, and let

$$M_t = \sup_{0 \leq s \leq t} X(s).$$

If $\mu < 0$, then

$$\lim_{t \to \infty} \mathsf{P}(M_t \geq x) = e^{(2\mu/\sigma^2)x}, \qquad x \geq 0;$$

otherwise,

$$\lim_{t \to \infty} \mathsf{P}(M_t \geq x) = 1, \qquad x \geq 0.$$

Proof. Note that $[X(t) - \mu t]/\sigma$ is a standard Brownian motion; hence,

$$e^{\theta X(t) - (\theta\mu + \frac{1}{2}\theta^2\sigma^2)t}$$

is a martingale for every value of the parameter θ (see Exercise 20). In particular, taking $\theta = -2\mu/\sigma^2$, we have the exponential martingale

$$U(t) := e^{-(2\mu/\sigma^2)X(t)}.$$

Let $T_{ax} := \inf\{t : X(t) \notin [a, x)\}$ for $a < 0$. (Note that $\mathsf{E}T_{ax} < \infty$; see Exercise 21.) Applying the optional sampling result— the version in (5.41) of Exercise 18—leads to the following:

$$1 = \mathsf{E}e^{-(2\mu/\sigma^2)X(T_{ax})}$$
$$= e^{-(2\mu/\sigma^2)a} \cdot \mathsf{P}[X(T_{ax}) = a] + e^{-(2\mu/\sigma^2)x} \cdot \mathsf{P}[X(T_{ax}) = x]. \qquad (5.8)$$

When $\mu < 0$, taking $a \to -\infty$, we have

$$\lim_{a \to -\infty} \mathsf{P}[X(T_{ax}) = x] = e^{(2\mu/\sigma^2)x}.$$

The left hand side above is the probability that X will ever reach x, i.e.,

$$\lim_{t \to \infty} \mathsf{P}(M_t \geq x) = \lim_{a \to -\infty} \mathsf{P}[X(T_{ax}) = x] = e^{(2\mu/\sigma^2)x}.$$

Next consider $\mu > 0$. Note that

$$\mathsf{P}[X(T_{ax}) = a] + \mathsf{P}[X(T_{ax}) = x] = 1;$$

from (5.8), we have

$$\mathsf{P}[X(T_{ax}) = x] = \frac{1 - e^{-2(\mu/\sigma^2)a}}{e^{-2(\mu/\sigma^2)x} - e^{-2(\mu/\sigma^2)a}}.$$

Letting $a \to -\infty$ in the above, we have

$$\lim_{a \to -\infty} \mathsf{P}(X(T_{ax}) = x) = 1;$$

hence,

$$\lim_{t \to \infty} \mathsf{P}(M_t \geq x) = 1.$$

Finally, consider $\mu = 0$. In this case, X, being a driftless Brownian motion, is itself a martingale. Applying (5.41) once again, we have

$$0 = \mathsf{E}X(T_{ax}) = a\mathsf{P}[X(T_{ax}) = a] + x\mathsf{P}[X(T_{ax}) = x].$$

Hence,

$$\mathsf{P}[X(T_{ax}) = x] = 1 - \mathsf{P}[X(T_{ax}) = a] = \frac{-a}{x - a}.$$

Letting $a \to -\infty$ yields

$$\lim_{t \to \infty} \mathsf{P}[M_t \geq x] = \lim_{a \to -\infty} \mathsf{P}(X(T_{ax}) = x) = 1.$$

$\square$

The above lemma is closely related to a single-server queueing model, in which the analogue of M_t in the lemma is the maximum of a random walk; and this maximum, in turn, is equal in distribution to the delay (waiting time) process. The negative drift case, $\mu < 0$, corresponds to the traffic intensity $\rho < 1$ in the queueing model, in which case, the limiting delay also follows an exponential distribution; see, for example, Section 6.5.

Next, we state a simple version of what is known as Ito's lemma/formula. Let $W = \{W(t), t \geq 0\}$ be a Wiener process (the standard Brownian motion), and let $A = \{A(t), t \geq 0\}$ be a continuous process with finite variation and $A(0) = 0$. (A process or a function is said to have finite variation if it can be expressed as the difference of two nondecreasing processes or functions.) Let $X(0)$ be a random variable independent of W. Then, we have the following result.

Theorem 5.6 Let f be a function having a continuous second derivative. Let

$$X(t) = X(0) + aW(t) + A(t),$$

where a is a constant. Then, for $t \geq 0$,

$$f(X(t)) = f(X(0)) + a \int_0^t f'(X(s))dW(s) + \int_0^t f'(X(s))dA(s)$$
$$+ \frac{1}{2}a^2 \int_0^t f''(X(s))ds, \tag{5.9}$$

where the first integral above is the so-called Ito's integral, whereas the last two integrals are the usual Riemann–Stieltjes and Riemann integrals.

We shall not attempt to present the formal definition of Ito's integral; interested readers may refer to, for example, Harrison [8], or Karatzas and Shreve [9]. Intuitively and symbolically, we write

$$dW(t) = Z \cdot \sqrt{dt}, \tag{5.10}$$

where Z is a standard normal variate. Note that since

$$\mathsf{E}[(dW(t))^2] = dt = \mathsf{E}[(Z\sqrt{dt})^2],$$

the equation in (5.10) (whose proof is beyond the scope of this book), can be viewed as a "natural" strengthening of the above equation in expectation to a probability-one (a.s.) relation. Hence, $[dW(t)]^2$ is of order dt. On the other hand, we note that $[dA(t)]^2$ is of order $(dt)^2$ or $o(dt)$.

Furthermore, we can express (5.9) in differential form as follows:

$$df(X(t)) = f'(X(t))[adW(t) + dA(t)] + \frac{a^2}{2}f''(X(t))dt. \qquad (5.11)$$

The first-derivative part on the right-hand side above is nothing but the usual "chain rule" in differentiation. The second-derivative part, which in regular calculus is of order $(dt)^2$ and hence vanishes under differentiation, follows from

$$[dX(t)]^2 = [adW(t) + dA(t)]^2 = a^2dt + o(dt),$$

taking into account (5.10), and replacing Z^2dt by its mean dt. (The variance of Z^2dt is of order $(dt)^2$. Hence, as $dt \to 0$, Z^2dt can be viewed as a deterministic quantity. As to the coefficient $\frac{1}{2}$ in (5.9) and (5.11), it is associated with the second derivative in the Taylor expansion of $f(x)$.)

In other words, in differentiating $f(X(t))$ when X is a Brownian motion, the second derivative cannot be ignored, as its leading term is of order dt. In this sense, (5.11) can be viewed as the chain rule for Ito's calculus.

Also note that since W is a martingale, the Ito's integral in (5.9),

$$\int_0^t f'(X(s))dW(s),$$

is also a martingale. To ensure (square) integrability, it is standard to require the condition

$$\mathsf{E}\int_0^t [f'(X(s))]^2 ds < \infty,$$

for each $t \geq 0$. Predictability of the process $f'(X(t))$ is guaranteed by the continuity of the process X and the function $f'(x)$. (Predictability is implied by left continuity, which of course is implied by continuity.)

Another important fact of the Brownian motion is that it is a *strong Markov process*. For a stochastic process, $X \in \mathcal{D}^J$, which is adapted to a filtration $\mathcal{F} = \{\mathcal{F}_t, t \geq 0\}$. It is called a strong Markov process if for any $\mathcal{F}$-stopping time T and any real number $s \geq 0$, we have

$$P[X(T+s) \in A \,|\, \mathcal{F}_T] = P[X(T+s) \in A \,|\, X(T)], \quad \forall A \in \mathcal{B}(\mathcal{D}^J). \quad (5.12)$$

Note that when the stopping time T degenerates to a constant t, the above is simply the usual Markov property. It is easily verified that a Brownian motion satisfies the Markov property. For a proof that the Brownian motion also satisfies the strong Markov property as stated above, refer to Karatzas and Shreve [9].

Finally, we state Donsker's theorem, which relates the limit of a scaled i.i.d. summation to Brownian motion.

Theorem 5.7 (Donsker's Theorem) Let $\{\xi_i, i \geq 1\}$ be a sequence of i.i.d. random variables. Assume that ξ_i has a finite mean m and a finite standard deviation σ. For each $n \geq 1$, define a centered summation process $X_n = \{X_n(t), t \geq 0\}$ by

$$X_n(t) = \frac{1}{\sqrt{n}\sigma} \sum_{i=1}^{\lfloor nt \rfloor} [\xi_i - m],$$

where the summation is understood to be zero when $nt < 1$. Then,

$$X_n \xrightarrow{d} W, \qquad \text{as } n \to \infty,$$

where W is the Wiener process (standard Brownian motion).

5.4   Two Fundamental Processes

Let $X \in \mathcal{D}$ be a nondecreasing process, and let $Y = \{Y(t), t \geq 0\}$ denote its inverse, defined as follows:

$$Y(t) = \sup\{s \geq 0 : X(s) \leq t\}, \qquad 0 \leq t < \infty. \tag{5.13}$$

The lemmas below establish the inverse relations between the limits associated with the two processes.

Lemma 5.8 Consider the (X, Y) pair introduced above. Suppose

$$\frac{X(t)}{t} \xrightarrow{a.s.} m, \qquad \text{as } t \to \infty, \tag{5.14}$$

for a positive constant m, and set $\mu := 1/m$. Then,

$$\frac{Y(t)}{t} \xrightarrow{a.s.} \mu, \qquad \text{as } t \to \infty. \tag{5.15}$$

Furthermore, as $n \to \infty$,

$$\bar{X}^n(t) := \frac{1}{n} X(nt) \xrightarrow{a.s.} mt, \quad \text{u.o.c.}, \tag{5.16}$$

$$\bar{Y}^n(t) := \frac{1}{n} Y(nt) \xrightarrow{a.s.} \mu t, \quad \text{u.o.c.} \tag{5.17}$$

Proof. First we establish (5.15). By the definition of Y, we have

$$X(Y(t)-) \leq t < X(Y(t) + 1), \qquad t \geq 0,$$

where we used the fact that X has a left limit. This, along with (5.14), implies $Y(t) \xrightarrow{a.s.} \infty$ as $t \to \infty$. The above inequality can be rewritten as

$$\frac{Y(t)}{X(Y(t)-)} \geq \frac{Y(t)}{t} > \frac{Y(t) + 1}{X(Y(t) + 1)} \times \frac{Y(t)}{Y(t) + 1},$$

which leads to (5.15) as $t \to \infty$. (Note that when t is large enough, $X(Y(t)-)$ is clearly positive, and also note that (5.14) clearly implies $X(t-)/t \xrightarrow{\text{a.s.}} m$ as $t \to \infty$.)

Next, we show that (5.14) implies (5.16) (and that (5.15) implies (5.17)k which is completely analogous). That the limit in (5.16) holds for each fixed $t > 0$ is clearly implied by (5.14), in view of

$$\bar{X}^n(t) = \frac{1}{nt} X(nt) \times t,$$

and the limit clearly holds for $t = 0$. Now we show that the convergence is uniform on any compact set. Fix $T > 0$. For any given $\epsilon > 0$, there exists a $t_0 > 0$ such that for all $t \geq t_0$,

$$\left| \frac{X(t)}{t} - m \right| < \frac{\epsilon}{T}, \tag{5.18}$$

because of (5.14). Since X is RCLL, X must be bounded on any bounded interval; hence,

$$M := \sup_{0 \leq t \leq t_0} |X(t)| < \infty.$$

So it suffices to show that for $n > (M + mT)/\epsilon$, we have

$$|\bar{X}^n(t) - mt| < \epsilon \qquad \text{for all } t \in [0, T]. \tag{5.19}$$

There are two cases. For any $t \in [0, T]$, if $nt \geq t_0$, then in view of (5.18), we have

$$|\bar{X}^n(t) - mt| = |t| \left| \frac{X(nt)}{nt} - m \right| < |T| \times \frac{\epsilon}{T} = \epsilon.$$

On the other hand, if $nt < t_0$, then

$$|\bar{X}^n(t) - mt| \leq \frac{1}{n} \left[\sup_{0 \leq u \leq t_0} |X(u)| + mT \right] = \frac{1}{n}(M + mT) < \epsilon.$$

$\square$

Lemma 5.9 Consider the (X, Y) pair introduced above. Suppose that as $T \to \infty$,

$$\|X - m\|_T := \sup_{0 \leq t \leq T} |X(t) - mt| \stackrel{\text{a.s.}}{=} O\left(\sqrt{T \log \log T}\right) \tag{5.20}$$

for a positive constant m. Set $\mu := 1/m$. Then as $T \to \infty$,

$$\|Y - \mu\|_T := \sup_{0 \leq t \leq T} |Y(t) - \mu t| \stackrel{\text{a.s.}}{=} O\left(\sqrt{T \log \log T}\right).$$

Proof. First note that

$$\mu t - Y(t) = \mu[X(Y(t)) - mY(t)] + \mu[t - X(Y(t))]. \tag{5.21}$$

It follows from the definitions of X and Y that

$$\sup_{0 \le t \le T} |t - X(Y(t))| \le \sup_{0 \le t \le Y(T)} |X(t) - X(t-)|. \tag{5.22}$$

It follows from (5.20) that $X(t)/t \xrightarrow{\text{a.s.}} m$ as $t \to \infty$; hence, by Lemma 5.8, we have $Y(t)/t \xrightarrow{\text{a.s.}} \mu$ as $t \to \infty$, implying for some $t_0 > 0$, $Y(t) \le (\mu + 1)t$ almost surely for $t \ge t_0$. The latter, combined with (5.20) and (5.22), in view of (5.21), proves the lemma.   □

In the next section we shall focus on a particular X process

$$X(t) := \sum_{i=1}^{\lfloor t \rfloor} \xi_i, \quad t \ge 0, \tag{5.23}$$

where $\xi = \{\xi_i, i = 1, 2, \dots\}$ is a sequence of nonnegative i.i.d. random variables in $\Re$, and for $0 \le t < 1$, the summation in (5.23) is understood to be zero. In this case, Y, still the inverse of X following (5.13), is called a renewal process.

5.5 Limit Theorems for the Two Fundamental Processes

Here we study various joint limit theorems for the i.i.d. summation process $X = \{X(t), t \ge 0\}$ and its associated renewal (counting) process $Y = \{Y(t), t \ge 0\}$ introduced at the end of the last section. Let us start with the following notation:

$$\bar{X}^n(t) := \frac{1}{n} X(nt) \quad \text{and} \quad \bar{Y}^n(t) := \frac{1}{n} Y(nt), \tag{5.24}$$

$$\hat{X}^n(t) := \sqrt{n} \left[\bar{X}^n(t) - \bar{X}(t) \right] \quad \text{and} \quad \hat{Y}^n(t) := \sqrt{n} \left[\bar{Y}^n(t) - \bar{Y}(t) \right], \tag{5.25}$$

where $\bar{X}(t) := mt$ ($m > 0$), and $\bar{Y}(t) := \mu t$ with $\mu = 1/m$.

Note that while both the "bar" and the "hat" processes above use the same time scaling (multiplying by n), they differ in space scaling, with the "bar" processes using division by n, while the "hat" processes uses division by $\sqrt{n}$.

We simulate a special case of the renewal process Y, namely a unit Poisson process, to illustrate the idea of the scaling and the limit theorems. Figure 5.1 shows four sample paths of a unit Poisson process for the first

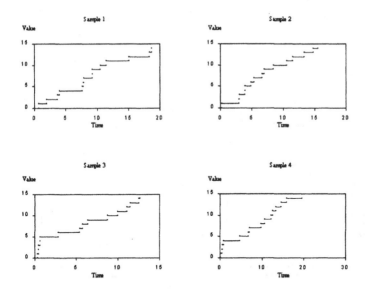

FIGURE 5.1. Four sample paths of a unit Poisson process

15 arrivals, whereas Figure 5.2 shows the same four sample paths of a unit Poisson process for a much longer time interval, 25×10^4. In order to plot the whole sample paths, both time and space have to be scaled. The plot actually shows the paths of $\bar{Y}^n$ with $n = 10^4$, and we note that all paths are almost the same straight line with a unit slope. Finally, Figure 5.3 shows the four sample paths of $\hat{Y}^n$ with $n = 10^4$, each of which shows the deviation between each sample path $\bar{Y}^n$ in Figure 5.2 and the straight line $\bar{Y}(t) = t$. In order to make the deviation visible in the plot, the space is scaled to 10^2 (instead of 10^4). We note that from the functional central limit theorem, each of these sample paths is approximately a sample path of a standard Brownian motion.

5.5.1 Functional Strong Law of Large Numbers

Theorem 5.10 (FSLLN) Suppose that ξ_i has a finite mean $m > 0$. Then, as $n \to \infty$,

$$(\bar{X}^n, \bar{Y}^n) \xrightarrow{\text{a.s.}} (\bar{X}, \bar{Y}) \qquad \text{u.o.c.} \tag{5.26}$$

Proof That $X(n)/n \to m$ almost surely follows from the standard SLLN. Hence, the desired limits follows from Lemma 5.8. $\square$

Note that when $m = 0$, $\xi_i = 0$ almost surely (since it is assumed that $\xi_i \geq 0$ almost surely). Hence, $\bar{X}^n = 0$ and $\bar{Y}^n = \infty$ almost surely.

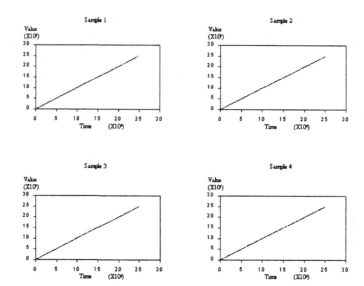

FIGURE 5.2. Four sample paths of a unit Poisson process: An illustration of the functional strong law of large numbers

5.5.2 Functional Central Limit Theorem

Theorem 5.11 (FCLT) Suppose ξ_i has a finite variance σ^2. Then,

$$(\hat{X}^n, \hat{Y}^n) \overset{d}{\to} (\hat{X}, \hat{Y}),$$

where $\hat{X}(t) = \sigma W(t)$, $\hat{Y}(t) = -\mu \hat{X}(\mu t)$, and W is a Wiener process, i.e., a one-dimensional standard Brownian motion.

Proof It follows from Donsker's theorem (Theorem 5.7) that

$$\hat{X}^n = \sqrt{n}[\bar{X}^n - \bar{X}] \overset{d}{\to} \hat{X} \quad \text{as } n \to \infty,$$

and it follows from Theorem 5.10 that $\bar{X}^n \overset{a.s.}{\longrightarrow} \bar{X}$ and $\bar{Y}^n \overset{a.s.}{\longrightarrow} \bar{Y}$ u.o.c. as $n \to \infty$. Then, from Theorem 5.3 (random time change theorem), we have

$$\hat{X}^n(\bar{Y}^n) = \sqrt{n}[\bar{X}^n(\bar{Y}^n) - m\bar{Y}^n] \overset{d}{\to} \hat{X}(\bar{Y})$$

as $n \to \infty$, where we note that $\bar{X}(t) = mt$. The above implies $\hat{Y}^n \overset{d}{\to} \hat{Y}$ as $n \to \infty$, if we have $\sqrt{n}[\bar{X}^n(\bar{Y}^n(t)) - t] \overset{a.s.}{\longrightarrow} 0$ u.o.c. as $n \to \infty$. The latter follows from $|\bar{X}^n(\bar{Y}^n(t)) - t| \leq 1/n$ for all $t \geq 0$, in view of (5.13) and (5.23). We also note that all of the above convergences hold jointly. □

Remark 5.12 Defining a mapping on $f: x \to x(1)$ on $\mathcal{D}^2$ and applying the continuous mapping theorem (Theorem 5.2) to the FCLT above, we

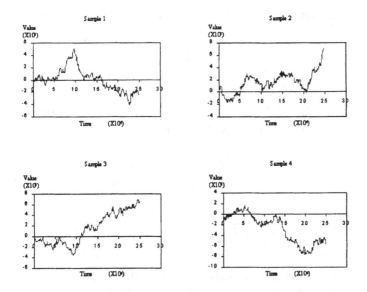

FIGURE 5.3. Four sample paths of a unit Poisson process: An illustration of the functional central limit theorem

have $(\hat{X}^n(1), \hat{Y}^n(1)) \xrightarrow{\mathrm{d}} (\hat{X}(1), \hat{Y}(1))$ as $n \to \infty$, which is the central limit theorem for an i.i.d. summation and its associated renewal process. In other words, the FCLT implies the CLT.

5.5.3 Functional Law of the Iterated Logarithm

Clearly, FCLT is a refinement of FSLLN in the sense of characterizing the deviation between the original stochastic processes (with scaling) and their FSLLN limits. The functional law of the iterated logarithm (FLIL) below is another refinement of FSLLN.

Theorem 5.13 (FLIL) Suppose that ξ has a finite variance. Then

$$\|X - m\|_T := \sup_{0 \le t \le T} |X(t) - mt| \stackrel{\text{a.s.}}{=} O\left(\sqrt{T \log \log T}\right), \qquad (5.27)$$

$$\|Y - \mu\|_T := \sup_{0 \le t \le T} |Y(t) - \mu t| \stackrel{\text{a.s.}}{=} O\left(\sqrt{T \log \log T}\right), \qquad (5.28)$$

as $T \to \infty$.

Proof. The FLIL bound (5.27) follows from the standard LIL (the version for the maximum of the summation process). Then the FLIL bound (5.28) follows from Lemma 5.9. $\qquad\square$

The standard version of the FLIL, following, for instance, Theorems 1.3.2 and 3.2.2 of Csörgő and Révész [4], can be stated as follows:

$$\limsup_{T\to\infty} \frac{\sup_{0\le t\le T} |X(t) - mt|}{\sqrt{T\log\log T}} = \sqrt{2}\sigma,$$

which clearly implies (5.27). On the other hand, (5.28) follows from Corollary 2.1.2 of Csörgő and Horváth [3], which gives

$$\limsup_{T\to\infty} \frac{\sup_{0\le t\le T} |Y(t) - \mu t|}{\sqrt{T\log\log T}} = \frac{\mu^{3/2}\sigma}{\sqrt{2}}$$

and

$$\liminf_{T\to\infty} \sqrt{\frac{\log\log T}{T}} \sup_{0\le t\le T} |Y(t) - \mu t| = \frac{\pi\mu^{3/2}\sigma}{\sqrt{8}}.$$

Therefore, Theorem 5.13 is a weaker version of the standard FLIL.

Replacing t by nt, and T by nT in (5.27) and (5.28), we have as $n \to \infty$,

$$\sup_{0\le t\le T} |\bar{X}^n(t) - \bar{X}(t)| = O\left(\sqrt{\frac{\log\log n}{n}}\right),$$

$$\sup_{0\le t\le T} |\bar{Y}^n(t) - \bar{Y}(t)| = O\left(\sqrt{\frac{\log\log n}{n}}\right),$$

for any fixed $T \ge 0$. Letting $n \to \infty$ yields the FSLLN limit (5.26), of course with the additional requirement that ξ have a finite variance.

5.5.4 Functional Strong Approximation

We have now observed that with a finite second moment of ξ, FSLLN can be refined via FCLT and FLIL. Bringing the moment condition of ξ to an even higher order leads to further refinement via the following result:

Theorem 5.14 (FSAT) Suppose that ξ_i has a finite moment of order $r > 2$. Then there exists a Wiener process $W = \{W(t), t \ge 0\}$ such that

$$\|X - \tilde{X}\|_T \equiv \sup_{0\le t\le T} |X(t) - \tilde{X}(t)| \stackrel{\text{a.s.}}{=} o(T^{1/r}), \tag{5.29}$$

$$\|Y - \tilde{Y}\|_T \equiv \sup_{0\le t\le T} |Y(t) - \tilde{Y}(t)| \stackrel{\text{a.s.}}{=} o(T^{1/r'}), \tag{5.30}$$

as $T \to \infty$, where $r' = r$ if $r < 4$, and $r' < 4$ if $r \ge 4$; and

$$\tilde{X}(t) = mt + \sigma W(t), \quad t \ge 0, \tag{5.31}$$

$$\tilde{Y}(t) = \mu t - \mu\sigma W(\mu t), \quad t \ge 0. \tag{5.32}$$

The proof for the above theorem can be found in Chapters 1 and 2 of Csörgő and Horváth [3].

Remark 5.15 If $r \geq 4$, the result in (5.30) can be improved to

$$\|Y - \tilde{Y}\|_T \stackrel{\text{a.s.}}{=} O((\log T)^{1/2}(T \log \log T)^{1/4}). \tag{5.33}$$

If $\mathsf{E}\left(e^{t\xi_i}\right) < \infty$ in a neighborhood of $t = 0$, then the approximation (5.29) can be improved to

$$\|X - \tilde{X}\|_T \stackrel{\text{a.s.}}{=} O(\log T). \tag{5.34}$$

Remark 5.16 The above approximations, as well as those in Theorem 5.14, are the best possible, in the sense that (5.29) holds if and only if $\mathsf{E}(\xi_i)^r < \infty$, and that if (5.34) holds with $o(\log T)$, then ξ_i must be a normal random variable. Though the approximation for Y can be improved upon, the approximations given by the theorem and the above remark are the best possible almost sure *simultaneous joint* approximation of X and Y. (By simultaneous joint approximation, we mean that the approximations $\tilde{X}$ and $\tilde{Y}$ for X and Y are related by the same Wiener process W as in (5.31) and (5.32). Refer to Section 2.2 of Csörgő and Horváth [3] for the best approximations for a renewal process Y.)

Remark 5.17 The strong approximation has a probabilistic counterpart. Suppose that $\mathsf{E}\left(e^{t\xi_i}\right) < \infty$ in a neighborhood of $t = 0$. Then, there exists a Wiener process $W = \{W(t), t \geq 0\}$ such that for all $T \geq 1$ and $x \geq 0$, we have

$$\mathsf{P}\left\{\|X - \tilde{X}\|_T > C_1 \log T + x\right\} \leq C_2 e^{-C_3 x},$$

where $\tilde{X}$ takes the form (5.31), and C_1, C_2, and C_3 are positive constants that do not depend on T and x. A similar form exists for the renewal process Y.

We now show that FSAT implies FSLLN, FCLT, and FLIL, given the higher-order moment condition. Start with FCLT. Fix $T > 0$, and replace T by nT and t by nt in FSAT (5.29). We have

$$\sup_{0 \leq t \leq T}\left|\hat{X}^n(t) - \frac{\sigma}{\sqrt{n}}W(nt)\right| \stackrel{\text{a.s.}}{=} \frac{o(1)}{n^{(1/2)-(1/r)}}, \quad \text{as } n \uparrow \infty. \tag{5.35}$$

From the scaling property of a Brownian motion, it is known that $W(nt)/\sqrt{n}$ is also a standard Brownian motion. This, along with the converging-together theorem, yields FCLT.

As for the FSLLN, divide (5.35) by $\sqrt{n}$ to get

$$\sup_{0 \leq t \leq T}\left|\bar{X}^n(t) - \bar{X}(t) - \frac{\sigma}{n}W(nt)\right| \stackrel{\text{a.s.}}{=} \frac{o(1)}{n^{(1/2)-(1/r)}}, \quad \text{as } n \uparrow \infty,$$

and apply the FSLLN for the Brownian motion, namely

$$\frac{1}{n}W(nt) \xrightarrow{\text{a.s.}} 0, \quad \text{u.o.c.,}$$

as $n \uparrow \infty$, to conclude the FSLLN. Finally, FSAT and the FLIL for the Brownian motion, namely

$$\|W\|_T = \sup_{0 \le t \le T} |W(t)| \stackrel{\text{a.s.}}{=} O\left(\sqrt{T \log \log T}\right),$$

yield

$$\|X - \bar{X}\|_T = \|(X - \tilde{X}) + \sigma W\|_T \stackrel{\text{a.s.}}{=} O\left(\sqrt{T \log \log T}\right).$$

5.6 Exponential Rate of Convergence

When the moment generating function of ξ exists, we can further refine the FSLLN in terms of characterizing the rate of convergence.

Theorem 5.18 Suppose that

$$\mathsf{E}\left(e^{t\xi_i}\right) < \infty$$

in a neighborhood of zero. Fix a finite $T > 0$. Then there exists a positive function ϕ on $(0, \infty)$ such that for all $n \ge 1$ and $\epsilon > 0$,

$$\mathsf{P}\{\|\bar{X}^n - \bar{X}\|_T \ge \epsilon\} \le Ce^{-\phi(\epsilon)n}, \tag{5.36}$$
$$\mathsf{P}\{\|\bar{Y}^n - \bar{Y}\|_T \ge \epsilon\} \le Ce^{-\phi(\epsilon)n}, \tag{5.37}$$

where C is a constant (which does not depend on n).

Let $\ell(\theta) = \log \mathsf{E}(e^{\theta\xi_1})$ be the log moment generating function of ξ_1; and let

$$g(x) := \sup_{\theta}(\theta x - \ell(\theta) \vee \ell(-\theta)). \tag{5.38}$$

Lemma 5.19 Suppose that the moment generating function of ξ_1 is finite in a neighborhood of zero. Then the function g in (5.38) is convex with $g(m) = 0$, and achieves the unique minimum at $x = m$, i.e., $g(x) > g(m) = 0$ for all $x \ne m$.

(The proof of the above elementary result is left as an exercise; see Exercise 29.)

Proof. (Theorem 5.18) By the inverse relationship between X and Y, we have

$$\{X(n) \leq t\} \equiv \{Y(t) \geq n\}.$$

Hence, it can be verified directly that

$$\{\|\bar{Y}^n - \bar{Y}\|_T > \epsilon\} \subseteq \{\|\bar{X}^n - \bar{X}\|_{\mu T + \epsilon} > \epsilon\}.$$

This, combined with (5.36), immediately yields (5.37). Therefore, it suffices to prove the inequality in (5.36).

Without loss of generality, assume $m = 0$; for otherwise, we can consider the i.i.d. sequence $\{\xi_i - m\}$. Since $\{X(n)\}$ is a martingale and $e^{\alpha x}$ is an increasing convex function $(\alpha > 0)$, the sequence $\{e^{\alpha X(n)}\}$ is a submartingale (Exercise 10). By the submartingale inequality (Exercise 11), we have

$$P\left\{\max_{1 \leq j \leq n} e^{\alpha X(j)} > z\right\} \leq \frac{1}{z} E\left(e^{\alpha X(n)}\right) = \frac{1}{z} e^{n\ell(\alpha)},$$

for $\alpha, z > 0$. The above immediately implies

$$P\left\{\max_{1 \leq j \leq n} X(j) > \frac{1}{\alpha} \log z\right\} \leq \frac{1}{z} e^{n\ell(\alpha)}.$$

Similarly, since $\{-X(n)\}$ is also a martingale, we have

$$P\left\{\max_{1 \leq j \leq n} e^{-\alpha X(j)} > z\right\} \leq \frac{1}{z} E\left(e^{-\alpha X(n)}\right) = \frac{1}{z} e^{n\ell(-\alpha)},$$

for $\alpha, z > 0$, and

$$P\left\{\max_{1 \leq j \leq n} (-X(j)) > \frac{1}{\alpha} \log z\right\} \leq \frac{1}{z} e^{n\ell(-\alpha)}.$$

Hence,

$$P\left\{\max_{1 \leq j \leq n} |X(j)| > \frac{1}{\alpha} \log z\right\} \leq P\left\{\max_{1 \leq j \leq n} (-X(j)) > \frac{1}{\alpha} \log z\right\}$$

$$+ P\left\{\max_{1 \leq j \leq n} X(j) > \frac{1}{\alpha} \log z\right\}$$

$$\leq \frac{1}{z}\left[e^{n\ell(\alpha)} + e^{n\ell(-\alpha)}\right]$$

$$\leq \frac{2}{z} e^{n[\ell(\alpha) \vee \ell(-\alpha)]}.$$

Now replacing n by nT and z by $e^{n\alpha\epsilon}$ yields

$$P\left\{\max_{1 \leq j \leq nT} |X(j)| > n\epsilon\right\} \leq 2 e^{-[\alpha\frac{\epsilon}{T} - \ell(\alpha) \vee \ell(-\alpha)]nT}.$$

Since the above holds for all α, it follows from the definition of the function g that

$$P\left\{\max_{1\leq j\leq nT}|X(j)| > n\epsilon\right\} \leq 2e^{-g(\epsilon/T)nT}.$$

Letting $C = 2$ and $\phi(\epsilon) = g(\epsilon/T)T$ proves the theorem.   □

The rate of convergence in Theorem 5.18 is often a topic in large deviations theory. However, the large deviations principle (LDP) for the i.i.d. summation process $\bar{X}^n$ has been proved only on a metric that is weaker than the Skorohod J_1 topology and hence weaker than the uniform topology (Puhalskii [11]). The LDP for the renewal process $\hat{Y}^n$ has been proved by Puhalskii [11] under the additional assumption that $P(\xi_1 > 0) > 0$. Thus, we have provided here a direct proof that does not make use of the LDP.

5.7   Notes

Billingsley [1] is a standard reference for results on weak convergence in $\mathcal{D}[0,1]$, and extensions to $\mathcal{D}^J[0,\infty)$ can be found in Whitt [13] and Ethier and Kurtz [6]. In this book, weak convergence is exclusively in $\mathcal{D}^J$, and all the stochastic processes that arise as the limits of weak convergence are exclusively in $\mathcal{C}^J$. Therefore, it is sufficient for us to use only the uniform topology, instead of Skorohod topologies; see Pollard [10] and Section 18 of Billingsley [1].

The Skorohod representation theorem is in Skorohod [12], and a more accessible proof can be found in Pollard [10] and Ethier and Kurtz [6]. The continuous mapping theorem and the random time-change theorem can be found in Whitt [13], where these theorems are established also for other topologies and other notions of convergence. The extended versions of these theorems in Whitt [13] are necessary if one deals with the weak convergence in $\mathcal{D}^J$ whose limiting process is only in $\mathcal{D}^J$ but not $\mathcal{C}^J$.

Most of the materials in the section on Brownian motion can be found in, for example, Harrison [8] and Karatzas and Shreve [9].

SLLNs and LIL can be found in most probability textbooks, for example, Chung [2]. Donsker's theorem can be found in Billingsley [1]. The proof of Lemma 5.19 is similar to the proof of Lemma 2.2.5 in Dembo and Zeitouni [5].

Strong approximation theorems can be found in Csörgő and Révész [4] and Csörgő and Horváth [3]. Csörgő and Révész [4] also contains the FLIL for i.i.d. summations. Csörgő and Horváth [3] also provides various refinements for the strong approximation of the i.i.d. summation and its associated renewal process, and extends the strong approximation to a more general process (other than the i.i.d. summation) and its associated inverse (not necessarily a renewal process).

Glynn [7] provides an elegant survey on weak convergence in $\mathcal{D}$, as well as strong approximations.

5.8   Exercises

1. Show that the definitions of convergence in distribution given by (5.4) and (5.5) are equivalent for the case of random vectors.

2. Let U be a uniform random variable in $[0,1]$. Let $X \equiv 0$, and let $Y = \{Y_t, 0 \le t \le 1\}$, where $Y_t = 0$ if $t \ne U$ and $Y_t = 1$ if $t = U$. Show that X and Y have the same finite-dimensional distributions, but X and Y have different distributions.

3. Give a counterexample to the equivalence mentioned in Exercise 1 for the case of stochastic processes.

4. Let $x \equiv 0$ and let $x_n = \{x_n(t), t \ge 0\}$ with

$$x_n(t) = \begin{cases} nt & \text{if } 0 \le t \le 1/n, \\ 2 - nt & \text{if } 1/n < t \le 2/n, \\ 0 & \text{if } 2/n < t \le 1. \end{cases}$$

Let X be a stochastic process with a unit mass at x (i.e., $\mathsf{P}(X = x) = 1$), and X_n be a stochastic process with a unit mass at x_n (i.e., $\mathsf{P}(X_n = x_n) = 1$), $n \ge 1$. Show that all finite-dimensional distributions of X_n converge weakly to X, but x_n does not converge weakly to x.

5. (Glynn [7]) Let U be a uniform random variable in $[0, 1]$ and let

$$X_n(t) = e^{-n(t-U)^2}$$

for $t \ge 0$. Let $X(t) = 0$ for all $t \ge 0$.

   (a) Show that the finite-dimensional distributions of X_n converge to those of X.

   (b) Show that X_n does not converge to X in distribution.

6. This exercise leads to a proof for a special case of Theorem 5.1.

   (a) Let X be a random variable and F be its distribution. Assume that F is continuous. Show that $F(X)$ is uniformly distributed in $[0, 1]$.

   (b) Let F be a distribution function and let U be a uniform random variable in $[0, 1]$. Set

$$F^{-1}(x) = \inf\{y : F(y) > x\}.$$

Show that the distribution function of $F^{-1}(U)$ is F. Do you need to assume that F is continuous?

(c) Prove Theorem 5.1 for the case where $S = \Re$.

7. Prove the continuous mapping theorem (Theorem 5.2).

8. Prove the random time-change theorem (Theorem 5.3).

9. Prove the convergence-together theorem (Theorem 5.4).

10. Let $\{Z_n, n \geq 1\}$ be a submartingale and ϕ be an increasing convex function defined on $\Re$. Suppose $E|\phi(Z_n)| < \infty$ for all $n \geq 1$. Show that $\{\phi(Z_n), n \geq 1\}$ is also a submartingale. When $\{Z_n, n \geq 1\}$ is a martingale, show that $\{\phi(Z_n), n \geq 1\}$ is a submartingale, if ϕ is a convex function.

11. *(Doob's Inequality)* Suppose that $\{X_n\}$ is a submartingale. Let $a \geq 0$,

$$A = \{ \max_{0 \leq k \leq n} X_k \geq a\},$$

and $\mathbf{1}_A := \mathbf{1}\{\omega \in A\}$. Then,

$$aP(A) \leq E[X_n \mathbf{1}_A] \leq E[X_n^+].$$

12. Let $X = \{X_t, t \in \mathcal{T}\}$ be a martingale. Show that

$$E[X_t] = E[X_0] \quad \text{for all } t \in \mathcal{T}.$$

13. *(Optional Sampling Theorem)* A random variable T is called a stopping time if $\{T = t\} \in \mathcal{F}_t$ for all $t \geq 0$. Suppose $\{X_t, t \in \mathcal{T}\}$ is a martingale, and T is a *bounded* stopping time, i.e., $0 \leq T \leq t_0$ (a.s.) for some constant t_0. Prove that $E[X_0] = E[X_T] = E[X_{t_0}]$.

14. A stochastic process $\{X_t, t \in \mathcal{T}\}$ is termed uniformly integrable if it satisfies the following condition:

$$\lim_{a \to \infty} \sup_t E[|X_t| \mathbf{1}\{|X_t| > a\}] = 0. \tag{5.39}$$

More specifically, for any given $\epsilon > 0$ there exists an a_0 such that for all $a \geq a_0$,

$$E[|X_t| \mathbf{1}\{|X_t| > a\}] \leq \epsilon, \quad \forall t \in \mathcal{T}. \tag{5.40}$$

(It is important to note that a_0 has to be *independent* of t.) Show that $\{X_t, t \in \mathcal{T}\}$ is uniformly integrable if any one of the following conditions holds:

- (boundedness) if $E[\sup_t |X_t|] < \infty$;

- (1+δ moment) if there exists a δ > 0 such that $E|X_t|^{1+\delta} \le K < \infty$ for all $t \in \mathcal{T}$.

15. Prove that if $\{X_n\}$ and $\{Y_n\}$ are both uniformly integrable, then $\{X_n + Y_n\}$ is uniformly integrable.

16. Show that $\{X_n\}$ being uniformly integrable implies $\sup_n E|X_n| < \infty$.

17. Suppose that $\{X_n\}$ is a submartingale and $\sup_n E|X_n| < \infty$. Show that as $n \to \infty$, $X_n \to X$ a.s. and $E|X| < \infty$.

18. Let $\{X_t, t \ge 0\}$ be a martingale and T a stopping time.

    (a) Show that $\{X_{T \wedge n}, n = 0, 1, 2, \dots\}$ is a martingale (known as a "stopped" martingale).

    (b) If $\{X_{T \wedge n}, n = 0, 1, 2, \dots\}$ is uniformly integrable, show that

    $$E[X_0] = \lim_n E[X_{T \wedge n}] = E[\lim_n X_{T \wedge n}] = E[X_T]. \qquad (5.41)$$

    (c) Show that if $\{X_t, t \ge 0\}$ is uniformly integrable, then $\{X_{T \wedge n}, n = 0, 1, 2, \dots\}$ is also uniformly integrable.

19. Let $W = \{W(t), t \ge 0\}$ be a standard Brownian motion. Define

    $$g(\theta, W(t)) := e^{\theta W(t) - \theta^2 t/2},$$

    where θ is a parameter.

    (a) Show that $g(\theta, W(t))$ is a martingale.

    (b) Let $g^{(r)}(\theta, W(t))$ be the rth-order derivative with respect to θ for $r = 1, 2, \dots$. Show that $g^{(r)}(\theta, W(t))$ is a martingale, and conclude that $W(t)$, $W^2(t) - t$, $W^3(t) - 3tW(t)$, and $W^4(t) - 6tW^2(t) + 3t^2$ are martingales.

    (c) Let $T = \inf\{t : W(t) \notin (-a, a)\}$ for some given $a > 0$, i.e., the first time the Brownian motion exits the interval $(-a, a)$. Show that

    $$E[T] \;\; = \;\; a^2 \quad \text{and} \quad E[T^2] = \frac{5}{3}a^4.$$

20. Let X be a Brownian motion with drift μ and standard deviation σ. Show that each of the following processes are martingales (with respect to the filtration generated by X):

    (a) $\{X(t) - \mu t, t \ge 0\}$;

    (b) $\{[X(t) - \mu t]^2 - \sigma^2 t, t \ge 0\}$;

    (c) $\{e^{\theta X(t) - (\theta \mu + \frac{1}{2}\theta^2 \sigma^2)t}, t \ge 0\}$ for any given $\theta \in \Re$.

21. Let X be a Brownian motion with drift μ and standard deviation σ starting at the origin.

   (a) Let $\tau_b = \inf\{t : X(t) \geq b\}$ $(b > 0)$. Show that $P\{\tau_b < \infty\} = 1$ if $\mu \geq 0$.

   (b) Let $\tau_{ab} = \inf\{t : X(t) \notin [a,b]\}$ $(a < 0 < b)$. Show that $E\tau_{ab} < \infty$. [Hint: For the case $\mu = 0$, apply the optional sampling theorem to the martingale $Z(t) := X(t)^2 - \sigma^2 t$ with stopping time $\tau_{ab} \wedge n$, and for the case $\mu \neq 0$, apply the optimal sampling theorem to the martingale
$$e^{\theta X(t) - (\theta\mu + \frac{1}{2}\theta^2\sigma^2)t}$$
   with stopping time $\tau_{ab} \wedge n$.]

   (c) Let $\tau_a = \infty\{t : X(t) \leq a\}$ $(a < 0)$. Show that $E\tau_a < \infty$ if $\mu < 0$.

22. A process is called a Gaussian process if its finite-dimensional joint distributions are multivariate normal. Let X be a standard Brownian motion.

   (a) Convince yourself that X must be a Gaussian process. Show that

$$E[X(t)] = 0 \quad \text{and} \quad \text{Cov}(X(t), X(s)) = t \wedge s.$$

   In fact, standard Brownian motion can also be defined as a Gaussian process with $E[X(t)] = 0$ and $\text{Cov}(X(t), X(s)) = t \wedge s$.

   (b) The conditional stochastic process $\{X(t), 0 \leq t \leq 1 | X(1) = 0\}$ is known as the *Brownian bridge*. Show that it is also a Gaussian process. Find

$$E[X(s)|X(1) = 0] \quad \text{and} \quad \text{Cov}[(X(s), X(t))|X(1) = 0].$$

23. Show by a counterexample that convergence (5.15) in Lemma 5.8 does *not* in general imply (5.14). Find a sufficient condition such that this implication does hold.

24. Let $\{X_n, n \geq 1\}$ be a nondecreasing sequence in $\mathcal{D}$ and $X \in \mathcal{C}$.

   (a) Prove that if as $n \to \infty$, $X_n(t) \to X(t)$ for rational $t \geq 0$, then $X_n \to X$, u.o.c., as $n \to \infty$.

   (b) Prove the above with the nondecreasing assumption replaced, so that the result can be applied to the proof in Lemma 5.8.

25. Consider a batch arrival process. Suppose that the interarrival times of the batches are i.i.d. with mean a_k^{-1} and SCV c_{0k}^2, and the batch sizes are i.i.d. with mean b_k and SCV c_{bk}^2. Assume that the batch sizes

are independent of interarrival times and that the arrivals among different classes are mutually independent. Specifically, let $N(t)$ denote the number of batches that have arrived during $[0, t]$, and let v_n be the size of the nth batch. Then the batch process $A = \{A(t), t \geq 0\}$ is defined by

$$A(t) = \sum_{n=1}^{N(t)} v_n, \qquad t \geq 0,$$

where the summation is understood to be zero if $N(t) = 0$.

(a) State and prove a functional central limit theorem for A.

(b) State and prove a functional strong approximation theorem for A.

26. Consider a model for "autonomous breakdowns." Let $\{(u(n), d(n)), n \geq 1\}$ be an i.i.d. sequence, where $u(n)$ and $d(n)$ denote the duration of the nth up-time and the nth down-time (including repair time) of a machine, respectively.

(a) For concreteness, assume that the machine starts in the first up period. Let $c(t)$ be the total time that the machine has been up during $[0, t]$. Represent $c = \{c(t), t \geq 0\}$ in terms of $\{(u(n), d(n)), n \geq 1\}$.

(b) Show that if both $u(n)$ and $d(n)$ have finite expectations, then as $n \to \infty$,

$$\frac{c(nt)}{n} \to \kappa := \frac{Eu(1)}{Eu(1) + Ed(1)} \qquad \text{u.o.c.}$$

(c) Assume that both $u(n)$ and $d(n)$ have finite second moments. Let d and c_d^2 (u and c_u^2) denote the mean and the squared coefficient of variation of $d_j(n)$ ($u_j(n)$). Furthermore, assume that $u_j(n)$ and $d_j(n)$ are independent. Show that

$$n^{-1/2}[c(n\cdot) - n\kappa\cdot] \overset{d}{\to} \hat{c}(\cdot),$$

where $\hat{c} = \{\hat{c}(t), t \geq 0\}$ is a driftless Brownian motion with variance $u^2 d^2 (c_u^2 + c_d^2)/(u + d)$.

(d) Repeat the above without assuming that $u(n)$ and $d(n)$ are independent.

27. Show that the FSAT bound (5.29) implies the FSAT bound (5.30).

28. Show that the probabilistic form of the strong approximation in Remark 5.17 implies the rate of convergence (5.36) in Theorem 5.18.

29. Prove Lemma 5.19.

30. Suppose that $\{\xi_i, i \geq 0\}$ is an i.i.d. sequence with exponential distribution.

   (a) Show that the rate of convergence given by Theorem 5.18 implies the FLIL bound. [Hint: Show that

$$\sum_n e^{-\phi\left(\sqrt{\frac{\log \log n}{n}}\right)n} < \infty,$$

   and then apply the Borel–Cantelli lemma.] (Note that the above result does not mean that Theorem 5.18 implies the FLIL theorem, since more assumptions are required for the former.)

   (b) Can you generalize the above to the nonexponential case?

References

[1] Billingsley, P. (1968). *Convergence of Probability Measures.* Wiley, New York.

[2] Chung, K.L. (1974). *A Course in Probability Theorem,* Academic Press, New York.

[3] Csörgő, M. and L. Horváth. (1993). *Weighted Approximations in Probability and Statistics,* Wiley, New York.

[4] Csörgő, M. and P. Révész. (1981). *Strong approximations in probability and statistics,* Academic Press, New York.

[5] Dembo, A. and O. Zeitouni. (1993). *Large Deviation Techniques and Applications,* Jones and Bartlett, Boston.

[6] Ethier, S.N. and T.G. Kurtz. (1986). *Markov Processes,* Wiley, New York.

[7] Glynn, P.W. (1990). Diffusion Approximations. *Handbooks in Operations Research and Management Science, Volume 2, Stochastic Models,* eds. D.P. Heyman and M.J. Sobel, 145–198.

[8] Harrison, J.M. (1985). *Brownian Motion and Stochastic Flow Systems,* Wiley.

[9] Karatzas, I. and S.E. Shreve (1991). *Brownian Motion and Stochastic Calculus,* Springer-Verlag, New York.

[10] Pollard, D. (1984). *Convergence of Stochastic Processes*. Springer-Verlag.

[11] Puhalskii, A.A. (1995). Large deviation analysis of the single server queue. *Queueing Systems, Theory and Applications*, **21**, 5–66.

[12] Skorohod, A.V. (1956). "Limit theorems for stochastic processes." *Theory of Probability and Its Applications*, **1**, 261–290.

[13] Whitt, W. (1980). Some useful functions for functional limit theorems. *Mathematics of Operations Research*, **5**, 67–85.

6

Single-Station Queues

The main subject of this chapter is limit theorems for the queue-length and the workload processes in the G/G/1 queue. The limit theorems include the functional strong law of large numbers (FSLLN), the functional law of the iterated logarithm (FLIL), the functional central limit theorem (FCLT) and the strong approximation. In addition, we also establish an exponential rate of convergence result for the fluid approximation. The limit of the FSLLN and the FLIL is a single station fluid model. Because of this, the FSLLN is often known as the *fluid approximation*. Similarly, the limit of the FCLT and the strong approximation takes the form of a one-dimensional reflected Brownian motion. Since this limit is a diffusion process, the FCLT has been conventionally known as *diffusion approximation*. Throughout this and the following chapters we shall follow the convention to use the terms "fluid approximation" and "diffusion approximation" interchangeably with their underlying limit theorems. (In contrast, in Chapter 10, where the proposed approximations are not necessarily supported by limit theorems, we shall use the term "Brownian approximation," in keeping with the Brownian network models in the research literature as approximations for queueing networks.)

We start in Section 6.1 with a pathwise construction of the queue-length process and the workload process, the key performance measures of a G/G/1 queue. The primitive data that we use to construct these processes are the i.i.d. summations of interarrival times and service times. In Section 6.2 an alternative representation is given for the queue-length process and the workload process, which leads to a one-dimensional reflection mapping. This one-dimensional reflection mapping links each limit theorem for the primitive data (studied in the previous chapter) to a corresponding limit theorem for the queue-length and the workload processes, respectively, in Sections 6.3–6.4 and in Sections 6.6–6.7. In Section 6.5 we define a one-dimensional reflected Brownian motion that arises as the limit of the diffusion approximation. We characterize the stationary distribution of the reflected Brownian motion, and propose an approximation for the stationary distribution of the queue-length and the workload processes.

6.1   Queue Length and Workload Processes: A Pathwise Construction

Consider a single-server queue in isolation. Initially, i.e., at $t = 0$, there are $Q(0)$ jobs in the system. In addition, there is an exogenous stream of jobs arriving at the system for service. Let u_1 be the arrival epoch of the first job arriving from outside; and let u_i be the interarrival time between the $(i-1)$st and the ith arrivals, $i = 2, 3, \ldots$. Let v_i be the required service time of the job that is the ith to be served, including the $Q(0)$ jobs that are present in the system at time zero. Let $u = \{u_i, i = 1, 2, \ldots\}$ and $v = \{v_i, i = 1, 2, \ldots\}$. Naturally, assume $Q(0)$, u_i and v_i, for all i, to be nonnegative.

Set

$$U(0) := 0, \qquad U(n) := \sum_{i=1}^{n} u_i, \quad n \geq 1,$$

$$V(0) := 0, \qquad V(n) := \sum_{i=1}^{n} v_i, \quad n \geq 1;$$

and

$$A(t) := \sup\{n : U(n) \leq t\}, \qquad (6.1)$$
$$S(t) := \sup\{n : V(n) \leq t\}. \qquad (6.2)$$

Clearly, $A(t)$ gives the number of jobs that have arrived during the time interval $(0, t]$, and $S(t)$, the number of jobs the server can *potentially* complete during the first t units of time (that is, provided that the server is busy all the time). We shall refer to the counting processes A and S as the *arrival process* and the *service process*, respectively.

The above $Q(0)$, A, and S constitute the primitive (given) data. In addition, we assume (in fact, throughout the book) that the service discipline is work-conserving, or nonidling, i.e., the server cannot stay idle if there are jobs present in the system.

Let $Q(t)$ denote the number of jobs in the system at time t, and $Z(t)$ denote the amount of time required to complete all jobs that are present at time t, in the system. These are two key performance measures, and we shall refer to $Q = \{Q(t), t \geq 0\}$ and $Z = \{Z(t), t \geq 0\}$ as the queue length process and the workload process, respectively.

With the above discussions, it should be clear that the sample path of the queue length process must satisfy the two relations

$$Q(t) = Q(0) + A(t) - S(B(t)), \tag{6.3}$$

$$B(t) = \int_0^t 1\{Q(s) > 0\}ds, \tag{6.4}$$

where $B(t)$ denotes the cumulative amount of time when the server is busy over the time interval $[0, t]$, and hence $S(B(t))$ is the number of jobs that have departed (after service completion) from the system in the same time interval. We shall refer to $B = \{B(t), t \geq 0\}$ as the *busy time process*.

The equation in (6.3) is simply a flow-balance relation, whereas (6.4) enforces the work-conserving assumption (also known as the nonidling assumption) that the server is busy if and only if the system is nonempty. By an inductive construction over time, it is clear that (6.3) and (6.4) uniquely determine the queue-length process Q and the busy time process B.

From the definition of $V(n)$ above, we know that $V(Q(0) + A(t))$ represents the total amount of work (measured in time units) over all jobs that have arrived up to time t, including those that are initially present in the system. Therefore, the workload at time t can be expressed as follows:

$$Z(t) = V(Q(0) + A(t)) - B(t). \tag{6.5}$$

Finally, another process of interest is the *idle time process* $I = \{I(t), t \geq 0\}$, defined as follows:

$$I(t) = t - B(t) = \int_0^t 1\{Q(s) = 0\}ds. \tag{6.6}$$

Clearly, $I(t)$ is the cumulative amount of time the server is idle during $[0, t]$.

In summary, the equations in (6.3) through (6.6) provide a pathwise construction of the dynamics of the single-server queue.

6.2   One-Dimensional Reflection Mapping

Suppose that both u_i and v_i have finite means, denoted by $1/\lambda$ and $m \equiv 1/\mu$, respectively. From the limit theorems of the last chapter, we know that (under certain conditions) $A(t)$ and $S(t)$ are asymptotically "close to" λt and μt, respectively, as $t \to \infty$. This motivates us to apply a "centering" operation to the queue length representation in (6.3), and rewrite it as follows:

$$Q(t) \quad = \quad X(t) + Y(t), \tag{6.7}$$

with

$$X(t) := Q(0) + (\lambda - \mu)t + [A(t) - \lambda t] - [S(B(t)) - \mu B(t)],$$
$$Y(t) := \mu I(t).$$

Furthermore, the following relations must hold: For all $t \geq 0$,

$$Q(t) \geq 0; \tag{6.8}$$
$$dY(t) \geq 0, \qquad Y(0) = 0; \tag{6.9}$$
$$Q(t)\, dY(t) = 0. \tag{6.10}$$

In words, $dY(t) \geq 0$ means that Y is nondecreasing, since the idle time process $I(t)$ is measured as a cumulation over time; and $Q(t)dY(t) = 0$ reflects the nonidling condition: The idle time cannot cumulate while the queue length is positive. Note that $Q(t)dY(t) = 0$ is equivalent to

$$\int_0^\infty Q(t)dY(t) = 0,$$

given (6.8) and (6.9).

In fact, given an appropriate X—not necessarily the "centered" process above—the relations in (6.7) through (6.10) uniquely determine Q and Y. This is stated in the theorem below, and the mapping from X to (Q, Y) defined by these relations is known as the (one-dimensional) *reflection mapping*.

Theorem 6.1 For any $x \in \mathcal{D}_0$, there exists a unique pair (y, z) in $\mathcal{D}^2$ satisfying

$$z = x + y \geq 0, \tag{6.11}$$
$$dy \geq 0 \text{ and } y(0) = 0, \tag{6.12}$$
$$z\, dy = 0. \tag{6.13}$$

In fact, the unique y and z can be expressed as

$$y(t) = \sup_{0 \leq s \leq t} [-x(s)]^+, \tag{6.14}$$
$$z(t) = x(t) + \sup_{0 \leq s \leq t} [-x(s)]^+. \tag{6.15}$$

We shall call z the reflected process of x and y the regulator of x. Denote the unique y and z by $y = \Psi(x)$ and $z = \Phi(x)$. Then, the mappings Ψ and Φ are Lipschitz continuous on $\mathcal{D}_0[0, t]$ under the uniform topology (refer to (5.6)) for any fixed t (including $t = \infty$).

Proof. That y and z given by (6.14) and (6.15) satisfy (6.11) and (6.12) is trivial. For (6.13), notice that for any given t, $dy(t) > 0$ implies that the sup in (6.14) is attained at $s = t$, and hence $z(t) = 0$.

Next we show the uniqueness of the two mappings Ψ and Φ. Let y' and z' be another pair satisfying the relations in (6.11) through (6.13). Note that $z - z' = y - y'$ is the difference of two nondecreasing functions, and $z(0) - z'(0) = 0$. We have

$$
\frac{1}{2}(z(t) - z'(t))^2 = \int_0^t (z(u) - z'(u))d(z(u) - z'(u))
$$

$$
= \int_0^t (z(u) - z'(u))d(y(u) - y'(u))
$$

$$
= -\int_0^t [z'(u)dy(u) + z(u)dy'(u)]
$$

$$
\leq 0,
$$

where the last equality follows from $z\,dy = 0$ and $z'\,dy' = 0$, and the inequality holds because z, z', dy, and dy' are all nonnegative. Therefore, we must have $z \equiv z'$, and hence $y \equiv y'$.

In view of (6.14), the Lipschitz continuity of Ψ follows from Exercise 1, and then in view of (6.11), the Lipschitz continuity of Φ follows from that of Ψ. □

Applying the above theorem to the relations in (6.7) through (6.10), we can write the queue-length process Q and the idle time process I as $Q = \Phi(X)$ and $I = m\Psi(X)$. As we shall demonstrate later, these expressions turn out to be extremely useful in establishing various limit theorems and approximations. Intuitively, for instance, if we can approximate X by another (usually simpler) process, say ξ, then we can approximate $Q = \Phi(X)$ by $\Phi(\xi)$ with a similar accuracy, thanks to the Lipschitz continuity of the mapping Φ.

When X is a one-dimensional Brownian motion that starts at $x = X(0) \geq 0$, and with drift θ and standard deviation σ, the process $Z = \{Z(t), t \geq 0\} := \Phi(X)$, with Φ being the one-dimensional reflection mapping defined in Theorem 6.1, is known as a one-dimensional *reflected Brownian motion*. In this case we shall write Z as $Z = RBM_x(\theta, \sigma^2)$. Figure 6.1 shows four sample paths of the reflected (standard) Brownian motion. (In fact, they are the reflected process paths of the four sample paths in Figure 5.3, which are approximately the paths of a standard Brownian motion.) (Note that when X is a driftless Brownian motion starting at the origin, it can be shown that the process $Z = \Phi(X)$ has the same distribution as $|X| = \{|X(t)|, t \geq 0\}$. Each sample path of the latter process $|X|$ can be seen as a reflection of the corresponding sample path of X around the horizontal axis. This may give rise to the name of the reflected Brownian motion, though in general, the sample path of a reflected Brownian motion may not be the reflection of a sample path from a Brownian motion.)

Next, we consider the stationary distribution of the reflected Brownian motion Z. Formally, the stationary distribution of Z is a probability mea-

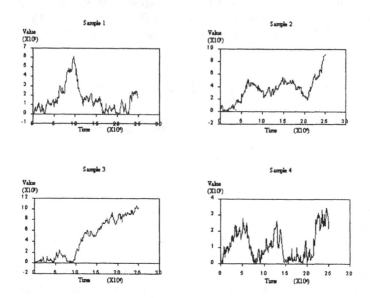

FIGURE 6.1. Four sample paths of a reflected Brownian motion

sure π (on $(\Re_+, \mathcal{B}(\Re_+))$) such that

$$\mathsf{E}_\pi[f(Z(t))] \equiv \int_{\Re_+} \mathsf{E}_z[f(Z(t)]d\pi(z) = \mathsf{E}_\pi[f(Z(0))]$$

for all bounded continuous functions f on $\Re^J$.

Theorem 6.2 Let $Z = RBM_x(\theta, \sigma^2)$ be the one-dimensional reflected Brownian motion. Then, the process Z has a stationary distribution if and only if $\theta < 0$, in which case the stationary distribution is exponential with rate

$$\eta := -\frac{2\theta}{\sigma^2}.$$

Proof. First suppose the Brownian motion starts at $x = X(0) = 0$. In this case,

$$\begin{aligned}
\mathsf{P}\{Z(t) \geq z\} &= \mathsf{P}\{X(t) + \sup_{0 \leq s \leq t} [-X(s)] \geq z\} \\
&= \mathsf{P}\{\sup_{0 \leq u \leq t} \tilde{X}_t(u) \geq z\},
\end{aligned}$$

where

$$\tilde{X}_t(u) = \begin{cases} X(t) - X(t-u), & 0 \leq u \leq t; \\ X(u), & t < u < \infty. \end{cases}$$

Clearly, for any fixed t, $\tilde{X}_t = \{\tilde{X}_t(u), u \geq 0\}$ is a Gaussian process with drift θ and covariance $\mathrm{Cov}(\tilde{X}_t(u), \tilde{X}_t(v)) = \sigma^2(u \wedge v)$. Hence, $\tilde{X}_t$ is a Brownian motion with drift θ and standard deviation σ (and also starts at the origin).

Let $\tau_z^t := \inf\{u \geq 0 : \tilde{X}_t(u) \geq z\}$. Then,

$$P\{Z(t) \geq z\} = P\{\tau_z^t \leq t\} = P\{\tau_z^0 \leq t\} = P\left\{\sup_{0 \leq u \leq t} X(u) \geq z\right\},$$

where the last equality follows from the fact that $\tilde{X}_t = \{\tilde{X}^t(u), u \geq 0\}$ and X are identical in distribution for all $t \geq 0$.

Therefore, following Lemma 5.5, we know that if $\theta \geq 0$, then

$$\lim_{t \to \infty} P\left\{\sup_{0 \leq u \leq t} X(u) \geq z\right\} = 1,$$

and if $\theta < 0$, then

$$\lim_{t \to \infty} P\left\{\sup_{0 \leq u \leq t} X(u) \geq z\right\} = e^{-\eta z}.$$

Next, suppose the Brownian motion starts at $X(0) = x > 0$. Let $\tau := \inf\{t \geq 0 : X(t) = 0\}$. If $\theta \leq 0$, then $\tau < \infty$ (Exercise 4). Clearly, $\{X(\tau + t), t \geq 0\}$ is a Brownian motion starting at the origin, and $\{Z(\tau + t), t \geq 0\}$ is a reflected Brownian motion also starting at the origin. Hence,

$$\lim_{t \to \infty} P\{Z(t) \geq z\} = \lim_{t \to \infty} P\{Z(\tau + t) \geq z\},$$

which equals $e^{-\eta z}$ if $\theta < 0$, and equals 1 if $\theta = 0$.

If $\theta > 0$, consider two cases: $\tau < \infty$ and $\tau = \infty$. In the first case, the same argument as above leads to

$$\lim_{t \to \infty} P\{Z(t) \geq z | \tau < \infty\} = 1.$$

In the second case, we must have

$$\lim_{t \to \infty} P\{Z(t) \geq z | \tau = \infty\} = \lim_{t \to \infty} P\{X(t) \geq z | \tau = \infty\} = 1.$$

(Refer to Exercise 4 for the last equality.)

Finally, suppose that Z (starting from $x = 0$) has the limit distribution π. Then for any bounded continuous function f on $\Re^J$ and $t \geq 0$,

$$\int_{\Re_+} f(z) d\pi(z) = \lim_{s \to \infty} E_0[f(Z(t + s))]$$

$$= \lim_{s \to \infty} E_0[E_{Z(t)}[f(Z(t + s))]]$$

$$= \int_{\Re_+} E_x[f(Z(t))],$$

where the second equality follows from the Markov property and the third equality follows from the Feller continuity (i.e., $E_x[f(Z(t))]$ is bounded and continuous in $x \in \Re_+$). This establishes that a limiting distribution must be a stationary distribution. It is left as an exercise to show that a stationary distribution must also be a limiting distribution.   □

Given the existence, we use an alternative way to obtain the stationary distribution π of the reflected Brownian motion $Z = RBM_x(\theta, \sigma^2)$. Note that we can write Z as

$$Z(t) = Z(0) + \theta t + \sigma W(t) + Y(t),$$

where W is a standard Brownian motion and Y is the regulator of the RBM. Then it follows from Ito's formula that for any twice continuously differentiable function f,

$$f(Z(t)) - f(Z(0)) = \sigma \int_0^t f'(Z(s)) dW(s)$$

$$+ \int_0^t \left[\frac{1}{2} \sigma^2 f''(Z(s)) + \theta f'(Z(s)) \right] ds$$

$$+ \int_0^t f'(Z(s)) dY(s).$$

Note that

$$\int_0^t f'(Z(s)) dW(s)$$

is a martingale, and that

$$\int_0^t f'(Z(s)) dY(s) = f'(0) Y(t),$$

due to the fact that $Y(\cdot)$ increases at t only when $Z(t) = 0$. Now taking the conditional expectation with respect to the stationary distribution π on both sides of Ito's formula, we have

$$0 = \mathsf{E}_\pi \int_0^t \left[\frac{1}{2} \sigma^2 f''(Z(s)) + \theta f'(Z(s)) \right] ds + f'(0) \mathsf{E}_\pi [Y(t)]$$

$$= t \int_0^\infty \left[\frac{1}{2} \sigma^2 f''(z) + \theta f'(z) \right] d\pi(z) + f'(0) \mathsf{E}_\pi [Y(t)].$$

Letting $f(z) = z$ in the above yields $\mathsf{E}_\pi[Y(t)] = -\theta t$. Hence, we can rewrite the above as

$$0 = \int_0^\infty \left[\frac{1}{2} \sigma^2 f''(z) + \theta f'(z) \right] d\pi(z) - \theta f'(0).$$

This is known as a *basic adjoint relation* (BAR). By taking $f(z) = e^{-\alpha z}$ in the BAR, we obtain a Laplace transform for the stationary distribution,

$$\pi^*(\alpha) = \int_0^\infty e^{-\alpha z} d\pi(z) = \frac{\theta}{\theta - \frac{1}{2}\sigma^2 \alpha} = \frac{\eta}{\alpha + \eta},$$

with $\eta = -2\theta/\sigma^2$ as given in the above theorem. Inverting the Laplace transformation, we conclude that π must have an exponential distribution with rate η.

The next theorem provides a pathwise bound on the reflected Brownian motion with a negative drift.

Theorem 6.3 Let $Z = RBM_x(\theta, \sigma^2)$ be the one-dimensional reflected Brownian motion. If $\theta < 0$, then

$$\sup_{0 \leq t \leq T} |Z(t)| \stackrel{a.s.}{=} O(\log T). \tag{6.16}$$

Proof. First consider $Z(0) = 0$. As shown in the proof of Theorem 6.2), we have

$$P\left(\sup_{0 \leq t \leq T} |Z(t)| \geq z\right) = P\left(\sup_{0 \leq t \leq T} \sup_{0 \leq s \leq t} [X(t) - X(s)] \geq z\right)$$

$$= P\left(\sup_{0 \leq s \leq T} X(s) \geq z\right)$$

$$\leq P\left(\sup_{0 \leq s \leq \infty} X(s) \geq z\right)$$

$$= e^{\theta z}$$

for any $z \geq 0$. Taking $z = (2 \log T)/\nu$ in the above yields

$$P\left(\sup_{0 \leq t \leq T} |Z(t)|/\log T \geq 2/\nu\right) \leq \frac{1}{T^2};$$

then by the Borel–Cantelli lemma we establish (6.16). If $Z(0) = X(0) = x \neq 0$, define $X^1(t) = X(t) - X(0)$. Define $Z^1 = \Phi(X^1)$. Since X^1 is a Brownian motion with a negative drift starting from the origin, the above proof establishes that the bound (6.16) holds for Z^1; then the Lipschitz continuity of the reflection mapping Φ establishes the bound (6.16) for $Z = \Phi(X)$. □

Lemma 6.4 Let $\{\theta_n, n \geq 1\}$ be a sequence of real numbers and let $\{x_n, n \geq 1\}$ be a sequence in $\mathcal{D}[0, \infty)$ with $x_n(0) \geq 0$. Let

$$z_n(t) = x_n(t) + \theta_n t + y_n(t), \tag{6.17}$$

$$y_n(t) = \sup_{0 \leq u \leq t} [-x_n(u) - \theta_n u]^+.$$

Suppose that as $n \to \infty$,

$$x_n \to x \qquad \text{u.o.c.} \tag{6.18}$$

with $x \in \mathcal{C}[0, \infty)$.

(i) If $\theta_n \to \theta$ as $n \to \infty$, then

$$(y_n, z_n) \to (y, z) \qquad \text{u.o.c.} \qquad \text{as } n \to \infty,$$

where

$$z(t) = x(t) + \theta t + y(t),$$
$$y(t) = \sup_{0 \le u \le t} [-x(u) - \theta u]^+.$$

(ii) If $\theta_n \to -\infty$ as $n \to \infty$ and $x(0) = 0$, then

$$(y_n(t) + \theta_n t, z_n(t)) \to (-x(t), 0) \qquad \text{u.o.c.} \qquad \text{as } n \to \infty.$$

(iii) If $\theta_n \to +\infty$ as $n \to \infty$, then

$$(y_n(t), z_n(t) - \theta_n t) \to (0, x(t)) \qquad \text{u.o.c.} \qquad \text{as } n \to \infty.$$

Proof. (i) This clearly follows from Theorem 6.1.

(ii) In view of (6.17) and (6.18), it suffices to show that $z_n \to 0$ u.o.c. as $n \to \infty$. Define

$$\tau_n(t) = \sup\{s : z_n(s) = 0, s \le t\}.$$

It is clear that $z_n(\tau_n(t)-) = 0$. Hence, we have

$$0 \le z_n(t) = z_n(t) - z_n(\tau_n(t)-) = x_n(t) - x_n(\tau_n(t)-) + \theta_n(t - \tau_n(t)),$$
$$(6.19)$$

where in obtaining the last equality, we used $y_n(t) = y_n(\tau_n(t)-)$. The latter clearly holds if $\tau_n(t) = t$; otherwise, it follows from the fact that $z_n(u) > 0$ for $u \in (\tau_n(t), t)$ and that $y_n(\cdot)$ increases at time t only when $z_n(t) = 0$ (Theorem 6.1).

Next, since $\theta_n \to -\infty$ as $n \to \infty$, for n large enough, $\theta_n < 0$, and the inequality in (6.19) implies

$$0 \le t - \tau_n(t) \le \frac{1}{-\theta_n}[x_n(t) - x_n(\tau_n(t))].$$

The above, in view of (6.18), implies that

$$\tau_n(t) \to t \qquad \text{u.o.c.} \qquad \text{as } n \to \infty, \qquad\qquad (6.20)$$

where we also used the fact that x is continuous and $0 \le \tau_n(t) \le t$.

Again, using (6.19) and noting that $\theta_n < 0$ for n large enough yields

$$0 \le z_n(t) = z_n(t) - z_n(\tau_n(t)) \le x_n(t) - x_n(\tau_n(t))$$

for n large enough; this, together with (6.18), (6.20) and the continuity of x, implies the convergence of z_n to zero u.o.c.

(iii) In view of (6.17) and (6.18), it is sufficient to show that $y_n \to 0$ u.o.c. as $n \to \infty$, or equivalently, for any $T > 0$ and $\epsilon > 0$, there exists an n_0 such that for $n \geq n_0$,

$$-x_n(u) - \theta_n u \leq \epsilon \qquad \text{for all } u \in [0, T].$$

The latter is clearly possible since as $n \to \infty$, $\theta_n \to \infty$ and $x_n \to x$ u.o.c. with x being continuous and $x(0) \geq 0$. □

6.3   Fluid Limit (FSLLN)

Here we study the fluid approximation for the single-server queue introduced in Section 6.1. The fluid approximation is concerned with the limits of $Q(nt)/n$, $Z(nt)/n$, and $B(nt)/n$ as $n \to \infty$; and "fluid limit" alludes to the fact (which will become evident below) that in the limit the queueing system behaves like a buffer that processes a deterministic, continuous stream of fluid.

Our starting point is to assume that the arrival and service processes, under the usual scaling, satisfy the FSLLN:

$$\bar{A}^n(t) \to \lambda t \quad \text{and} \quad \bar{S}^n(t) \to \mu t, \qquad \text{u.o.c.,} \quad \text{as } n \to \infty, \qquad (6.21)$$

where

$$\bar{A}^n(t) = \frac{1}{n} A(nt) \quad \text{and} \quad \bar{S}^n(t) = \frac{1}{n} S(nt).$$

Following Theorem 5.10, we know that a sufficient condition for the above to hold is that the arrival and service processes both be renewal processes with finite means: $\mathsf{E}(u_i) = 1/\lambda$ and $\mathsf{E}(v_i) = 1/\mu$.

Furthermore, as we take the limit $n \to \infty$, we shall allow the initial queue length to depend on n, denoted by $Q^n(0)$, and assume that almost surely,

$$\bar{Q}^n(0) := \frac{1}{n} Q^n(0) \to \bar{Q}(0), \qquad \text{as } n \to \infty, \qquad (6.22)$$

with $\bar{Q}(0) \geq 0$ being a constant. This means, in particular, that the initial queue length (as $n \to \infty$) could be either finite (implying $\bar{Q}(0) = 0$) or infinite (possibly $\bar{Q}(0) > 0$). As we let $n \to \infty$, it then becomes necessary to envision a sequence of queues, indexed by n, with the nth having an initial queue length $Q^n(0)$. Accordingly, we shall index the performance of the sequence of queues by a superscript n, denoted by Q^n, Z^n, and B^n, etc. And, specifically, we are concerned with the limits of

$$\bar{Q}^n(t) := \frac{1}{n} Q^n(nt), \quad \bar{Z}^n(t) := \frac{1}{n} Z^n(nt), \quad \text{and} \quad \bar{B}^n(t) := \frac{1}{n} B^n(nt)$$

as n approaches infinity.

Theorem 6.5 Suppose that (6.22) and (6.21) hold. Then as $n \to \infty$,

$$(\bar{Q}^n, \bar{Z}^n, \bar{B}^n) \to (\bar{Q}, \bar{Z}, \bar{B}), \quad \text{u.o.c.,} \quad \text{as } n \to \infty, \tag{6.23}$$

where

$$\bar{Q} = \bar{X} + \bar{Y},$$
$$\bar{X}(t) = \bar{Q}(0) + (\lambda - \mu)t,$$
$$\bar{Y} = \Psi(\bar{X}),$$
$$\bar{Z} = m\bar{Q},$$
$$\bar{B}(t) = t - m\bar{Y}(t),$$

and $m = 1/\mu$. Furthermore,

$$\bar{Q}(t) = [\bar{Q}(0) + (\lambda - \mu)t]^+, \tag{6.24}$$
$$\bar{Y}(t) = [-\bar{Q}(0) - (\lambda - \mu)t]^+. \tag{6.25}$$

Proof. Applying scaling to (6.7) (i.e., replacing t by nt and dividing both sides by n), we have

$$\bar{Q}^n(t) = \bar{X}^n(t) + \bar{Y}^n(t),$$

where

$$\bar{X}^n(t) = \bar{Q}^n(0) + (\lambda - \mu)t + [\bar{A}^n(t) - \lambda t]$$
$$- [\bar{S}^n(\bar{B}^n(t)) - \mu\bar{B}^n(t)],$$
$$\bar{Y}^n(t) = \mu(t - \bar{B}^n(t)) = \mu \int_0^t 1\{\bar{Q}^n(s) = 0\}ds.$$

Clearly, $\bar{X}^n$, $\bar{Y}^n$, and $\bar{Q}^n$ jointly satisfy the relations in (6.11) through (6.13) that define the one-dimensional reflection mapping. Hence, $\bar{Q}^n = \Phi(\bar{X}^n)$ and $\bar{Y}^n = \mu \bar{I}^n = \Psi(\bar{X}^n)$; and by Theorem 6.1, to establish the convergence of $\bar{Q}^n$ and $\bar{Y}^n$ in (6.23), it suffices to show the convergence of $\bar{X}^n$. But this is implied by (6.22) and (6.21), if we note that $0 \leq \bar{B}^n(t) \leq t$ for all $n \geq 1$ and $t \geq 0$. Next, the convergence of $\bar{Y}^n$ and the equality $\bar{B}^n(t) = t - m\bar{Y}^n(t)$ implies the convergence of $\bar{B}^n$.

The equalities in (6.24) and (6.25) follow from Theorem 6.1, taking into account $x + [-x]^+ = [x]^+$ for any real number x, and

$$\bar{Y}(t) = \sup_{0 \leq s \leq t} [-\bar{X}(s)]^+ = [-\bar{X}(t)]^+,$$

where the last equality holds because $\bar{X}$ is linear.

Finally, consider the convergence of $\bar{Z}^n$. Recall relation (6.5); we have

$$Z^n(t) = mQ^n(0) + (\rho - 1)t + [V(Q^n(0) + A(t)) - m(Q^n(0) + A(t))]$$
$$+ m[A(t) - \lambda t] + I^n(t), \tag{6.26}$$

where $\rho = \lambda m$. Scaling to the above yields

$$\bar{Z}^n(t) = m\bar{Q}^n(0) + (\rho - 1)t + [\bar{V}^n(\bar{Q}^n(0) + \bar{A}^n(t)) - m(\bar{Q}^n(0) + \bar{A}^n(t))]$$
$$+ m[\bar{A}^n(t) - \lambda t] + m\bar{Y}^n(t).$$

Next, by Theorem 5.10, (6.21) implies

$$\bar{V}^n(t) = \frac{1}{n}V(\lfloor nt \rfloor) \to mt, \quad \text{u.o.c.,} \quad \text{as } n \to \infty.$$

Then, the convergence of $\bar{Q}^n(0)$, $\bar{V}^n$, $\bar{A}^n$, and $\bar{Y}^n$ clearly implies the convergence of $\bar{Z}^n$, following Theorem 5.3 (the time-change theorem); and the limit is

$$\bar{Z}(t) = m\bar{Q}(0) + (\rho - 1)t + m\bar{Y}(t) = m\bar{Q}(t).$$

□

The limiting queue-length process $\bar{Q}$, as we briefly mentioned above, can be interpreted as the fluid level of a deterministic, continuous flow system as follows: There is a single buffer, with fluid continuously flowing in at a constant rate λ and flowing out at a rate μ (unless the buffer is empty, in which case the outflow rate is the smaller of the inflow rate λ and the processing rate μ). Let $\bar{Q}(0)$ be the initial fluid level. Then, the fluid level at time t is

$$\bar{Q}(t) = \bar{Q}(0) + \lambda t - \mu t + \bar{Y}(t), \tag{6.27}$$

where $\bar{Y}(t)$ indicates the cumulative amount of lost outflow capacity (at rate μ) due to an empty buffer, during the time interval $[0, t]$. Clearly, we must have

$$\bar{Q}(t) \geq 0 \qquad \text{for all } t \geq 0, \tag{6.28}$$
$$\bar{Y}(\cdot) \text{ is nondecreasing with } \bar{Y}(0) = 0. \tag{6.29}$$

Furthermore, the following condition must hold:

$$\bar{Y}(\cdot) \text{ can increase at time } t \text{ only when } \bar{Q}(0) = 0. \tag{6.30}$$

In view of Theorem 6.1, the relations in (6.27) through (6.30) imply that $\bar{Q}$ must be the same as the limiting queue length in (6.24), and $\bar{Y}$ must be the same as given by (6.25). In other words, the fluid approximation limit of the single-server queue gives rise to a fluid model where the fluid level is the limiting queue-length process. Therefore, we shall also refer to this fluid approximation limit as the fluid limit.

Remark 6.6 When $\bar{Q}(0) = 0$, the limits in Theorem 6.5 simplify to the following (with $\rho := \lambda/\mu$):

$$\bar{Q}(t) = (\rho - 1)^+ \mu t,$$
$$\bar{Z}(t) = (\rho - 1)^+ t,$$
$$\bar{B}(t) = (\rho \wedge 1)t.$$

Remark 6.7 From (6.24), it is clear that if the traffic intensity satisfies $\rho < 1$ (or equivalently, $\lambda < \mu$), then, regardless of the initial fluid level $\bar{Q}(0)$, the fluid level process $\bar{Q}$ will reach zero in a finite time and stay zero afterwards. More precisely, $\bar{Q}(t) = 0$ for all $t \geq \tau$ with $\tau = \bar{Q}(0)/(\mu - \lambda)$. On the other hand, when $\rho > 1$, $\bar{Q}(t)$ grows linearly, at a rate $\lambda - \mu$, over time t; and when $\rho = 1$, it stays constant at the initial level. This is analogous to the M/M/1 queue, where $\rho < 1$, $\rho > 1$ and $\rho = 1$ correspond, respectively, to the cases in which the queue-length process is positive recurrent, transient and null recurrent.

6.4   Diffusion Limit (FCLT)

Consider the queueing model described in Section 6.2. Our concern here is the weak convergence of the following processes:

$$\hat{Q}^n(t) := \sqrt{n}[\bar{Q}^n(t) - (\lambda - \mu)^+ t] = \frac{Q(nt) - (\lambda - \mu)^+ nt}{\sqrt{n}},$$

$$\hat{Z}^n(t) := \sqrt{n}[\bar{Z}^n(t) - (\rho - 1)^+ t] = \frac{Z(nt) - (\rho - 1)^+ nt}{\sqrt{n}},$$

$$\hat{B}^n(t) := \sqrt{n}[(\rho \wedge 1)t - \bar{B}^n(t)] = \frac{(\rho \wedge 1)nt - B(nt)}{\sqrt{n}}.$$

Note that as with the FCLT limit in the last chapter, the centering of each process above is around its fluid limit, and the scaling is by a factor of $\sqrt{n}$.

Apply the same scaling to the primitives (the arrival and service processes):

$$\hat{A}^n(t) := \sqrt{n}\left[\bar{A}^n(t) - \lambda t\right] = \frac{A(nt) - n\lambda t}{\sqrt{n}},$$

$$\hat{S}^n(t) := \sqrt{n}\left[\bar{S}^n(t) - \mu t\right] = \frac{S(nt) - n\mu t}{\sqrt{n}},$$

$$\hat{V}^n(t) := \sqrt{n}\left[\bar{V}^n(t) - mt\right] = \frac{V(\lfloor nt \rfloor) - nmt}{\sqrt{n}}.$$

Our starting point is the following (joint) weak convergence:

$$\left(\hat{A}^n, \hat{S}^n, \hat{V}^n\right) \xrightarrow{d} \left(\hat{A}, \hat{S}, \hat{V}\right), \qquad \text{as } n \to \infty. \tag{6.31}$$

From the FCLT (Theorem 5.11), we know that the above holds if the sequence of interarrival times (u) and the sequence of service times (v) are independent, and each is an i.i.d. sequence, with a finite second moment. In this case, $\hat{A}$ and $\hat{S}$ are two independent driftless Brownian motions with variances λc_a^2 and μc_s^2, respectively, and $\hat{V}(t) = -m\hat{S}(mt)$. Here c_a and c_s denote the coefficients of variation of u_i and v_i, respectively. (The coefficient of variation of a random variable is the ratio of its standard deviation to its mean.)

Theorem 6.8 Assume the weak convergence in (6.31). Then,

$$(\hat{Q}^n, \hat{Z}^n, \hat{B}^n) \xrightarrow{d} (\hat{Q}, \hat{Z}, \hat{B}), \qquad \text{as } n \to \infty, \qquad (6.32)$$

with the limit $(\hat{Q}, \hat{Z}, \hat{B})$ taking the following forms, depending on the traffic intensity ρ:

(i) $\rho < 1$:

$$\hat{Q} = 0,$$
$$\hat{Z} = 0,$$
$$\hat{B} = m\left[\hat{S}(\rho t) - \hat{A}(t)\right],$$

(ii) $\rho = 1$:

$$\hat{Q} = \hat{X} + \hat{Y} \geq 0,$$
$$\hat{X} = \hat{A} - \hat{S},$$
$$\hat{Y} = \Psi(\hat{X}),$$
$$\hat{Z} = m\hat{Q},$$
$$\hat{B} = m\hat{Y}$$

(iii) $\rho > 1$:

$$\hat{Q} = \hat{A} - \hat{S},$$
$$\hat{Z} = m\hat{Q},$$
$$\hat{B} = 0.$$

Proof. Applying scaling to (6.7) (replacing t by nt and dividing both sides by $\sqrt{n}$) yields

$$\tilde{Q}^n(t) := \frac{1}{\sqrt{n}}Q(nt) = \hat{X}^n(t) + (\lambda - \mu)\sqrt{n}t + \hat{Y}^n(t), \qquad (6.33)$$

where

$$\hat{X}^n(t) = \tilde{Q}^n(0) + \hat{A}^n(t) - \hat{S}^n(\bar{B}^n(t)),$$
$$\hat{Y}^n(t) = \sqrt{n}\bar{Y}^n(t) = \mu\sqrt{n}[t - \bar{B}^n(t)].$$

Clearly, $\hat{X}^n(t) + (\lambda - \mu)\sqrt{n}t$, $\hat{Y}^n$, and $\tilde{Q}^n$ jointly satisfy the relations in (6.11) through (6.13) in Theorem 6.1 that define the one-dimensional reflection mapping. (Note that the reason why we use $\tilde{Q}^n$ instead of $\hat{Q}^n$ is that when $\lambda > \mu$, $\hat{Q}^n$ is *not* nonnegative and we cannot use the reflection mapping.)

From the Skorohod representation theorem (Theorem 5.1) we can replace the weak convergence in (6.31) by the following:

$$(\hat{A}^n, \hat{S}^n, \hat{V}^n) \to (\hat{A}, \hat{S}, \hat{V}), \quad \text{u.o.c.,} \quad \text{as } n \to \infty. \tag{6.34}$$

From Theorem 6.5, we have

$$\bar{B}^n(t) := \frac{1}{n}B(nt) \to (\rho \wedge 1)t, \quad \text{u.o.c.} \tag{6.35}$$

Also note that $\hat{Q}^n(0) = Q(0)/\sqrt{n} \to 0$. Hence, from the limits in (6.34) and (6.35), and the time-change theorem (Theorem 5.3), we have

$$\hat{X}^n \to \hat{X}, \quad \text{u.o.c.,}$$

where $\hat{X}(t) = \hat{A}(t) - \hat{S}((\rho \wedge 1)t)$.

Applying Lemma 6.4 to (6.33) with $\theta_n = (\lambda - \mu)\sqrt{n}$, we have the convergence of $\hat{Q}^n$ and $\hat{B}^n$ by noting that $\hat{Q}^n(t) = \tilde{Q}^n(t) - (\theta_n)^+ t$ and $\hat{B}^n(t) = m\left[\hat{Y}^n(t) - (\mu - \lambda)^+\sqrt{n}t\right]$, where the three cases $\theta_n \to -\infty$, $\theta_n \to \theta(= 0)$, and $\theta_n \to +\infty$ correspond to $\rho < 1$, $\rho = 1$, and $\rho < 1$, respectively.

For the convergence of $\hat{Z}^n$, we rewrite (6.26) as follows:

$$\hat{Z}^n(t) = m\hat{Q}^n(0) + \hat{V}^n(\bar{Q}^n(0) + \bar{A}^n(t)) + m\hat{A}^n(t) + \hat{B}^n(t).$$

The time-change theorem, along with the limits of $\hat{Q}^n(0)$, $\hat{V}^n$, $\bar{Q}^n$, $\bar{A}^n$, $\hat{A}^n$, and $\hat{Y}^n$, then implies the convergence of $\hat{Z}^n$, with the limit

$$\hat{Z}(t) = \hat{V}(\lambda t) + m\hat{A}(t) + \hat{B}(t) = m[\hat{A}(t) - \hat{S}(\rho t)] + \hat{B}(t),$$

where the last equality follows from $\hat{V}(t) = -m\hat{S}(mt)$. Hence, $\hat{Z}(t) = 0$ when $\rho < 1$, since $\hat{B}(t) = m[\hat{S}(\rho t) - \hat{A}(t)]$; and $\hat{Z}(t) = m\hat{Q}(t)$ when $\rho \geq 1$, since $\hat{Q}(t) = \hat{X}(t) + \hat{Y}(t)$, $\hat{X}(t) = \hat{A}(t) - \hat{S}(\rho t)$, and $\hat{B}(t) = m\hat{Y}(t)$. $\quad\square$

Remark 6.9 It is clear from the proof that the theorem holds as long as the weak convergence in (6.31) does. In particular, for the theorem to hold, it is necessary to assume neither that the arrival process and the service process are mutually independent, nor that they are renewal processes (i.e., constructed from i.i.d. sequences).

Remark 6.10 For the case $\rho < 1$, by Theorem 6.5, the scaled process $\bar{Q}^n$ converges to zero u.o.c. as $n \to \infty$, whereas Theorem 6.8 shows that $\hat{Q}^n = \sqrt{n}\bar{Q}^n$ converges weakly to zero, providing a lower bound on the rate at which $\bar{Q}^n$ converges to zero. On the other hand, the limit of the centered idle time process is a Brownian motion.

For the case $\rho = 1$, $\hat{X} = \hat{A} - \hat{S}$. When the arrival process and the service process are mutually independent renewal processes, $\bar{X}$ (the difference of the two Brownian motions) is a Brownian motion, with a zero drift and a variance equal to $\lambda c_a^2 + \mu c_s^2 = \lambda(c_a^2 + c_s^2)$ (since $\lambda = \mu$ in this case). The limiting process $\hat{Q}$ in this case is a reflected Brownian motion as defined in Section 6.2.

For the case $\rho > 1$, note that by Theorem 6.5, $\bar{Q}^n(t)$ converges u.o.c. to a linear function $(\lambda - \mu)t$. Hence, our scaling involves centering $\bar{Q}^n(t)$ at $(\lambda - \mu)t$. Consequently, unlike the first two cases, $\hat{Q}^n(t) = \sqrt{n}[\bar{Q}^n(t) - (\lambda - \mu)t]$ may not be nonnegative. Also note that $\hat{B} = 0$ indicates that the cumulative idle time converges to zero at a rate higher than the one indicated by the fluid limit in Theorem 6.5.

Traditionally, the diffusion limit theorem is established for a sequence of queues indexed by n, under a heavy traffic condition. Hence, it is also known as the *heavy traffic limit theorem* or *heavy traffic approximation*. Let λ_n and μ_n be, respectively, the arrival rate and the service rate for the nth network. The heavy traffic condition is specified by

$$\sqrt{n}[\lambda_n - \mu_n] \to \theta, \qquad \text{as } n \to \infty.$$

In this case, the above limit theorem still holds after minor modifications. Let us focus on the queue-length process. (The modification for the other processes is similar.) First, the queue-length process for the nth queue is scaled as

$$\hat{Q}^n(t) = \frac{1}{\sqrt{n}}Q^n(nt),$$

instead of

$$\hat{Q}^n(t) = \frac{1}{\sqrt{n}}[Q^n(nt) - (\lambda_n - \mu_n)]$$

(though the limit theorem may also be established for the latter). The processes $\hat{A}^n$ and $\hat{S}^n$ are redefined as

$$\hat{A}^n(t) = \frac{A^n(nt) - n\lambda_n t}{\sqrt{n}},$$

$$\hat{S}^n(t) = \frac{S^n(nt) - n\mu_n t}{\sqrt{n}}.$$

Under the weak convergence in (6.31) (it is not necessary to assume that the arrival process A^n and the service process S^n are renewal processes), the weak convergence in (6.32) still holds, and the limiting process $\hat{Q}$ is the same as the case for $\rho = 1$, except that a drift term θ should be added to the Brownian motion $\hat{X}$.

The primary motivation in considering a sequence of queues is to help identify the appropriate reflected Brownian motion to approximate the original queueing processes. As we know from Chapter 1, the queue-length process and the workload process have stationary distributions only when the traffic intensity ρ is strictly less than one. On the other hand, as we showed in the theorem, when $\rho < 1$, the diffusion limit of the queue-length process is zero. Clearly, a value zero would not provide us with a good approximation for the stationary queue-length process. An alternative way to formulate the problem is to assume that a particular queue under consideration is among a sequence of queues, whose traffic intensities approach one. Once the limit is identified, we can use it to approximate the queue in question. For example, we can replace the drift of the Brownian motion θ by $\sqrt{n}(\lambda - \mu)$. Then, inverting the scaling, we can approximate the queue-length process Q by a reflected Brownian motion

$$\hat{Q}(t) = \hat{X}(t) + \hat{Y}(t),$$

where $\hat{X}(t) = Q(0) + \hat{A}(t) - \hat{S}(\rho t) + (\lambda - \mu)t$ and $Y = \Psi(\hat{X})$.

6.5   Approximating the G/G/1 Queue

We propose here an explicit approximation for the stationary distributions of the queue length and the workload process of a G/G/1 queue. To do so, we give an intuitive derivation of the approximation. We first rewrite the queue-length process as

$$Q(t) = Q(0) + A(t) - S(B(t)) = X(t) + Y(t),$$

where

$$X(t) = Q(0) + [A(t) - \lambda t] - [S(B(t)) - \mu B(t)] + (\lambda - \mu)t,$$
$$Y(t) = \mu[t - B(t)].$$

Recall that B is the busy time process and Y/μ gives the cumulative idle time. Under the work-conserving service discipline, it must be that $Y(\cdot)$ increases at time t only when $Q(t) = 0$; therefore,

$$Q(t) = \Phi(X)(t) = X(t) + \sup_{0 \leq s \leq t} [-X(s)]^+.$$

Consequently, if X can be approximated by another process, say $\hat{X}$, then Q can be approximated by

$$\hat{Q}(t) := \Phi(\hat{X})(t) = \hat{X}(t) + \sup_{0 \leq s \leq t} [-\hat{X}(s)]^+.$$

Now we identify an approximation for X. It follows from the convergence (6.34) that

$$\hat{A}^n(t) \equiv \sqrt{n}\left[\frac{1}{n}A(nt) - \lambda t\right] \overset{\text{d}}{\approx} \hat{A}(t),$$

where $\overset{\text{d}}{\approx}$ means approximately equal in distribution. If in the above we replace nt by t and multiply both side by $\sqrt{n}$, we can obtain

$$[A(t) - \lambda t] \overset{\text{d}}{\approx} \sqrt{n}\hat{A}(t/n) \overset{\text{d}}{=} \hat{A}(t),$$

where the equality in distribution is due to the scaling property of a driftless Brownian motion. Similarly, we also have

$$[S(t) - \mu t] \overset{\text{d}}{\approx} \hat{S}(t).$$

Replacing t by $B(t)$ in the above yields

$$[S(B(t)) - \mu B(t)] \overset{\text{d}}{\approx} \hat{S}(B(t)).$$

Note that the cumulative busy time should be proportional to $(\rho \wedge 1)t$, where $\rho = \lambda/\mu$. Hence, it is plausible that the above approximation can be replaced by

$$[S(B(t)) - \mu B(t)] \overset{\text{d}}{\approx} \hat{S}((\rho \wedge 1)t).$$

Combining the above approximations for $A(t)$ and $S(B(t))$ yields the following approximation for X:

$$\hat{X}(t) \quad = \quad Q(0) + \hat{A}(t) - \hat{S}((\rho \wedge 1)t) + (\lambda - \mu)t.$$

Since $\hat{A}$ and $\hat{S}$ are two independent driftless Brownian motions, $\hat{X}$ is also a Brownian motion starting at $Q(0)$ with drift $(\lambda - \mu)$ and variance

$$\sigma^2 = \lambda c_a^2 + \mu(\rho \wedge 1)c_s^2 = \lambda c_a^2 + (\lambda \wedge \mu)c_s^2.$$

Therefore, we propose the reflected Brownian motion

$$\hat{Q} = \Phi(\hat{X}) = RBM_{Q(0)}(\lambda - \mu, \lambda c_a^2 + (\lambda \wedge \mu)c_s^2)$$

as an approximation for the queue-length process Q. When $\rho > 1$, an alternative approximation for the queue-length process Q is the Brownian motion

$$\hat{X} = BM_{Q(0)}(\lambda - \mu, \lambda c_a^2 + \mu c_s^2),$$

as suggested directly by Theorem 6.8. The disadvantage of this second approximation is that while the queue-length process is nonnegative, the Brownian motion is not.

A similar argument leads to an approximation to the workload process W by the reflected Brownian motion $\hat{Z} := m\hat{Q}$ if $\rho \leq 1$. This approximation gives an approximate version of Little's law. When $\rho > 1$, the above argument would lead to an approximating reflected Brownian motion

$$\hat{Z} = m \times RBM_{mQ(0)}(\lambda - \mu, \lambda(c_a^2 + c_s^2)),$$

or an alternative Brownian motion approximation

$$\hat{X} = m \times BM_{Q(0)}(\lambda - \mu, \lambda c_a^2 + \mu c_s^2).$$

It follows from Theorem 6.2 that when $\lambda < \mu$, the reflected Brownian motions that respectively approximate the queue length and the workload process have stationary distributions. Specifically, we may approximate the stationary distribution of the queue-length process by an exponential distribution with mean

$$\frac{\rho(c_a^2 + c_s^2)}{2(1 - \rho)},$$

and approximate the stationary distribution of the workload process by an exponential distribution with mean

$$\frac{\lambda(c_a^2 + c_s^2)}{2(1 - \rho)}.$$

6.6   Functional Law of the Iterated Logarithm

Here we establish a version of the functional law of the iterated logarithm (FLIL) for the G/G/1 queue as described in Section 6.1. To this end, we assume that both the arrival sequence u and the service sequence v have finite second moments; hence, by Theorem 5.13, their summation processes and their associated renewal processes jointly satisfy the FLIL:

$$\sup_{0 \leq t \leq T} |A(t) - \lambda t| \overset{\text{a.s.}}{=} O\left(\sqrt{T \log \log T}\right), \tag{6.36}$$

$$\sup_{0 \leq t \leq T} |S(t) - \mu t| \overset{\text{a.s.}}{=} O\left(\sqrt{T \log \log T}\right), \tag{6.37}$$

$$\sup_{0 \leq t \leq T} |V(t) - mt| \overset{\text{a.s.}}{=} O\left(\sqrt{T \log \log T}\right), \tag{6.38}$$

as $T \to \infty$.

Theorem 6.11 Suppose that (6.36)–(6.38) hold. Then, as $T \to \infty$,

$$\sup_{0 \le t \le T} |Q(t) - \bar{Q}(t)| \stackrel{\text{a.s.}}{=} O\left(\sqrt{T \log \log T}\right), \tag{6.39}$$

$$\sup_{0 \le t \le T} |Z(t) - \bar{Z}(t)| \stackrel{\text{a.s.}}{=} O\left(\sqrt{T \log \log T}\right), \tag{6.40}$$

$$\sup_{0 \le t \le T} |B(t) - (\rho \wedge e)t| \stackrel{\text{a.s.}}{=} O\left(\sqrt{T \log \log T}\right), \tag{6.41}$$

where Q, Z, and B are, respectively, the queue, the workload and the busy time processes,

$$\bar{Q} = \bar{X} + \bar{Y}, \tag{6.42}$$
$$\bar{X}(t) = Q(0) + (\lambda - \mu)t, \tag{6.43}$$
$$\bar{Y} = \Psi(\bar{X}), \tag{6.44}$$
$$\bar{Z} = m\bar{Q}, \tag{6.45}$$

and Φ is the oblique reflection mapping as defined in Theorem 6.1.

Remark 6.12 The deterministic process $\bar{Q}$ can be interpreted as the fluid level process of a fluid model as described in Section 6.3 with inflow rate λ, outflow rate μ, and initial fluid level $Q(0)$.

The approximations in (6.39)–(6.40) still hold with $\bar{Q}(t) = (\lambda - \mu)^+ t$ and $\bar{Z}(t) = (\rho - e)^+ t$. This is equivalent to replacing $Q(0)$ by 0 in (6.43). (See Exercise 7.)

It will be clear from the proof that there is no need to assume that the arrival and the service processes are renewal processes. The theorem holds as long as the FLIL equalities (6.36)–(6.38) hold.

Remark 6.13 The theorem can be generalized to the case where the initial queue length varies with T, and we leave this as an exercise. The same holds for the summation and the counting processes: The FLIL equalities in (6.39)–(6.41) imply the fluid limit in (6.23) (see Exercise 8). However, this does not mean that Theorem 6.5 is a consequence of the above theorem, since the latter requires more assumptions on the primitive data.

Remark 6.14 It would be interesting to know what the following limit is:

$$\limsup_{T \to \infty} \frac{\sup_{0 \le t \le T} |Q(t) - \bar{Q}(t)|}{\sqrt{T \log \log T}}.$$

Such a limit theorem would be a stronger form of the FLIL (consistent with the classical FLIL for the i.i.d. summation). The same questions may be asked for the workload process and the busy time process.

Proof (of Theorem 6.11). First, rewrite the relation in (6.7) as

$$Q(t) = X(t) + Y(t), \tag{6.46}$$

where

$$X(t) = Q(0) + (\lambda - \mu)t + [A(t) - \lambda t] - [S(B(t)) - \mu B(t)], \tag{6.47}$$

$$Y(t) = \mu[t - B(t)]. \tag{6.48}$$

In view of $0 \le B(t) \le t$, applying the FLIL equalities in (6.36)–(6.37) to (6.47) yields

$$\|X - \bar{X}\|_T = \sup_{0 \le t \le T} |X(t) - \bar{X}(t)| = O(\sqrt{T \log \log T}), \quad T \uparrow \infty,$$

where $\bar{X}$ follows from (6.43). The above, together with the Lipschitz continuity of the reflection mapping (Theorem 6.1), leads to the equality in (6.39), and

$$\|Y - \bar{Y}\|_T = \sup_{0 \le t \le T} |Y(t) - \bar{Y}(t)| = O(\sqrt{T \log \log T}). \tag{6.49}$$

To derive the equality in (6.40), we rewrite (6.5) as

$$Z(t) = mQ(t) + [V(Q(t) + S(B(t))) - m(Q(t) + S(B(t)))]$$
$$+ m[S(B(t)) - \mu B(t)].$$

Then the equality follows from (6.39), (6.37), and (6.38).

Finally, the proof of (6.41) is completed by combining (6.48) and (6.49) with (6.50) below, which is established in the following lemma. □

Lemma 6.15 Let $\bar{Y} = \Psi(\bar{X})$ with $\bar{X}$ as in (6.43). Then, there exists an $M > 0$ such that

$$\sup_{0 \le t < \infty} |\bar{Y}(t) - (\mu - \lambda)^+ t| \le M. \tag{6.50}$$

Proof. Let $\bar{X}^*(t) = (\lambda - \mu)t$. Then, it can be directly verified that $\Psi(\bar{X}^*)(t) \equiv (\mu - \lambda)^+ t$. By the Lipschitz continuity of the reflection mapping, there exists an $M' > 0$ such that

$$\|\Psi(\bar{X}) - \Psi(\bar{X}^*)\| \le M'\|\bar{X} - \bar{X}^*\| = M'\|Q(0)\| \equiv M.$$

□

6.7 Strong Approximation

Suppose that both the arrival sequence u and the service sequence v have finite rth moments with $r > 2$. As in Section 6.4, let c_a and c_s denote the coefficients of variation for u_j and v_j, respectively. By the FSAT (Theorem 5.14), we can assume without loss of generality that there exist two independent standard Brownian motion processes W^0 and W^1 such that as $T \to \infty$,

$$\sup_{0 \le t \le T} |A(t) - \tilde{A}(t)| \stackrel{\text{a.s.}}{=} o(T^{1/r}), \tag{6.51}$$

$$\sup_{0 \le t \le T} |S(t) - \tilde{S}(t)| \stackrel{\text{a.s.}}{=} o(T^{1/r}), \tag{6.52}$$

$$\sup_{0 \le t \le T} |V(t) - \tilde{V}(t)| \stackrel{\text{a.s.}}{=} o(T^{1/r'}), \tag{6.53}$$

where

$$\tilde{A}(t) = \lambda t + \lambda^{1/2} c_a W^0(t),$$
$$\tilde{S}(t) = \mu t + \mu^{1/2} c_s W^1(t),$$
$$\tilde{V}(t) = mt - c_s m^{1/2} W^1(mt),$$

and in (6.53), we can choose $r' = r$ when $r < 4$ and any $r' < 4$ when $r \ge 4$.

Theorem 6.16 Suppose that the FSAT (6.51)–(6.53) hold with $r > 2$. Then as $T \to \infty$,

$$\sup_{0 \le t \le T} |Q(t) - \tilde{Q}(t)| \stackrel{\text{a.s.}}{=} o(T^{1/r'}), \tag{6.54}$$

$$\sup_{0 \le t \le T} |Z(t) - \tilde{Z}(t)| \stackrel{\text{a.s.}}{=} o(T^{1/r'}), \tag{6.55}$$

$$\sup_{0 \le t \le T} |B(t) - \tilde{B}(t)| \stackrel{\text{a.s.}}{=} o(T^{1/r'}), \tag{6.56}$$

where Q, Z, and B are the queue length, the workload, and the busy time processes,

$$\tilde{Q} = \tilde{X} + \tilde{Y}, \tag{6.57}$$
$$\tilde{X}(t) = Q(0) + (\lambda - \mu)t + \lambda^{1/2} c_a W^0(t)$$
$$\qquad - \mu^{1/2} c_s W^1((\rho \wedge 1)t), \tag{6.58}$$
$$\tilde{Y} = \Psi(\tilde{X}),$$
$$\tilde{Z}(t) = m\tilde{Q}(t) + m^{1/2} c_s [W^1((\rho \wedge 1)t) - W^1(\rho t)], \tag{6.59}$$
$$\tilde{B}(t) = t - \mu \tilde{Y}(t),$$

and Ψ is the reflection mapping as defined in Theorem 6.1. In (6.54)–(6.56), we can choose $r' = r$ when $r < 4$ and any $r' < 4$ when $r \ge 4$.

Remark 6.17 For $r \geq 4$, the FSAT in (6.54) and (6.55) can be improved as follows:

$$\sup_{0 \leq t \leq T} |Q(t) - \tilde{Q}(t)| = O((T \log\log T)^{1/4}(\log T)^{1/2}),$$

$$\sup_{0 \leq t \leq T} |Z(t) - \tilde{Z}(t)| = O((T \log\log T)^{1/4}(\log T)^{1/2}),$$

as $T \to \infty$. [This follows from the proof of the theorem in view of Lemma 6.21.] A stronger result would be that (6.54) holds with $r' = r$, $r > 2$ (regardless of whether $r < 4$ or $r \geq 4$). However, it is not clear whether such a result will hold.

Remark 6.18 The process $\tilde{X}$ is a Brownian motion starting at $Q(0)$ with drift $(\lambda - \mu)$. The process $\tilde{Q}$ is a reflected Brownian motion as defined in Section 6.5. The theorem holds, even when the vector $Q(0)$ and the processes A, S and R are not independent. When they are independent, the variance of the Brownian motion $\tilde{X}$ is given by

$$\Gamma = \lambda c_a^2 + (\lambda \wedge \mu) c_s^2.$$

When $\rho \leq 1$, the equality in (6.59) is Little's formula. This relation fails, however, when $\rho > 1$.

Remark 6.19 When $\rho < 1$ (corresponding to the case of a stable system), the limit $\tilde{Q}$ is consistent with the Brownian approximation for the queue-length process proposed in Section 6.5. In other words, the strong approximation theorem provides the theoretic support for the proposed approximation. Note that previously the approximation was proposed based on the diffusion approximation, where heuristics must be used by introducing a sequence of networks that approach a heavy traffic condition. In addition, the diffusion approximation involves scaling in time and space; when applying it (heuristically) to approximating a queue, we have to invert the scaling process. Therefore, it is our view that whenever it exists, a strong approximation is superior to a diffusion approximation as a tool for approximating a queueing system. However, we caution that Theorem 6.3 suggests that any constant (or even any process with a slower than $T^{1/r}$ bound) can be the strong approximation limit. Therefore, it remains a challenge to discover the best approximation and the best bound for the queue length process.

Remark 6.20 The strong approximations in (6.54)–(6.56) imply the FLIL in (6.39)–(6.41) and the FCLT in (6.32); and the latter two imply the fluid approximation in (6.23). Of course, these do not mean that the corresponding theorems of the former imply those of the latter, since the former require stronger conditions on the primitive data.

We first state and prove a lemma. The role this lemma plays in deriving the strong approximation is comparable to the role the random time-change theorem (Theorem 5.3) plays in deriving the diffusion approximation. We call $\tau = \{\tau(t), t \geq 0\}$ a random time-change process if τ is a nondecreasing process with $\tau(0) = 0$ and $\tau(t) - \tau(s) \leq t - s$ for all $t \geq s \geq 0$.

Lemma 6.21 Let $X = \{X(t), t \geq 0\}$ be a standard Brownian motion and let $\tau = \{\tau(t), t \geq 0\}$ be a random time-change process defined on the same probability space. Assume that τ satisfies an FLIL bound; namely, for any given $T > 0$, there exists an $M > 0$ such that

$$\sup_{0 \leq t \leq T} |\tau(t) - \alpha t| \leq M \sqrt{T \log \log T},$$

where α is a nonnegative constant. Then,

$$\sup_{0 \leq t \leq T} |X(\tau(t)) - X(\alpha t)| \overset{\text{a.s.}}{=} O((T \log \log T)^{1/4} (\log T)^{1/2}), \qquad \text{as } T \uparrow \infty,$$

which implies

$$\sup_{0 \leq t \leq T} |X(\tau(t)) - X(\alpha t)| \overset{\text{a.s.}}{=} o(T^{1/r}), \qquad \text{as } T \uparrow \infty,$$

for any $r > 2$.

Proof. For the time being, let us assume that for any $\delta > 0$ there exists a constant $C = C(\delta) > 0$ such that the inequality

$$P \left\{ \sup_{\substack{0 \leq u,v \leq T \\ |u-v| \leq h}} |X(u) - X(v)| \geq y\sqrt{h} \right\} \leq C \left(1 + \frac{T}{h} \right) e^{-y^2/(2+\delta)} \qquad (6.60)$$

holds for every $T > 0$ and $0 < h < T$.

Taking $h = h(T)$ and $y = g(T)$ in the above inequality and then applying the Borel–Cantelli lemma yields the equality

$$\sup_{\substack{0 \leq u,v \leq T \\ |u-v| \leq h(T)}} |X(u) - X(v)| \overset{\text{a.s.}}{=} O((T \log \log T)^{1/4} (\log T)^{1/2}), \qquad \text{as } T \uparrow \infty,$$

under the condition that

$$\frac{g(T) \sqrt{h(T)}}{(T \log \log T)^{1/4} (\log T)^{1/2}}$$

is bounded for large T and

$$\sum_{T=10}^{\infty} \left(1 + \frac{T}{h(T)} \right) e^{-[g(T)]^2/(2+\delta)} < \infty.$$

In the above, taking

$$h(T) = M\sqrt{T \log \log T} \qquad \text{and} \qquad g(T) = 5(\log T)^{1/2}$$

proves the lemma.

Now, we return to the inequality (6.60). Noting that

$$\{(u,v) : |u - v| \le h, \quad 0 \le u \le T, \quad 0 \le v \le T\}$$
$$\subseteq \{(u,v) : 0 \le u \le T, 0 \le v - u \le h\}$$
$$\cup \{(u,v) : 0 \le v \le T, 0 \le u - v \le h\},$$

we can use Lemma 1.2.1 of Csörgő and Révész [1981] with T replaced by $T + h$ and C replaced by $C/2$ to obtain the desired result. $\square$

Proof (of Theorem 6.16). If (6.51)–(6.53) hold for $r \ge 4$, they clearly hold for $r < 4$. Therefore, it suffices for us to restrict our attention to the case $r < 4$.

First rewrite (6.3) as

$$Q(t) = X(t) + Y(t),$$

where

$$X(t) = \tilde{X}(t) + [A(t) - \tilde{A}(t)] - [S(B(t)) - \tilde{S}(B(t))]$$
$$\quad - \mu^{1/2} c_s [W^1(B(t)) - W^1((\rho \wedge 1)t)],$$
$$Y(t) = \mu[t - B(t)],$$

and $\tilde{X}$ follows (6.58). If we can prove that

$$\sup_{0 \le t \le T} |X(t) - \tilde{X}(t)| \overset{\text{a.s.}}{=} o(T^{1/r}), \tag{6.61}$$

then we immediately deduce (6.54) and

$$\sup_{0 \le t \le T} |Y(t) - \tilde{Y}(t)| \overset{\text{a.s.}}{=} o(T^{1/r})$$

from the Lipschitz continuity of the mappings Ψ and Φ; and the latter implies (6.56).

Note that $0 \le B(t) \le t$; hence, it follows from the FSAT assumptions in (6.51)–(6.52) that to prove (6.61) it suffices to show that

$$\sup_{0 \le t \le T} |W^1(B(t)) - W^1((\rho \wedge 1)t)| \overset{\text{a.s.}}{=} o(T^{1/r}), \tag{6.62}$$

for all $\ell, k = 1, \dots, K$. This is implied by Lemma 6.21, if we can show that there exists an $M > 0$ such that

$$\sup_{0 \le t \le T} |B(t) - (\rho \wedge 1)t| \le M\sqrt{T \log \log T}.$$

Note that the FSAT assumptions in (6.51)–(6.53) imply the FLIL assumptions in (6.36)–(6.38); hence, the FLIL equality in (6.41) implies the above inequality.

The proof for (6.55) is similar, if we observe that there exists an $M > 0$ such that

$$\sup_{0 \le t \le T} |S(B(t)) - (\lambda \wedge \mu)t| \le M\sqrt{T \log \log T},$$

which follows from the above bound for B and (6.37). Specifically, rewriting (6.5), in view of $Q(t) + S(B(t)) = Q(0) + A(t)$, we have

$$
\begin{aligned}
Z(t) = {}& \tilde{Z}(t) + [V(Q(t) + S(B(t))) - \tilde{V}(Q(t) + S(B(t)))] \\
& + m[Q(t) - \tilde{Q}(t)] + m[S(B(t)) - \tilde{S}(B(t))] \\
& + m^{1/2}c_s[W^1(B(t)) - W^1((\rho \wedge 1)t)] \\
& + m^{1/2}c_s[W^1(m[Q(0) + A(t)]) - W^1(\rho t)].
\end{aligned}
$$

Making use of (6.52)-(6.53) and (6.54) and following an argument similar to the above completes the proof.   □

6.8   Exponential Rate of Convergence for the Fluid Approximation

Suppose that the interarrival time sequence u and the service time sequence v have finite moment generating functions in a neighborhood of zero. Theorem 5.18 implies the exponential rates of convergence of the arrival process and service process.

To allow more generality, we let the initial queue length vary with a parameter n ($n = 1, 2, \ldots$), and denote it by $Q^n(0)$. We shall use a superscript n to emphasize the dependence of the performance measures on the initial state. Thus, Q^n, Z^n, and B^n denote, respectively, the queue length, the workload, and the busy time processes associated with the initial queue length $Q^n(0)$.

Theorem 6.22 Fix a finite $T > 0$. In addition to the above assumptions, assume that the initial queue length satisfies

$$P\{\|\bar{Q}^n(0) - \bar{Q}(0)\| \ge \epsilon\} \le C_1 e^{-\phi_1(\epsilon)n}, \tag{6.63}$$

for a nonnegative J-dimensional vector $\bar{Q}(0)$, a constant C_1, and a positive function ϕ_1 on $(0, \infty)$. Then, there exist a constant C and a positive function ϕ on $(0, \infty)$ such that for all $\epsilon > 0$ and $n \ge 1$,

$$P\{\|\bar{Q}^n - \bar{Q}\|_T \ge \epsilon\} \le C e^{-\phi(\epsilon)n}, \tag{6.64}$$

$$P\{\|\bar{Z}^n - \bar{Z}\|_T \ge \epsilon\} \le C e^{-\phi(\epsilon)n}, \tag{6.65}$$

$$P\{\|\bar{B}^n - \bar{B}\|_T \ge \epsilon\} \le C e^{-\phi(\epsilon)n}, \tag{6.66}$$

where $\bar{Q}$, $\bar{Z}$, and $\bar{B}$ are defined the same as in (6.42)–(6.45) in Theorem 6.11.

Proof. First, by Theorems 5.18, we have the following bounds on the rate of convergence for the scaled arrival process $\bar{A}^n$, service process $\bar{S}^n$, and service time sequence $\bar{V}^n$:

$$P\{\|\bar{A}^n - \lambda\|_T \geq \epsilon\} \leq C_2 e^{-\phi_2(\epsilon)n}, \tag{6.67}$$

$$P\{\|\bar{S}^n - \mu\|_T \geq \epsilon\} \leq C_3 e^{-\phi_3(\epsilon)n}, \tag{6.68}$$

and

$$P\{\|\bar{V}^n - m\|_{|\bar{Q}(0)| + (\lambda+\mu)T+\epsilon} \geq \epsilon\} \leq C_4 e^{-\phi_4(\epsilon)n}, \tag{6.69}$$

for all $\epsilon > 0$ and $n \geq 1$, where λ, μ, and m are used to denote the linear functions $\lambda(t) = \lambda t$, $\mu(t) = \mu t$, and $m(t) = mt$. In (6.67)–(6.69), C_i, $i = 2, 3, 4$, are positive constants, and ϕ_i, $i = 2, 3, 4$, are positive functions on $(0, \infty)$.

Next, we rewrite the scaled version of the queue length expression (6.7) as

$$\bar{Q}^n(t) = \bar{X}^n(t) + \bar{Y}^n(t),$$

where

$$\begin{aligned}\bar{X}^n(t) &= \bar{X}(t) + [\bar{Q}^n(0) - \bar{Q}(0)] + [\bar{A}^n(t) - \lambda t] \\ &\quad - [\bar{S}^n(\bar{B}^n(t)) - \mu\bar{B}^n(t)], \\ \bar{Y}^n(t) &= \mu(t - \bar{B}^n(t)),\end{aligned} \tag{6.70}$$

with $\bar{X}$ in (6.70) being defined by (6.43). Clearly, we have

$$\bar{Q}^n = \Phi(\bar{X}^n) \qquad \text{and} \qquad \bar{Y}^n = \Psi(\bar{X}^n),$$

where Ψ and Φ are the one-dimensional reflection mappings.

If we can prove that there exists a constant C_5 and a positive function ϕ_5 on $(0, \infty)$ such that for all $\epsilon > 0$ and $n \geq 1$,

$$P\{\|\bar{X}^n - \bar{X}\|_T \geq \epsilon\} \leq C_5 e^{-\phi_5(\epsilon)n} \tag{6.71}$$

holds, then (6.64) and

$$P\{\|\bar{Y}^n - \bar{Y}\|_T \geq \epsilon\} \leq C e^{-\phi(\epsilon)n}$$

with $\bar{Q} = \Phi(\bar{X})$ and $\bar{Y} = \Psi(\bar{X})$ will follow from the Lipschitz continuity of the reflection mapping (Theorem 6.1). Noting that $\bar{B}^n(t) = t - \bar{Y}^n(t)/\mu_j$ leads to the bound in (6.66).

We shall return to the proof of the bound in (6.65), after establishing the bound in (6.71). To this end, we define the following sets:

$$E(\epsilon) = \{\omega : \|\bar{X}^n - \bar{X}\|_T \geq \epsilon\},$$
$$E_1(\epsilon) = \{\omega : \|\bar{Q}^n(0) - \bar{Q}(0)\| \geq \epsilon\},$$
$$E_2(\epsilon) = \{\omega : \|\bar{A}^n - \lambda\|_T \geq \epsilon\},$$
$$E_3(\epsilon) = \{\omega : \|\bar{S}^n - \mu\|_T \geq \epsilon\},$$

It is clear from the definition of $\bar{X}^n$ in (6.70) that

$$E(\epsilon) \subseteq E_1\left(\frac{\epsilon}{3}\right) \cup E_2\left(\frac{\epsilon}{3}\right) \cup E_3\left(\frac{\epsilon}{3}\right),$$

where we used the fact that $0 \leq \bar{B}^n(t) \leq t$. Therefore, by (6.63) and (6.67)–(6.68),

$$P(E(\epsilon)) \leq \sum_{k=1}^{3} P\left(E_k\left(\frac{\epsilon}{3}\right)\right) \leq \sum_{k=1}^{3} C_k e^{-\phi_k(\epsilon/3)n}.$$

Letting

$$\phi_5(\epsilon) = \min_{1 \leq k \leq 3} \phi_k(\epsilon/3) \quad \text{and} \quad C_5 = \sum_{k=1}^{3} C_k$$

establishes the bound in (6.71).

Finally, we prove (6.65). First, rewrite (6.26) as

$$\bar{Z}^n(t) = [\bar{V}^n(\bar{Q}^n(t) + \bar{S}^n(\bar{B}^n(t))) - m(\bar{Q}^n(t) + \bar{S}^n(\bar{B}^n(t)))]$$
$$+ m[\bar{Q}^n(t) - \bar{Q}(t)] + \bar{Z}(t) + m[\bar{S}^n(\bar{B}^n(t)) - m^{-1}\bar{B}^n(t)],$$

where we used the fact that $\bar{Z} = m\bar{Q}$. Note that for all $t \geq 0$,

$$\|\bar{Q}(t) + \bar{S}(\bar{B}(t))\| \leq |\bar{Q}(0)| + (\lambda + \mu)t.$$

We have

$$\{\|\bar{Z}^n - \bar{Z}\| > \epsilon\} \subseteq \left\{\|\bar{V}^n - m\|_{|\bar{Q}(0)|+(\lambda+\mu)T} > \frac{\epsilon}{3}\right\}$$
$$\cup \left\{\|\bar{Q}^n - \bar{Q}\|_T > \frac{1}{3m}\right\} \cup E_3\left(\frac{1}{3m}\right).$$

Hence, the equality in (6.65) is proved in view of (6.69), (6.64), and (6.68). $\square$

6.9 Notes

The presentation of this chapter is largely based on Chen and Mandelbaum [3]. Theorem 6.1 can be found in Harrison [9]. Part (ii) of Lemma 6.4 is from

Lemma 2 of Chen and Zhang [4], but has its origin (in a less explicit form) from Johnson [8] and Chen and Mandelbaum [2]. Part (iii) of Lemma 6.4 is from Theorem 6.1 of Whitt [16]. The section on the exponential rate of convergence is based on Chen [1].

Some of the pioneering work in diffusion approximation (also known as heavy traffic approximation) includes Kingman [13] and Iglehart and Whitt [11, 12]. In particular, the limit theorems such as FSLLNs, FCLT, and FLIL presented in this chapter are the special cases studied in Iglehart and Whitt [11, 12], Iglehart [10], and Whitt [14]. Whitt [15] provides an excellent survey on the early development of diffusion approximations. Csörgő, et al. [5], Glynn and Whitt [6, 7], and Zhang et al. [17] are among the first to study the strong approximation for a queueing system.

6.10   Exercises

1. Let both x and x' be in $\mathcal{D}$. For any fixed $t > 0$ (including $t = \infty$), show that

$$\left| \sup_{0 \le u \le t} [x(u)]^+ - \sup_{0 \le u \le t} [x'(u)]^+ \right| \le \sup_{0 \le u \le t} |x(u) - x'(u)|.$$

2. Let Φ and Ψ be the mappings defined in Theorem 6.1. Show that the Lipschitz constant for Ψ is one, that is, for any fixed $t > 0$,

$$\|\Psi(x) - \Psi(x')\|_t \le \|x - x'\|_t \qquad \text{for all } x, x' \in \mathcal{D}_0.$$

Next show that the Lipschitz constant for Φ is larger than one, by finding a pair of $x, x' \in \mathcal{D}_0$ such that for some $t > 0$,

$$\|\Phi(x) - \Phi(x')\|_t > \|x - x'\|_t.$$

3. Let $x \in \mathcal{D}_0$ and $z = \Phi(x)$. Show that for any $t \ge s \ge 0$,

$$|z(t) - z(s)| \le \sup_{s \le u \le t} |x(t) - x(u)|.$$

4. Let X be a Brownian motion with drift θ and standard deviation σ starting at $X(0) = x$. Let $\tau = \inf\{t \ge 0 : X(t) = 0\}$.

   (a) Suppose that $\theta \le 0$. Show that $\mathsf{P}\{\tau < \infty\} = 1$.

   (b) Suppose that $\theta > 0$. Show that

$$\lim_{t \to \infty} \mathsf{P} \left\{ \sup_{0 \le s \le t} X(s) \ge x | \tau = \infty \right\} = 1.$$

5. Prove that the strong approximation bounds (6.54)-(6.56) imply the FLIL bounds (6.39)-(6.41).

6. Prove that the strong approximations (6.54)–(6.56) imply the FCLT weak convergence (6.32).

7. Show that the approximations in (6.39)–(6.40) hold with $\bar{Q}(t) = (\lambda - \mu)^+ t$ and $\bar{Z}(t) = (\rho - e)^+ t$.

8. Prove that either the FLIL bounds (6.39)–(6.41) or the FCLT weak convergence (6.32) imply the FSLLN convergence (6.23).

9. State and prove a weak convergence theorem for a sequence of queues.

10. State and prove a strong approximation theorem for a sequence of queues.

11. Show the strong approximation implies the weak convergence for a sequence of queues.

References

[1] Chen, H. (1996). Rate of convergence of the fluid approximation for generalized Jackson networks. *Journal of Applied Probability*, **33**, 3, 804–814.

[2] Chen, H. and A. Mandelbaum. (1991). Stochastic discrete flow networks: diffusion approximations and bottlenecks. *Annals of Probability*, **19**, 4, 1463–1519.

[3] Chen, H. and A. Mandelbaum. (1994). Hierarchical modelling of stochastic networks:, Part II: strong approximations. In *Stochastic Modelling and Analysis of Manufacturing Systems*, ed. by D.D. Yao, 107–131, Springer-Verlag.

[4] Chen, H. and H. Zhang. (1996). Diffusion approximations for multi-class re-entrant lines under a first-buffer-first-served discipline. *Queueing Systems, Theory and Applications*, **23**, 177–195.

[5] Csörgő, M., P. Deheuvels, and L. Horváth. (1987). An approximation of stopped sums with applications in queueing theory. *Advances in Applied Probability*. **19**, 674–690.

[6] Glynn, P.W. and W. Whitt. (1991a). A new view of the heavy-traffic limit theorem for infinite-server queues. *Advances in Applied Probability*, **23**, 188–209.

[7] Glynn, P.W. and W. Whitt. (1991b). Departures from many queues in series. *Annals of Applied Probability*, **1**, 546–572.

[8] Johnson, D.P. (1983). *Diffusion approximations for optimal filtering of jump processes and for queueing networks*, Ph.D. dissertation, University of Wisconsin.

[9] Harrison, J.M. (1985). *Brownian Motion and Stochastic Flow Systems*, Wiley.

[10] Iglehart, D.L. (1971). Multiple channel queues in heavy traffic, IV: Laws of the iterated logarithm. *Zeitschrift Wahrscheinlichkeitstheorie*, 17, 168–180.

[11] Iglehart, D.L. and W. Whitt. (1970a). Multiple channel queues in heavy traffic, I. *Advances in Applied Probability*, **2**, 150–177.

[12] Iglehart, D.L. and W. Whitt. (1970b). Multiple channel queues in heavy traffic, II. *Advances in Applied Probability*, **2**, 355–364.

[13] Kingman, J.F.C. (1965). The heavy traffic approximation in the theory of queues. In *Proceedings of Symposium on Congestion Theory*, W. Smith and W. Wilkinson (eds.), University of North Carolina Press, Chapel Hill, 137–159.

[14] Whitt, W. (1971). Weak convergence theorems for priority queues: preemptive-resume discipline. *Journal of Applied Probability*, **8**, 74–94.

[15] Whitt, W. (1974). Heavy traffic theorems for queues: a survey. In *Mathematical Methods in Queueing Theory*, ed. A.B. Clarke, Springer-Verlag.

[16] Whitt, W. (1980). Some useful functions for functional limit theorems. *Mathematics of Operations Research*, **5**, 67–85.

[17] Zhang, H., G. Hsu, and R. Wang. (1990). Strong approximations for multiple channel queues in heavy traffic. *Journal of Applied Probability*, **28**, 658–670.

7

Generalized Jackson Networks

In this chapter we consider a queueing network that generalizes the Jackson network studied in Chapter 2, by allowing renewal arrival processes that need not be Poisson and i.i.d. service times that need not follow exponential distributions. (However, we do not allow the service times of the network to be state-dependent; in this regard, the network is more restrictive than the Jackson network. Nevertheless, this network has been conventionally referred to as the *generalized* Jackson network.) Unlike the Jackson network, the stationary distribution of a generalized Jackson network usually does not have an explicit analytical form. Therefore, approximations and limit theorems that support such approximations are usually sought for the generalized Jackson networks.

Most of this chapter focuses on open networks, whereas the last section summarizes corresponding results for closed networks. We start with a description of the network model in Section 7.1. In Section 7.2 we introduce an *oblique reflection mapping*, which plays a key role in various limit theorems for a generalized Jackson network. This is analogous to the role played by the one-dimensional reflection mapping in a single-station queue. This is followed by two sections where we present two special processes defined through the reflection mapping. The first one is related to a *(homogeneous) fluid network*, which characterizes the FLIL and fluid approximation limit of a generalized Jackson network, and the second one is known as a *reflected Brownian motion*, which describes the FSAT and diffusion approximation limit of a generalized Jackson network. Upon a first reading, the reader may want to skip these two sections and proceed directly to Section 7.5, where we derive heuristically a Brownian approximation for the generalized open Jackson network.

In addition, Section 7.5 motivates the study of the fluid network in Section 7.3 and the reflected Brownian motion in Section 7.4. This section also provides a motivation for the more rigorous analysis in the later sections: The functional law of the iterated logarithm in Section 7.6 and the functional strong approximation theorem in Section 7.7 provide bounds for the proposed first-order and second-order approximations in Section 7.5. The fluid limit in Section 7.8 and the diffusion limit in Section 7.9 give an alternative justification for the approximations in Section 7.5. The corresponding results for the closed networks are presented in Section 7.10.

7.1   The Queueing Network Model

The network consists of J single-server stations, indexed by $j = 1, \ldots, J$. Each station has an infinite waiting room. Jobs arrive exogenously at station j following a renewal process with rate $\alpha_j \geq 0$. Jobs are served in the order of their arrival; that is, a first-in-first-out (FIFO) service discipline is in force. The service times of jobs at station j form an i.i.d. sequence with mean $m_j > 0$, or equivalently, a *service rate* $\mu_j = 1/m_j$. A job completing service at station i may be routed to station j (for additional service) with probability p_{ij}, whereas with probability $1 - \sum_{j=1}^{J} p_{ij}$ it will leave the network. The $J \times J$ matrix $P = (p_{ij})$ is called a *routing matrix*. It is clear that matrix P must be a substochastic matrix, namely, all of its components are nonnegative and each of its row sums is less than or equal to 1. When matrix P is an (irreducible) stochastic matrix and the exogenous arrival rate (vector) $\alpha = (\alpha_j) = 0$, we call the network an *(irreducible) closed network*. In this case, there are neither arrivals nor departures from the network; in other words, a fixed number of jobs that are initially present in the network will circulate among the J stations and remain in the network forever. When the matrix P has a spectral radius (or equivalently, its largest real eigenvalue) less than one, we call the network an *open network*. (In Lemma 7.1 of the next section, a number of equivalent conditions are listed for a nonnegative matrix to have a spectral radius less than unity.) In this case, any job in the network will leave the network after visiting a finite number of stations. Except in Section 7.10, we shall focus on the open network. Hence, for convenience, whenever we say a generalized Jackson network, we mean a generalized open Jackson network. Figure 7.1 is an example of a three-station generalized Jackson network. For this network, $\alpha_3 = 0$ and the routing matrix takes the form

$$
P = \begin{pmatrix} 0 & 0.5 & 0.5 \\ 0.5 & 0 & 0 \\ 0.5 & 0 & 0 \end{pmatrix}.
$$

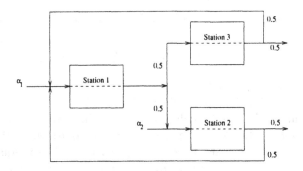

FIGURE 7.1. A three-station generalized Jackson network

Next, as was done for the single-station queue, we provide a pathwise construction of the queue length and the workload processes. To this end, we first introduce the primitive (given) data. Let $\mathcal{E} = \{j : \alpha_j > 0\}$ be the set of indices for those stations that have exogenous arrivals. Let $Q_j(0)$ be a nonnegative integer-valued random variable, representing the number of jobs at station j at time $t = 0$, $j = 1, \ldots, J$. For $j \in \mathcal{E}$, let $A_j = \{A_j(t), t \geq 0\}$ be the renewal process with $A_j(t)$ indicating the number of jobs that have arrived at station j exogenously during $(0, t]$. Let $A_j \equiv 0$ for $j \notin \mathcal{E}$. For $j = 1, \ldots, J$, let $v_j = \{v_j(\ell), \ell \geq 1\}$ be an i.i.d. sequence with $v_j(\ell)$ indicating the service time of the ℓth job served at station j, $\ell = 1, 2, \ldots$. We assume that for $j \in \mathcal{E}$, the mean interarrival time of A_j is finite and equal to $1/\alpha_j$, and that for $j = 1, \ldots, J$, the mean service time at station j (i.e., the mean of $v_j(1)$) is finite and equal to $m_j = 1/\mu_j$. We describe the routing of the job at station j by a sequence of J-dimensional i.i.d. random vectors, $\xi^j = \{\xi^j(\ell), \ell \geq 1\}$ taking values in $\{e^1, \ldots, e^J, 0\}$ (where e^i is the ith unit vector in $\Re^J$). The event $\xi^j(\ell) = e^i$ indicates that the ℓth job that completes service at station j will go next to station i, whereas the event $\xi^j(\ell) = 0$ indicates that the ℓth job that completes service at station j will leave the network. We assume that $Q(0) = (Q_j(0))$, A_j, $j \in \mathcal{E}$, v_j and ξ^j, $j = 1, \ldots, J$, are defined on the same probability space and are mutually independent.

For convenience, define the renewal process $S_j = \{S_j(t), t \geq 0\}$ associated with the service sequence v_j by

$$S_j(t) = \sup\{n : V_j(n) \leq t\},$$

where

$$V_j(0) = 0 \quad \text{and} \quad V_j(n) = \sum_{\ell=1}^{n} v_j(\ell), \quad n \geq 1,$$

$j = 1, \ldots, J$. The process $S = (S_j)$ will be called the *service process*. Let

$$R^j(0) = 0 \qquad \text{and} \qquad R^j(n) = \sum_{\ell=1}^{n} \xi^j(\ell), \quad n \geq 1;$$

we denote the ith component of R^j by R_i^j, $i, j = 1, \ldots, J$.

To recapitulate, the primitive data for a generalized Jackson network are the *arrival process* A_j, the *service process* S_j (or equivalently, the *service sequence* v_j), the *routing sequence* $R^j = \{R^j(n), n \geq 0\}$ (or equivalently, ξ^j), and the initial queue length $Q_j(0)$, $j = 1, \ldots, J$, all defined on the same probability space. From these data, we now construct some of the most important performance measures for the generalized Jackson network: the queue-length process $Q = \{Q(t), t \geq 0\}$, the workload process $Z = \{Z(t), t \geq 0\}$ and the busy time process $B = \{B(t), t \geq 0\}$. The jth component of $Q(t)$, $Q_j(t)$, indicates the number of jobs in station j (either in service or in waiting) at time t; the jth component of $Z(t)$, $Z_j(t)$, indicates the amount of time required at time t to serve all jobs currently in station j; and the jth component of $B(t)$, $B_j(t)$, indicates the amount of time the server at station j is busy (serving jobs).

It follows from their definitions that $S_j(B_j(t))$ indicates the number of jobs that have completed service at station j during the time interval $[0, t]$, and that $R_i^j(S_j(B_j(t)))$ indicates the number of jobs that enter station i from station j during the time interval $[0, t]$. Hence, we have the following *flow-balance* relation:

$$Q_j(t) = Q_j(0) + A_j(t) + \sum_{i=1}^{J} R_j^i(S_i(B_i(t))) - S_j(B_j(t)), \qquad (7.1)$$

for all $t \geq 0$ and $j = 1, \ldots, J$. Assume that a work-conserving rule is in force; we must have

$$B_j(t) = \int_0^t 1\{Q_j(s) > 0\}ds, \qquad (7.2)$$

for all $t \geq 0$ and $j = 1, \ldots, J$. By an inductive argument over the time, we can show that relations (7.1) and (7.2) uniquely determine the queue-length process Q and the busy time process B (assuming a tie-break rule when simultaneous events exist).

We next construct the workload process Z. Recall that $V_j(n)$ indicates the total service time for the first n jobs served at station j. Thus, we have

$$Z_j(t) = V_j\left(Q_j(0) + A_j(t) + \sum_{i=1}^{J} R_j^i(S_i(B_i(t)))\right) - B_j(t)$$
$$= V_j(Q_j(t) + S_j(B_j(t))) - B_j(t), \qquad (7.3)$$

for all $t \geq 0$ and $j = 1, \ldots, J$, where the first equality simply states that the total work (measured in time) in station j at time t equals the difference between the total arrived work and the total finished work (all measured in time), and the second equality follows from (7.1).

Another process often of interest is the (cumulative) *idle time* process $I = \{I(t), t \geq 0\}$, whose jth component is defined by

$$I_j(t) = t - B_j(t) = \int_0^t 1\{Q_j(s) = 0\}ds, \qquad (7.4)$$

for all $t \geq 0$ and $j = 1, \ldots$.

To close this section, we provide an alternative representation for the queue-length process and the workload process in terms of the oblique reflection mapping. Not only does this provide motivation for the oblique reflection mapping studied in the following section, it is also essential in the analysis throughout the remaining part of the chapter. It follows from the strong law of large numbers that with probability one,

$$\frac{1}{t}A_j(t) \to \alpha_j, \qquad \text{as } t \to \infty, \qquad (7.5)$$

$$\frac{1}{t}S_j(t) \to \mu_j, \qquad \text{as } t \to \infty, \qquad (7.6)$$

$$\frac{1}{n}R_i^j(n) \to p_{ji}, \qquad \text{as } n \to \infty. \qquad (7.7)$$

With the idea of "centering" as used in Section 6.2, we can rewrite (7.1) as

$$Q(t) = X(t) + (I - P')Y(t), \qquad (7.8)$$

where the jth components of $X(t)$ and $Y(t)$ are given by

$$X_j(t) = Q_j(0) + (\alpha_j + \sum_{i=1}^J \mu_i p_{ij} - \mu_j)t + [A_j(t) - \alpha_j t]$$

$$+ \sum_{i=1}^J p_{ij}[S_i(B_i(t)) - \mu_i B_i(t)] - [S_j(B_j(t)) - \mu_j B_j(t)]$$

$$+ \sum_{i=1}^J [R_j^i(S_i(B_i(t))) - p_{ij}S_i(B_i(t))], \qquad (7.9)$$

$$Y_j(t) = \mu_j[t - B_j(t)] = \mu_j I_j(t) = \mu_j \int_0^t 1\{Q_j(s) = 0\}ds, \qquad (7.10)$$

$j = 1, \ldots, J$. It is clear that for all $t \geq 0$,

$$Q(t) \geq 0, \qquad (7.11)$$

$$dY(t) \geq 0 \quad \text{and} \quad Y(0) = 0, \qquad (7.12)$$

$$Q_j(t)dY_j(t) = 0, \qquad j = 1, \ldots, J. \qquad (7.13)$$

As we shall see in the next section, relations (7.8) and (7.11)–(7.13) uniquely determine a pair of mappings from X to $Q := \Phi(X)$ and $Y := \Psi(X)$.

Similar to the derivation in the single-station queue, suppose that we can approximate X by a simpler process, say η; then we can approximate the queue-length process $Q = \Phi(X)$ by $\Phi(\eta)$, using the Lipschitz continuity of the reflection mapping Φ. In other words, to obtain an approximation or a limit theorem for the queue-length process Q, it suffices to obtain an approximation or a limit theorem for the process X.

7.2   Oblique Reflection Mapping

Let R be a $J \times J$ matrix and recall that $\mathcal{D}_0^J = \{x \in \mathcal{D}^J : x(0) \geq 0\}$. As a generalization of the one-dimensional reflection mapping, it is natural to consider the following problem: For an $x \in \mathcal{D}_0^J$, find a pair (y, z) in $\mathcal{D}^{2J}$ such that

$$z = x + Ry \geq 0, \tag{7.14}$$

$$dy \geq 0 \text{ and } y(0) = 0, \tag{7.15}$$

$$z_j dy_j = 0, \quad j = 1, \ldots, J. \tag{7.16}$$

This is also known as the Skorohod problem. If the pair exists and is unique for the given x, then we write $y = \Psi(x)$ and $z = \Phi(x)$, and call (Ψ, Φ) (sometimes just Φ) an *(oblique) reflection mapping*. The matrix R is called a *reflection matrix*.

Here we consider a special case of the above problem. Specifically, we assume that matrix R belongs to a class of matrices, known as M-matrices. A square matrix is an M-matrix if it has positive diagonal elements, nonpositive off-diagonal elements, and it has a nonnegative inverse. For a $J \times J$ M-matrix $R = (r_{ij})$, let D be a $J \times J$ diagonal matrix whose jth diagonal element is r_{jj}, $j = 1, \ldots, J$ (i.e., the diagonal elements of R). Then, we can write $R = (I - G)D$, where I is the identity matrix and $G := I - RD^{-1}$ is a nonnegative matrix. It is clear that $(I - G)$ is an M-matrix if and only if R is an M-matrix. We summarize some equivalent properties of a nonnegative matrix as follows. (See Exercise 2.)

Lemma 7.1 Suppose that G is a nonnegative. Then each of the following conditions implies all others:

1. matrix $(I - G)$ is an M-matrix;

2. matrix G is convergent, namely, $G^n \to 0$ as $n \to \infty$;

3. $(I - G)^{-1}$ exists;

4. $(I - G)^{-1}$ exists and is nonnegative.

5. matrix G has a spectral radius (or equivalently its largest real eigenvalue) less than unity.

6. All principal minors of $(I - G)$ are M-matrices and have nonnegative inverses;

7. $(I - G')^{-1}$ exists;

When applied to the generalized Jackson network, in view of (7.8) in the previous section, the reflection matrix takes the form $R = I - P'$, where $P = (p_{ij})$ is a substochastic matrix whose (i, j)th component represents the probability that a job after service completion at station i goes next to station j. The assumption that matrix P is convergent signifies that the network is an open network.

Theorem 7.2 (Oblique Reflection Mapping Theorem) Suppose that $J \times J$ matrix R is an M-matrix. Then for every $x \in \mathcal{D}_0^J$, there exists a unique pair (y, z) in $\mathcal{D}^{2J}$ satisfying

$$z = x + Ry \geq 0, \tag{7.17}$$

$$dy \geq 0 \text{ and } y(0) = 0, \tag{7.18}$$

$$z_j dy_j = 0, \quad j = 1, \ldots, J. \tag{7.19}$$

In other words, the reflection mapping (Φ, Ψ) is well defined on $\mathcal{D}_0^J$. In addition, the mappings Φ and Ψ are Lipschitz continuous mappings from $\mathcal{D}_0^J$ to $\mathcal{D}^J$ under the uniform norm.

The process $z = \Phi(x)$ and the process $y = \Psi(x)$ are referred to as the *reflected process* (of x) and the *regulator* (of x), respectively. The mapping (Φ, Ψ) in fact depends on the matrix R; to ease the notation, we shall omit reference to its dependence on the matrix R whenever this is obvious from context.

Before proving this theorem, we give an account of some useful mathematical properties of the reflection mapping.

1. (Least Element Characterization)   Fix $x \in \mathcal{D}_0^J$. If $y \in \mathcal{D}_0^J$ is the least element satisfying (7.17) and (7.18) (i.e., for any $\tilde{y}$ satisfying (7.17) and (7.18), $y(t) \leq \tilde{y}(t)$ all $t \geq 0$), then y (jointly with $z := x + Ry$) must satisfy (7.19), and hence, $y = \Psi(x)$. Conversely, let $y = \Psi(x)$. Suppose that $\tilde{y} \in \mathcal{D}^J$ satisfies (7.17) and (7.18). Then, it must be that $y(t) \leq \tilde{y}(t)$ for all $t \geq 0$.

2. (Nonanticipation)   For any $t > 0$, the restrictions of $\Psi(x)$ and $\Phi(x)$ to $[0, t]$ depend only on the restriction of x to $[0, t]$.

3. (Scale) For any real numbers $\alpha > 0$, $\beta > 0$ and $t \geq 0$, we have

$$\Psi\big(\alpha x(\beta \cdot)\big)(t) = \alpha \Psi(x)(\beta t),$$

$$\Phi\big(\alpha x(\beta \cdot)\big)(t) = \alpha \Phi(x)(\beta t),$$

4. (Shift) For any $s, t \geq 0$,

$$
\begin{aligned}
y(s + t) &= y(s) + \Psi\big(z(s) + x(s + \cdot) - x(s)\big)(t), \\
z(s + t) &= \Phi\big(z(s) + x(s + \cdot) - x(s)\big)(t),
\end{aligned}
$$

where $y = \Psi(x)$ and $z = \Phi(x)$.

5. (Path continuity) If $x \in \mathcal{D}_0^J$ is upper semicontinuous, then $\Phi(x)$ is continuous. (A function $x \in \mathcal{D}^J$ is said to be upper-semi-continuous if $\lim_{t \to t_0-} x(t) \leq x(t_0)$ for all $t_0 > 0$.)

The scale and shift properties follow immediately from the definition of the reflection mapping. The nonanticipation property should be clear from the proof of Theorem 7.2. The proofs for the other properties are outlined in the exercises at the end of this chapter.

Proof (of Theorem 7.2).   Note that we can write R as $R = (I - G)D$, where D is a diagonal matrix with positive diagonal elements and G is a nonnegative matrix with a spectral radius less than unity. Hence, we can assume without loss of generality that D is an identity matrix, i.e., $R = I - G$; otherwise, we can replace Dy by y. In view of Exercise 1, we can further assume that each row sum of matrix G is less than unity.

Define a mapping π from $\mathcal{D}_0^J$ to $\mathcal{D}_0^J$ by

$$
\pi(y)(t) = \sup_{0 \leq u \leq t} [-x(u) + Gy(u)]^+ \qquad \text{for all } t \geq 0. \tag{7.20}
$$

(Note that if $G = 0$, it follows from the one-dimensional reflection mapping that $\pi(y)$ as defined gives the unique solution to (7.17)–(7.19).) Momentarily we shall show that the mapping π has a unique fixed point. Let y be the fixed point. Then it is direct to verify that y, together with $z := x + (I - G)y$, satisfies (7.17)–(7.19). Conversely, suppose that (y, z) satisfies (7.17)–(7.19). Let $\tilde{y} = \pi(y)$. If we can show that $\tilde{y} = y$, then the uniqueness in the theorem is implied by the uniqueness of the fixed point for π. It follows from (7.17)–(7.18) that $\tilde{y} \leq y$. If $\tilde{y} \neq y$, then there must be $\tilde{y}_j(t_0) < y_j(t_0)$ for some j and t_0 and $dy_j(t_0) > 0$. On the other hand, it follows from (7.17) and $\tilde{y}_j(t_0) < y_j(t_0)$ that $z_j(t_0) > 0$, contradicting (7.19).

Now we show that for any fixed $T > 0$, π is a contraction mapping on $\mathcal{D}_0^J[0, T]$ (viewed as a Banach space with the uniform norm on compact sets). It is easy to verify that

$$
\|\pi(y^1) - \pi(y^2)\|_T \leq \theta \times \|y^1 - y^2\|_T
$$

for all $y^1, y^2 \in \mathcal{D}_0^J$, where θ is the largest row sum of matrix G, and by our assumption, $\theta < 1$. Since $T > 0$ is arbitrary, we have proved the existence and uniqueness of a fixed point for π on $\mathcal{D}_0^J$.

Finally, we prove the Lipschitz continuity of Ψ (and hence Φ). Given any $x^1, x^2 \in \mathcal{D}_0^J$, let

$$\pi_i(y)(t) = \sup_{0 \le u \le t} \left[-x^i(u) + Gy(u) \right]^+, \quad i = 1, 2.$$

Let $y^{i,0} = 0$ and $y^{i,n+1} = \pi_i(y^{i,n})$, $n \ge 0$ and $i = 1, 2$. It is clear that $y^{i,n} \to y^i = \Psi(x^i)$ as $n \to \infty$, $i = 1, 2$. On the other hand,

$$\|y^{1,n+1} - y^{2,n+1}\|_T \le \|x^1 - x^2\|_T + \theta \|y^{1,n} - y^{2,n}\|_T.$$

Letting $n \to \infty$ in the above yields

$$\|\Psi(x^1) - \Psi(x^2)\|_T \le \frac{1}{1-\theta} \|x^1 - x^2\|_T.$$

$\square$

7.3 A Homogeneous Fluid Network

The network consists of J buffers, indexed by $j = 1, \ldots, J$. Each buffer has an infinite storage capacity, and the buffers are interconnected by pipes to form a network within which fluid is circulating. (We assume that the pipes have infinite capacity and it takes no time for fluid to pass through the pipes.) The vectors α and μ are referred to as the *exogenous inflow rate (vector)* and the *potential outflow rate (vector)*, respectively; and the matrix P is referred to as the *flow-transfer matrix*. The quantity $\alpha_j t$ indicates the cumulative fluid that flows into buffer j exogenously (namely from the outside of the network) during the time interval $[0, t]$, and $\mu_j t$ indicates the cumulative potential outflow (of fluid) from buffer j during the time interval $[0, t]$ (namely the maximum outflow that can be realized during $[0, t]$ if buffer j never empties). The component p_{ij} of matrix P indicates the fraction of the outflow from buffer i that flows directly to buffer j, and $1 - \sum_{j=1}^{J} p_{ij}$ indicates the fraction of the outflow that leaves the network. Such a fluid network is referred to as fluid network (α, μ, P) with initial fluid level $z(0) = x(0)$.

Let $y_j(t)$ denote the cumulative amount of potential outflow that is lost due to emptiness at buffer j during $[0, t]$. Then the actual outflow from buffer j during $[0, t]$ is $\mu_j t - y_j(t)$, and the fluid level in buffer j at time t equals

$$z_j(t) = z_j(0) + \alpha_j t + \sum_{i=1}^{J} [\mu_j t - y_j(t)] p_{ij} - [\mu_j t - y_j(t)].$$

Rewriting the above in vector form yields

$$z(t) = x(t) + (I - P')y(t), \qquad t \ge 0, \tag{7.21}$$

where

$$x(t) = x(0) + [\alpha - (I - P')\mu]t, \qquad t \geq 0.$$

It is clear that the fluid level process must satisfy

$$z(t) \geq 0, \qquad t \geq 0, \tag{7.22}$$

and the cumulative loss process must satisfy

$$dy \geq 0 \text{ and } y(0) = 0. \tag{7.23}$$

We assume that the operations at the buffers observe a *work-conserving condition*, namely, no potential outflow at a buffer should be lost when the fluid level at that buffer is positive. Mathematically, this can be expressed as

$$z_j(t)dy_j(t) = 0 \qquad \text{for all } t \geq 0 \text{ and } j = 1, \ldots, J. \tag{7.24}$$

It follows from Theorem 7.2 that the relations in (7.21)–(7.24) uniquely determine the fluid level process $z = \{z(t), t \geq 0\}$ and the cumulative loss process $y = \{y(t), t \geq 0\}$. That is, $z = \Phi(x)$ and $y = \Psi(x)$.

Next we characterize the fluid network, particularly its asymptotic behavior as time evolves. To this end, we introduce an *effective* inflow rate (vector) λ, whose jth coordinate λ_j denotes the sum of exogenous and endogenous inflow rate to station j. Note that the effective outflow rate from a station is the minimum of its effective inflow rate and its potential outflow rate; hence, we must have

$$\lambda_j = \alpha_j + \sum_{i=1}^{J}(\lambda_i \wedge \mu_i)p_{ij}, \qquad j = 1, \ldots, J,$$

or in vector form,

$$\lambda = \alpha + P'(\lambda \wedge \mu). \tag{7.25}$$

The above equation is known as the *traffic equation*. Formally, the effective inflow rate is defined by the traffic equation. This is justified by the following theorem.

Theorem 7.3 The traffic equation (7.25) has a unique solution λ.

Proof. Define a mapping π from $\Re_+^J$ to $\Re_+^J$ by $\pi(x) = \alpha + P'(x \wedge \mu)$. Then λ is a solution to (7.25) if and only if it is a fixed point of π. Since $P^n \to 0$ and hence $(P')^n \to 0$ as $n \to \infty$, a high enough power of π is a contraction mapping. By Banach's contraction mapping theorem, π has a unique fixed point. $\qquad\square$

Let

$$\rho_j = \frac{\lambda_j}{\mu_j}$$

and call it the *traffic intensity* of station j, $j = 1, \ldots, J$. When $\rho \le e$, the traffic equation (7.25) simplifies to $\lambda = \alpha + P'\lambda$, a better-known form. In this case, the effective inflow rate has an explicit solution: $\lambda = (I - P')^{-1}\alpha$, and the condition $\rho \le e$ is the same as $\lambda = (I - P')^{-1}\alpha \le \mu$.

In general, buffer j is called a *nonbottleneck* if $\rho_j < 1$, a *bottleneck* if $\rho_j \ge 1$, a *balanced bottleneck* if $\rho_j = 1$, and a *strict bottleneck* if $\rho_j > 1$.

Theorem 7.4 Let y and z be, respectively, the cumulative loss process and the fluid level process in the fluid network (α, μ, P) with initial fluid level $z(0)$. Let λ and ρ be, respectively, its effective inflow rate and traffic intensity. Then, there exists a finite time τ such that
(i) if $\rho_j < 1$ (i.e., buffer j is a nonbottleneck), then

$$
\begin{aligned}
z_j(t) &= 0, & \text{for } t \ge \tau, \\
y_j(t) &= y_j(\tau) + (\mu_j - \lambda_j)(t - \tau), & \text{for } t \ge \tau,
\end{aligned}
$$

(ii) if $\rho_j \ge 1$ (i.e., buffer j is a bottleneck), then

$$
\begin{aligned}
z_j(t) &= z_j(\tau) + (\lambda_j - \mu_j)(t - \tau), & \text{for } t \ge \tau, \\
y_j(t) &= 0, & \text{for } t \ge 0.
\end{aligned}
$$

In particular, if $z(0) = 0$, then we can choose $\tau = 0$ and we have $z(t) = (\lambda - \mu)^+ t$ and $y(t) = (\mu - \lambda)^+ t$.

As a preparation for the proof of the theorem, we state the following lemma, whose proof is outlined in Exercise 20.

Lemma 7.5 Assume that P is a $J \times J$ substochastic matrix with a spectral radius less than unity. Let a and b be a partition of $\{1, \ldots, J\}$. Then

(i) P_a is a substochastic matrix with a spectral radius less than unity. (In particular, this implies that $(I - P'_a)^{-1}$ exists).

(ii) The matrix

$$\hat{P}_b := P_b + P_{ba}(I - P_a)^{-1}P_{ab}$$

is also a substochastic matrix with a spectral radius less than unity.

Proof (of Theorem 7.4). When $z(0) = 0$, it can be directly checked that $z(t) = (\lambda - \mu)^+ t$ and $y(t) = (\mu - \lambda)^+ t$ satisfy relations (7.21)–(7.24), and then the proof follows from the uniqueness of the reflection mapping.

Consider $z(0) \ne 0$. Let $a = \{j : \rho_j < 1\}$ be the set of nonbottleneck buffers and let $b = \{j : \rho_j \ge 1\}$ be the complement of a, i.e., the set of bottleneck buffers.

Let $f(t) = e'(I - P'_a)^{-1} z_a(t)$. We note that z is Lipschitz continuous and hence absolutely continuous, and so is f. If we can show that for some $\epsilon > 0$, $\dot{f}(t) < -\epsilon$ whenever $\dot{f}(t)$ exists and $f(t) > 0$, then there exists a finite τ such that for all $t \geq \tau$, $f(t) = 0$ and hence $z_a(t) = 0$. To this end, first rewrite the traffic equation (7.25) in block form,

$$\lambda_a = \alpha_a + P'_a \lambda_a + P'_{ba} \mu_b,$$
$$\lambda_b = \alpha_b + P'_{ab} \lambda_a + P'_b \mu_b.$$

Solving the above yields

$$\lambda_a = (I - P'_a)^{-1}(\alpha_a + P'_{ba}\mu_b), \tag{7.26}$$
$$\lambda_b = \alpha_b + P'_{ab}(I - P'_a)^{-1}\alpha_a + [P_b + P_{ba}(I - P_a)^{-1}P_{ab}]'\mu_b. \tag{7.27}$$

Next, rewrite (7.21) in block form,

$$z_a(t) = z_a(0) + [\alpha_a - (I - P'_a)\mu_a + P'_{ba}\mu_b]t$$
$$+ (I - P'_a)y_a(t) - P'_{ba}y_b(t), \tag{7.28}$$
$$z_b(t) = z_b(0) + [\alpha_b + P'_{ab}\mu_a - (I - P'_b)\mu_b]t$$
$$- P'_{ab}y_a(t) + (I - P'_b)y_b(t). \tag{7.29}$$

It follows from (7.28) and (7.26) that

$$f(t) = e'(I - P'_a)^{-1} z_a(t)$$
$$= f(0) + e'[(\lambda_a - \mu_a)t + y_a(t)] - e'(I - P'_a)^{-1}y_b(t).$$

Since $\dot{y}_b(t) \geq 0$, we have

$$\dot{f}(t) \leq e'[\dot{y}_a(t) - (\mu_a - \lambda_a)].$$

(Here and below, the existence of all derivatives is implicitly assumed.) If $f(t) > 0$, then $z_i(t) > 0$ for some $i \in a$, and hence $\dot{y}_i(t) = 0$ in view of (7.24). In addition, note that $\dot{y}_a(t) \leq (\mu_a - \lambda_a)$ (see Exercise 19); we have

$$\dot{f}(t) \leq -\min_{i \in a}(\mu_i - \lambda_i).$$

Thus, we have proved that there exists a finite τ such that $z_a(t) = 0$ for $t \geq \tau$.

Next, consider $t \geq \tau$. Solving $y_a(t)$ from (7.28) and substituting it into (7.29) yields

$$z_b(t) = z_b(0) + P'_{ab}(I - P'_a)^{-1} z_a(0) + (\lambda_b - \mu_b)t + (I - \hat{P}'_b)y_b(t),$$

where $\hat{P}_b := P_b + P_{ba}(I - P_a)^{-1}P_{ab}$ is a substochastic matrix with a spectral radius less than unity (by Lemma 7.5). Note that $\lambda_b \geq \mu_b$; then (ii) follows from the shift property of the reflection mapping (shifting the above process to τ).

Finally, since $z_a(t) = 0$ and $y_b(t) = y_b(\tau)$ for $t \geq \tau$, it follows that $y_a(t) = y_a(\tau) + (\mu_a - \lambda_a)(t - \tau)$ in view of (7.28) and the shift property (of the reflection mapping). $\qquad\qquad\square$

Theorem 7.4 characterizes the long-run behavior of bottleneck buffers and nonbottleneck buffers; however, it does not tell us how to identify bottleneck buffers and nonbottleneck buffers. We close this section with an algorithm that identifies bottleneck buffers:

Step 0 Set $a := \emptyset$ and $b := \{1, \ldots, J\}$;

Step 1 Compute

$$\hat{\alpha}_b := \alpha_b + P'_{ab}[I - P'_a]^{-1}\alpha_a, \qquad (7.30)$$

$$\hat{P}_b := P_b + P_{ba}[I - P_a]^{-1}P_{ab}, \qquad (7.31)$$

$$\hat{\theta}_b := \hat{\alpha}_b - [I - \hat{P}_b]\mu_b; \qquad (7.32)$$

Step 2 If $\hat{\theta}_b \geq 0$, return a as the set of nonbottlenecks, b as the set of bottlenecks, then STOP; otherwise, set $a := a \cup \{k : \hat{\theta}_k < 0\}$ and $b := a^c$, then GOTO Step 1.

The proof that the above algorithm does identify the set of nonbottlenecks (and hence bottlenecks) can be outlined as follows. The key idea is that we first identify some "obvious" nonbottleneck buffers, named *subcritical* buffers (defined below, also see Lemma 7.6 below). We then eliminate these buffers to form a subnetwork of all other buffers, and we do it in a way such that the bottlenecks and the nonbottlenecks remain invariant (see Lemma 7.7). This procedure is iterated until all buffers left (if any) in the subnetwork are bottlenecks.

Let $\theta = \alpha - (I - P')\mu$. We call buffer i *subcritical* if $\theta_i < 0$, and *supercritical* if $\theta_i \geq 0$. The following facts are immediate.

Lemma 7.6 Subcritical buffers are nonbottleneck buffers, but the converse may not hold. If all buffers are supercritical, then they are all bottlenecks.

For any partition a and b of $\{1, \ldots, J\}$, (7.30) and (7.31) actually define a transformation from a fluid network (α, μ, P) with buffers $\{1, \ldots, J\}$ to a subnetwork $(\hat{\alpha}_b, \mu_b, \hat{P}_b)$ with buffers b. Intuitively, the flows (both exogenous and endogenous) of the subnetwork are modified from the original network as though the passage flow through buffers in a were instantaneous. The following property of this transformation further justifies our iterative algorithm. (The proof of the lemma is left as an exercise.)

Lemma 7.7 If the set a does not include any strict bottleneck buffers, then the transformation (7.30) leaves invariant the effective inflow rate and the traffic intensity for buffers b. In other words, if $\rho_a \leq e$, then $\lambda_b = \hat{\lambda}_b$ and $\rho_b = \hat{\rho}_b$, where λ and ρ are respectively the effective inflow rate and the traffic intensity of the network (α, μ, P), and $\hat{\lambda}_b$ and $\hat{\rho}_b$ are respectively the effective inflow rate and the traffic intensity of the network $(\hat{\alpha}_b, \mu_b, \hat{P}_b)$.

7.4   A Reflected Brownian Motion

We assume throughout this section that R is an M-matrix, so that the reflection mapping is well-defined and continuous. Let $X = \{X(t), t \geq 0\}$ be a J-dimensional Brownian motion, starting at $x = X(0) \geq 0$, with drift vector θ and covariance matrix Γ. The process $Z = \{Z(t), t \geq 0\}$ with sample paths $Z = \Phi(X)$ is known as a reflected Brownian motion on the nonnegative orthant, and the process $Y = \{Y(t), t \geq 0\}$ with sample paths $Y = \Psi(X)$ is known as the regulator of the reflected Brownian motion. We shall denote the reflected Brownian motion by $Z = RBM_x(\theta, \Gamma; R)$, where R is the reflection matrix. Sometimes, we use $RBM(\theta, \Gamma)$ as a shorthand for $Z = RBM_x(\theta, \Gamma; R)$. From the definition of the reflection mapping, the process Z evolves like some Brownian motion in the interior of the nonnegative orthant $\Re_+^J$. When it reaches the jth face, $F_j := \{x = (x_i) \in \Re_+^J : x_j = 0\}$, of the nonnegative orthant, it is reflected (or pushed back) instantaneously toward the interior of the nonnegative orthant in the direction determined by the jth column of matrix R, and the jth component of the regulator $Y_j(t)$ indicates the cumulative amount of such effort in pushing it back.

Theorem 7.8 The reflected Brownian motion $Z = RBM_x(\theta, \Gamma; R)$ is a diffusion process (a strong Markov process with continuous sample paths).

Proof. Let $\mathcal{F}_t = \sigma(X(s); 0 \leq s \leq t)$ for $t \geq 0$ be the filtration generated by X. First it follows from the nonanticipation property and the continuity of the reflection mapping that both Z and Y are adapted to $\mathcal{F} = \{\mathcal{F}_t, t \geq 0\}$. That Z has a continuous sample path follows from the sample path continuity of a Brownian motion and the path continuity property of the reflection mapping. That Z is a strong Markov process follows from the shift property of the reflection mapping and the fact that a Brownian motion is a strong Markov process. $\square$

   We shall demonstrate in Section 7.5 that the RBM may arise as an approximation for the queue-length process of a generalized Jackson network. Hence, it is natural to approximate the stationary distribution of the queue-length process by the stationary distribution of the RBM. Formally, a stationary distribution of Z is a probability measure π (on $(\Re_+^J, \mathcal{B}(\Re_+^J))$)

such that

$$E_\pi[f(Z(t))] \equiv \int_{\Re_+^J} E_z[f(Z(t)]d\pi(z) = E_\pi[f(Z(0))]$$

for all bounded continuous functions f on $\Re_+^J$. We state the following theorem without proof.

Theorem 7.9 The reflected Brownian motion $Z = RBM_x(\theta, \Gamma; R)$ has a stationary distribution if and only if $R^{-1}\theta < 0$. When it exists, the stationary distribution is unique.

We now interpret the above necessary and sufficient condition in the context of approximating the generalized Jackson network. Let α be the exogenous arrival rate (vector), μ the service rate (vector), and P be the routing matrix of the generalized Jackson network. Let $Z = RBM_x(\theta, \Gamma; R)$ be a reflected Brownian motion that approximates the queue-length process of the network. In the next section we will show that $R = I - P'$, and $\theta = \alpha - (I - P')\mu$. (Note that the covariance matrix Γ does not have any impact on the existence of the stationary distribution.) In this case, the condition $R^{-1}\theta < 0$ is the same as $\lambda := (I - P')^{-1}\alpha < \mu$, or the traffic intensity being less than unity. Obviously, this is consistent with the existence of the stationary distribution for the generalized Jackson network.

Given the existence of a stationary distribution, we characterize the stationary distribution by a basic adjoint relation (BAR) as we did in Section 6.2 for the one-dimensional reflected Brownian motion. To this end, define the following operators:

$$\mathcal{L} = \frac{1}{2} \sum_{i,j=1}^J \Gamma_{ij} \frac{\partial^2}{\partial z_i \partial z_j} + \sum_{j=1}^J \theta_j \frac{\partial}{\partial z_j},$$

$$\mathcal{D}_j = \sum_{i=1}^J r_{ij} \frac{\partial}{\partial z_i} \equiv r_j' \nabla,$$

where r_j is the jth column of R. Write $Z(t) = X(0) + \theta t + \Gamma^{1/2}W(t)$, where $W(t) = X(t) - X(0) - \theta t$ is a J-dimensional standard Brownian motion. Then it follows from Ito's formula that for any function f defined on $\Re_+^J$ whose derivatives up to order 2 are all continuous and bounded,

$$f(Z(t)) - f(Z(0)) = \int_0^t \Delta f(Z(s))dW(s) + \int_0^t \mathcal{L}f(Z(s))ds$$

$$+ \sum_{j=1}^J \int_0^t \mathcal{D}_j f(Z(s))dY_j(s),$$

where $\Delta f(z) = (\partial f/\partial z_1, \dots, \partial f/\partial z_J)$. Now taking the conditional expectation with respect to the stationary distribution π on both sides of the above and noting that the first integral on the right-hand side is a martingale, we obtain

$$0 = \mathsf{E}_\pi \int_0^t \mathcal{L}f(Z(s))ds + \sum_{j=1}^J \mathsf{E}_\pi \int_0^t \mathcal{D}_j f(Z(s))dY_j(s)$$

$$= t\mathsf{E}_\pi[\mathcal{L}f(Z(0))] + \sum_{j=1}^J \mathsf{E}_\pi \int_0^t \mathcal{D}_j f(Z(s))dY_j(s)$$

$$= t\int_{\Re_+^J} \mathcal{L}f(z)d\pi(z) + \sum_{j=1}^J \mathsf{E}_\pi \int_0^t \mathcal{D}_j f(Z(s))dY_j(s).$$

It can be shown that a version of Y_j is a continuous additive function of Z with support on F_j (where we note that Y_j increases at time t only when $Z_j(t) = 0$); hence,

$$\mathsf{E}_\pi \int_0^t \mathcal{D}_j f(Z(s))dY_j(s) = t\int_{F_j} \mathcal{D}_j f(z)d\nu_j(z),$$

where ν_j is a finite Borel measure. (See Harrison and Williams [13] for a more rigorous treatment.) In summary, we have

Theorem 7.10 Let π be a stationary distribution of $Z = RBM_x(\theta, \Gamma; R)$. Then for each $j = 1, \ldots, J$, there exists a finite Borel measure ν_j on F_j (which is mutually absolutely continuous with the Lebesgue measure on the face F_j) such that a basic adjoint relation (BAR),

$$\int_{\Re_+^J} \mathcal{L}f(z)d\pi(z) + \sum_{j=1}^J \int_{F_j} \mathcal{D}_j f(z)d\nu_j(z) = 0, \tag{7.33}$$

holds for all continuous and bounded functions f defined on $\Re_+^J$ whose derivatives up to order 2 are also continuous and bounded.

Remark 7.11 We note that

$$\nu_j(F_j) = \frac{1}{t}\mathsf{E}_\pi[Y_j(t)], \qquad \text{for all } t > 0,$$

gives the average (over time) expected amount of push-back at the face F_j to keep the reflected Brownian motion within its domain, the nonnegative orthant.

Finally, we show how to use the basic adjoint relation to derive the stationary distribution in a special case where the stationary distribution has a product form. Let D be a J-dimensional diagonal matrix whose jth diagonal element is r_{jj} (the jth diagonal element of matrix R), $j = 1, \ldots, J$, and let Λ be a J-dimensional diagonal matrix whose jth diagonal element is Γ_{jj} (the jth diagonal element of matrix Γ), $j = 1, \ldots, J$.

Theorem 7.12 Suppose that $R^{-1}\theta < 0$. If

$$2\Gamma = RD^{-1}\Lambda + \Lambda D^{-1}R',\qquad(7.34)$$

then the stationary distribution of $Z = RBM_x(\theta,\Gamma;R)$ has a product form. Specifically, the density function of the stationary distribution takes the form

$$f(z_1,\ldots,z_J) = \prod_{j=1}^{J} f_j(z_j),\qquad(7.35)$$

with

$$f_j(z) = \begin{cases} \eta_j e^{-\eta_j z}, & z \geq 0, \\ 0, & z < 0, \end{cases} \qquad j = 1,\ldots,J,$$

where

$$\eta = -2\Lambda^{-1}DR^{-1}\theta.$$

Conversely, if the stationary density function takes the product form (7.35) (it is not necessary to assume the exponential form), then the condition (7.34) must hold and f_j must be of the exponential form as given above.

In Exercise 23 a proof is outlined to show that the BAR always holds under the condition (7.34) with the product form density function given in the above theorem. This would establish the sufficiency of Theorem 7.12, if the BAR could be shown to be sufficient for π to be a stationary distribution. The proof of the necessity part of Theorem 7.12 is outlined in Exercise 24.

7.5 Approximating the Network

Recall the relations in (7.8) and (7.11)–(7.13), and the oblique reflection theorem. In order to find an approximation for the queue-length process $Q = \Phi(X)$, it suffices to find an approximation for the process X, as defined in (7.9). In this section we argue heuristically that under some mild conditions, the process X can be approximated by a Brownian motion, and hence the queue-length process Q can be approximated by a reflected Brownian motion as defined in Section 7.4. We call such an approximation the Brownian approximation. (It would be useful to first review Section 6.5 on approximating the G/G/1 queue before proceeding further.) In addition, we present a numerical example at the end of this section to provide some evidence on the quality of the Brownian approximation.

First, it follows from the strong law of large numbers that

$$A(t) \approx \alpha t, \tag{7.36}$$

$$S(t) \approx \mu t, \quad \text{and} \tag{7.37}$$

$$R(t) \approx P't, \tag{7.38}$$

where $R(t) := (R^1(\lfloor t \rfloor), \ldots, R^J(\lfloor t \rfloor))$. Hence,

$$X(t) \approx \bar{X}(t) := Q(0) + \theta t,$$

where X is as defined in (7.9),

$$\theta = \alpha - (I - P')\mu. \tag{7.39}$$

It follows from the continuity of the oblique reflection mapping that $Q = \Phi(X)$ and $Y = \Psi(X)$ can be approximated by $\bar{Q} = \Phi(\bar{X})$ and $\bar{Y} = \Psi(\bar{X})$. In view of the special linear form of $\bar{X}$, it is clear that $\bar{Q}$ and $\bar{Y}$ can be interpreted, respectively, as the fluid level and the cumulative loss of the fluid network (α, μ, P), as described in Section 7.3. Let λ be the effective arrival rate determined by the traffic equation (7.25), and let $\rho_j = \lambda_j/\mu_j$, $j = 1, \ldots, J$, be the traffic intensity. Then it follows from Theorem 7.4 that

$$Y(t) \approx \bar{Y}(t) \approx (\mu - \lambda)^+ t.$$

In view of $Y_i(t) = \mu_i[t - B_i(t)]$, the above approximation implies the following approximation for the busy time process

$$B_j(t) \approx (1 \wedge \rho_j)t. \tag{7.40}$$

Combining the above with (7.37) yields,

$$S_j(B_j(t)) \approx \mu_j(1 \wedge \rho_j)t = (\lambda_j \wedge \mu_j)t. \tag{7.41}$$

All of the above are first-order approximations in the sense that the approximations involves only the means, the first-order information of the data.

Next, assume that both the interarrival times of the renewal (exogenous) arrival process A_j ($j \in \mathcal{E}$) and the service times $v_j(1)$ ($j = 1, \ldots, J$) have finite second moments. Let $c_{0,j}$ be the coefficient of variation of the interarrival times of A_j for $j \in \mathcal{E}$, $c_{0,j} = 0$ for $j \notin \mathcal{E}$, and let c_j be the coefficient of variation of $v_j(1)$ for $j = 1, \ldots, J$. Then, from the FCLT of Section 6.5, we have

$$A(t) - \alpha t \stackrel{d}{\approx} \hat{A}(t), \tag{7.42}$$

$$S(t) - \mu t \stackrel{d}{\approx} \hat{S}(t), \tag{7.43}$$

$$R(t) - P't \stackrel{d}{\approx} \hat{R}(t), \tag{7.44}$$

where $\hat{A}$, $\hat{S}$, and $\hat{R}^j$ (the jth coordinate of $\hat{R}$), $j = 1, \ldots, J$, are $J + 2$ independent J-dimensional driftless Brownian motions, whose covariance matrices are as follows:

$$\Gamma^0 = (\Gamma^0_{jk}) \qquad \text{with} \qquad \Gamma^0_{jk} = \alpha_j c^2_{0,j} \delta_{jk}, \qquad (7.45)$$

$$\Gamma^{J+1} = (\Gamma^{J+1}_{jk}) \qquad \text{with} \qquad \Gamma^{J+1}_{jk} = \mu_j c^2_j \delta_{jk}, \qquad (7.46)$$

$$\Gamma^\ell = (\Gamma^\ell_{jk}) \qquad \text{with} \qquad \Gamma^\ell_{jk} = p_{\ell j}(\delta_{jk} - p_{\ell k}). \qquad (7.47)$$

Combining the first-order approximations in (7.40) and (7.41) with the second-order approximations in (7.43) and (7.44), we have

$$S_j(B_j(t)) - \mu_j B_j(t) \stackrel{\mathrm{d}}{\approx} \hat{S}_j((1 \wedge \rho_j)t), \qquad (7.48)$$

$$R^i_j(S_i(B_i(t))) - p_{ij}S_i(B_i(t)) \stackrel{\mathrm{d}}{\approx} \hat{R}^i_j((\lambda_i \wedge \mu_i)t),, \qquad (7.49)$$

$i, j = 1, \ldots, J$. From (7.42)–(7.44) and (7.48)–(7.49), we can approximate $X_j(t)$ defined in (7.9) by

$$\tilde{X}_j(t) = Q_j(t) + \theta_j t + \hat{A}_j(t) - \hat{S}_j((1 \wedge \rho_j)t)$$
$$+ \sum_{i=1}^J \left[\hat{R}^i_j((\lambda_i \wedge \mu_i)t) + p_{ij}\hat{S}_i((1 \wedge \rho_i)t) \right],$$

with θ_j being the jth component of θ defined in (7.39). In vector form, we have

$$\tilde{X}(t) = Q(0) + \theta t + \hat{A}(t) + \sum_{i=1}^J \hat{R}^i((\lambda_i \wedge \mu_i)t) - (I - P')\hat{S}((e \wedge \rho)t).$$

The process $\tilde{X} = \{\tilde{X}(t), t \geq 0\}$ is a J-dimensional Brownian motion starting at $Q(0)$ with drift vector θ and covariance matrix $\Gamma = (\Gamma_{k\ell})$:

$$\Gamma_{k\ell} = \sum_{j=1}^J (\lambda_j \wedge \mu_j) \left[p_{jk}(\delta_{k\ell} - p_{j\ell}) + c^2_j(p_{jk} - \delta_{jk})(p_{j\ell} - \delta_{j\ell}) \right]$$
$$+ \alpha_k c^2_{0,k} \delta_{k\ell}, \qquad k, \ell = 1, \ldots, J. \qquad (7.50)$$

Therefore, the reflected Brownian motion $\Phi(\tilde{X}) = RBM_{Q(0)}(\theta, \Gamma; I - P')$ provides an approximation to the queue-length process Q.

To recapitulate, for a generalized Jackson network, with a given routing matrix (P), service rate vector (μ), and exogenous arrival rate vector (α), and the coefficients of variation of the interarrival and service times ($c_{0,j}$ and c_j, $j = 1, \ldots, J$), its queue-length process can be approximated by a reflected Brownian motion $RBM_{Q(0)}(\theta, \Gamma; I - P')$, where the drift vector θ follows (7.39) and the covariance matrix Γ follows (7.50).

When the network does not have any bottleneck station, i.e., $\lambda = (I - P')^{-1}\alpha < \mu$, the reflected Brownian motion has a stationary distribution, which is characterized by the basic adjoint relation (BAR) (7.33) in Theorem 7.10. Numerical solutions are possible based on the BAR, which shall be discussed in Chapter 10.

In this case, the product form condition (7.34) can be simplified to

$$2\Gamma_{jk} = -(p_{jk}\Gamma_{jj} + p_{kj}\Gamma_{kk}) \qquad \text{for all } j \neq k. \tag{7.51}$$

In other words, when the covariance matrix $\Gamma = (\Gamma_{jk})$ satisfies the above condition (7.51), the density function of the stationary distribution takes a product form,

$$f(x_1, \ldots, x_J) = \prod_{j=1}^{J} f_j(x_j),$$

with

$$f_j(x) = \eta_j e^{-\eta_j x}, \qquad x \geq 0,$$

and $f_j(x) = 0$ for $x < 0$, where

$$\eta_j = \frac{2(\mu_j - \lambda_j)}{\Gamma_{jj}},$$

$j = 1, \ldots, J$. It can be directly verified that when $c_{0,j} = 1$ and $c_j = 1$ for all $j = 1, \ldots, J$, the condition in (7.51) holds. A Jackson network with constant service rates at all stations is one such special case.

We note that for the Jackson network, the above product form stationary distribution for the approximation RBM is exactly the same as the stationary distribution for the original Jackson network. However, this is not true in general. Note that the above heuristic deviation suggests only that the reflected Brownian motion is an approximation of the queue-length process for each finite time t or a bounded time interval. (Indeed, this can be justified by the limit theorems as we shall do in the following sections.) However, it has not been established that the stationary distribution of the RBM would arise as some kind of limit of the stationary distribution of the queue-length process. On the other hand, this should not prevent us from proposing the stationary distribution of the RBM as an approximation for the stationary distribution of the corresponding queueing network. Numerical experiments have indicated that such an approximation provides quite impressive results. While the specifics of the numeric algorithm are not discussed till Chapter 10, we next present some numerical results.

Consider the three-station network given by Figure 7.1. The traffic intensity of this network is

$$\rho = \begin{pmatrix} (2\alpha_1 + \alpha_2)m_1 \\ (\alpha_1 + 1.5\alpha_2)m_2 \\ (\alpha_1 + 0.5\alpha_2)m_3 \end{pmatrix}.$$

In this network, jobs arrive to stations 1 and 2 following two independent Poisson processes with the same rate $\alpha_1 = \alpha_2$ respectively. There are no exogenous arrivals to station 3 (i.e., $\alpha_3 = 0$). Jobs that finish service at station 1 proceed to either station 2 or station 3, with equal probability. Jobs completing service at station 2 either join station 1 or exit the network, each with probability 0.5. Jobs completing service station 3 either join station 1 or exit the network, with equal probability. All service times are i.i.d. at each station, independent among stations, and independent of arrivals. The mean and SCV of the service times at station j are m_j and c_j^2, $j = 1, 2, 3$. We consider 11 versions of the system, with parameters shown by Table 7.1. In Cases 1 through 5, station 1 is heavily loaded, while stations 2 and 3 are moderately loaded; in Cases 6 and 7, all stations are lightly loaded; in Cases 8 and 9, all stations are moderately loaded; and Case 10 and 11, all stations are heavily loaded.

Table 7.2 shows the simulation estimates and SRBM approximations of the mean sojourn time at each station. The mean sojourn time is the sum of the mean waiting time and the mean service time, and the mean waiting time is approximated by the workload, which relates to the queue length by Little's law (see Theorem 7.19 below). Specifically, we approximate the mean sojourn time at station j by $m_j \mathsf{E}_\pi[\tilde{Q}_j] + m_j$, $j = 1, \ldots, J$. In simulation, the random variables are fitted with Erlang distributions, exponential distributions, or Gamma distributions depending on the SCV being less than 1, equal to 1, or larger than 1, respectively. In this table, the number in parentheses after each simulation result is the half-width of the 95% confidence interval, expressed as a percentage of the simulation estimate. The number in parentheses after each SRBM estimate represents the percentage error from the simulation result. The mean sojourn time from the SRBM model is computed by the BNA/FM algorithm developed by Chen and Shen in [7]. We note that Case 2 corresponds to the case where the network is a Jackson network; in this case, the exact average sojourn time is given by

$$\mathsf{E}[\mathcal{S}_j] = \frac{m_j}{1 - \rho_j}, \qquad j = 1, 2, 3;$$

hence, $\mathsf{E}[\mathcal{S}_1] = 6$, $\mathsf{E}[\mathcal{S}_2] = 2.4$, and $\mathsf{E}[\mathcal{S}_3] = 2$.

7.6 Functional Law of the Iterated Logarithm

In this section we establish a version of the functional law of the iterated logarithm (FLIL) for the generalized Jackson network as described in Section 7.1. To this end, we assume that the interarrival times and the service times have finite second moments. It then follows from Theorem 5.13 that the arrival process, the service process, and the routing process satisfy the

Case	α_1	α_2	m_1	m_2	m_3	c_1^2	c_2^2	c_3^2	ρ_1	ρ_2	ρ_3
1	0.50	0.50	0.60	0.60	0.80	0.25	0.25	0.25	0.90	0.75	0.60
2	0.50	0.50	0.60	0.60	0.80	1.00	1.00	1.00	0.90	0.75	0.60
3	0.50	0.50	0.60	0.60	0.80	2.00	2.00	2.00	0.90	0.75	0.60
4	0.50	0.50	0.60	0.60	0.80	2.00	0.25	0.25	0.90	0.75	0.60
5	0.50	0.50	0.60	0.60	0.80	0.25	2.00	2.00	0.90	0.75	0.60
6	0.50	0.50	0.20	0.20	0.40	0.25	0.25	0.25	0.30	0.25	0.30
7	0.50	0.50	0.20	0.20	0.40	0.25	0.25	0.25	0.30	0.25	0.30
8	0.50	0.50	0.40	0.50	0.80	0.25	0.25	0.25	0.60	0.63	0.60
9	0.50	0.50	0.40	0.50	0.80	2.00	2.00	2.00	0.60	0.63	0.60
10	0.75	0.75	0.40	0.50	0.80	0.25	0.25	0.25	0.90	0.94	0.90
11	0.75	0.75	0.40	0.50	0.80	2.00	2.00	2.00	0.90	0.94	0.90

TABLE 7.1. Parameters of the generalized Jackson queueing network

System No.	Method	$E(\mathcal{S}_1)$	$E(\mathcal{S}_2)$	$E(\mathcal{S}_3)$
1	Simulation	4.879 (3.9%)	1.572 (0.8%)	1.280 (0.4%)
	SRBM	4.544 (6.9%)	1.578 (0.4%)	1.303 (1.8%)
2	Simulation	6.019 (4.1%)	2.399 (1.4%)	2.000 (0.8%)
	SRBM	5.957 (1.0%)	2.385 (0.6%)	2.004 (0.2%)
3	Simulation	7.610 (4.1%)	3.405 (1.4%)	2.880 (1.2%)
	SRBM	7.668 (0.8%)	3.394 (0.3%)	2.839 (1.4%)
4	Simulation	7.402 (5.8%)	2.020 (1.1%)	1.902 (0.6%)
	SRBM	7.426 (0.3%)	2.029 (0.5%)	1.857 (2.4%)
5	Simulation	5.009 (5.1%)	3.029 (1.6%)	2.296 (1.0%)
	SRBM	4.976 (0.7%)	3.054 (0.8%)	2.385 (3.9%)
6	Simulation	0.255 (0.2%)	0.239 (0.1%)	0.504 (0.2%)
	SRBM	0.257 (0.8%)	0.242 (1.3%)	0.500 (0.8%)
7	Simulation	0.318 (0.5%)	0.297 (0.4%)	0.649 (0.6%)
	SRBM	0.317 (0.3%)	0.297 (0.0%)	0.629 (3.1%)
8	Simulation	0.777 (0.6%)	1.024 (0.5%)	1.522 (0.8%)
	SRBM	0.780 (0.4%)	1.028 (0.4%)	1.504 (1.2%)
9	Simulation	1.262 (0.8%)	1.714 (1.0%)	2.596 (1.3%)
	SRBM	1.290 (2.2%)	1.743 (1.7%)	2.609 (0.5%)
10	Simulation	2.620 (1.9%)	5.550 (4.1%)	5.191 (3.4%)
	SRBM	2.602 (0.7%)	5.416 (2.4%)	5.015 (3.4%)
11	Simulation	5.738 (3.6%)	11.091 (3.1%)	11.677 (4.1%)
	SRBM	5.808 (1.2%)	11.128 (0.3%)	11.669 (0.1%)

TABLE 7.2. Average sojourn time of the generalized Jackson queueing network

FLIL. That is, as $T \to \infty$,

$$\sup_{0 \le t \le T} \|A(t) - \alpha t\| \overset{\text{a.s.}}{=} O(\sqrt{T \log \log T}), \tag{7.52}$$

$$\sup_{0 \le t \le T} \|S(t) - \mu t\| \overset{\text{a.s.}}{=} O(\sqrt{T \log \log T}), \tag{7.53}$$

$$\sup_{0 \le t \le T} \|R(t) - P't\| \overset{\text{a.s.}}{=} O(\sqrt{T \log \log T}), \tag{7.54}$$

$$\sup_{0 \le t \le T} \|V(t) - mt\| \overset{\text{a.s.}}{=} O(\sqrt{T \log \log T}), \tag{7.55}$$

where $\alpha = (\alpha_j)$, $\mu = (\mu_j)$ and P are the exogenous arrival rate (vector), the service rate (vector) and the routing matrix, respectively; and $m = (m_j)$, with $m_j = 1/\mu_j$ being the expected service time for a job at station j. In (7.55), $V(t) := V(\lfloor t \rfloor)$.

As we shall see, the FLIL limit of the generalized Jackson network can be described by the fluid network introduced in Section 7.3. In particular, the exogenous arrival rate, the service rate, and the routing matrix correspond respectively to the exogenous inflow rate, the potential outflow rate, and the flow-transfer matrix. Corresponding to the effective inflow rate, we define an *effective* arrival rate (vector) λ, whose jth coordinate λ_j denotes the sum of exogenous and endogenous arrival rates to station j. The effective arrival rate must satisfy (and, in fact, is uniquely determined by) the traffic equation (7.25) (see Theorem 7.3). Similarly, let

$$\rho_j = \frac{\lambda_j}{\mu_j}$$

and call it the *traffic intensity* of station j, $j = 1, \ldots, J$.

Theorem 7.13 Let Q, Z, and B be, respectively, the queue, the workload, and the busy time processes. Suppose that (7.52)-(7.55) hold. Then, as $T \to \infty$,

$$\sup_{0 \le t \le T} \|Q(t) - \bar{Q}(t)\| \stackrel{\text{a.s.}}{=} O(\sqrt{T \log \log T}), \tag{7.56}$$

$$\sup_{0 \le t \le T} \|Z(t) - \bar{Z}(t)\| \stackrel{\text{a.s.}}{=} O(\sqrt{T \log \log T}), \tag{7.57}$$

$$\sup_{0 \le t \le T} \|B(t) - (\rho \wedge e)t\| \stackrel{\text{a.s.}}{=} O(\sqrt{T \log \log T}), \tag{7.58}$$

where

$$\bar{Q} = \bar{X} + (I - P')\bar{Y}, \tag{7.59}$$

$$\bar{X}(t) = Q(0) + \theta t, \tag{7.60}$$

$$\theta = \alpha - (I - P')\mu, \tag{7.61}$$

$$\bar{Y} = \Psi(\bar{X}), \tag{7.62}$$

$$\bar{Z} = M\bar{Q}, \tag{7.63}$$

$M = \text{diag}(m)$ is a J-dimensional diagonal matrix whose jth diagonal element equals m_j, $j = 1, \ldots, J$, and Ψ is the oblique reflection mapping defined in Theorem 7.2 with $R = I - P'$.

Remark 7.14 The deterministic process $\bar{Q}$ can be interpreted as the fluid level process of a fluid network (α, μ, P) with initial state $Q(0)$ as described in Section 7.3. Refer to that section, in particular Theorem 7.4, for a characterization of the limit process $\bar{Q}$.

Remark 7.15 The approximations in (7.56)–(7.58) still prevail with $\bar{Q}(t)$ $= (\lambda - \mu)^+ t$ and $\bar{W}(t) = (\rho - e)^+ t$. This is equivalent to replacing $Q(0)$ by 0 in (7.60).

Remark 7.16 Note the above theorem provides a justification (in terms of an approximate bound) for the first-order approximation proposed in Section 7.5.

Remark 7.17 The proof of the theorem is similar to the proof of Theorem 6.11. In particular, as it is clear from the proof (of the latter) that there is no need to assume that the arrival and the service processes are renewal processes, the routing process forms i.i.d. sequences, and all of them are mutually independent. The theorem holds as long as the FLIL equalities (7.52)–(7.55) hold.

Remark 7.18 As remarked after Theorem 6.11, it would be interesting to know what the limit

$$\limsup_{T \to \infty} \frac{\sup_{0 \le t \le T} |Q_j(t) - \bar{Q}_j(t)|}{\sqrt{T \log \log T}},$$

is for each $j = 1, \ldots, J$. Such a limit theorem would be a stronger form of the FLIL (consistent with the classical FLIL for the i.i.d. summation). The same question can be asked about the workload process and the busy time process.

Proof (of Theorem 7.13). Recall the representation of the queue-length process by $Q = \Phi(X)$ with X following (7.9) in the last section. In view of $0 \le B_j(t) \le t$ for all $t \ge 0$ and $j = 1, \ldots, J$, applying the FLIL equalities in (7.52)–(7.55) to (7.9) yields

$$\|X - \bar{X}\|_T = \sup_{0 \le t \le T} \|X(t) - \bar{X}(t)\| \stackrel{\text{a.s.}}{=} O(\sqrt{T \log \log T}), \quad T \to \infty,$$

where $\bar{X}$ follows (7.60). The above, together with the Lipschitz continuity of the reflection mapping (Theorem 7.2), leads to the equality in (7.56), and

$$\|Y - \bar{Y}\|_T = \sup_{0 \le t \le T} \|Y(t) - \bar{Y}(t)\| \stackrel{\text{a.s.}}{=} O(\sqrt{T \log \log T}), \quad T \to \infty. \quad (7.64)$$

The proof of the two equalities in (7.57) and (7.58) are similar to the proof of Theorem 6.11. The first equality involves rewriting (7.3) by centering, and the second one uses the relations in (7.64) and (7.10) and a modification of Lemma 6.15. $\qquad \square$

7.7 Strong Approximation

We establish here a strong approximation theorem for the generalized Jackson network. We shall assume that the interarrival times of the arrival processes A_j, $j \in \mathcal{E}$, and the service processes S_j, $j = 1, \ldots, J$, have finite rth moments with $r > 2$. Based on the FSAT (Theorem 5.14), we can assume that all processes are defined on a probability space such that there exist $J + 2$ independent J-dimensional Wiener process W^j, $j = 0, 1, \ldots, J, J+1$, so that the following is fulfilled: as $T \to \infty$,

$$\sup_{0 \le t \le T} |A(t) - \alpha t - (\Gamma^0)^{1/2} W^0(t)| \stackrel{\text{a.s.}}{=} o(T^{1/r}), \qquad (7.65)$$

$$\sup_{0 \le t \le T} |S(t) - \mu t - (\Gamma^{J+1})^{1/2} W^{J+1}(t)| \stackrel{\text{a.s.}}{=} o(T^{1/r}), \qquad (7.66)$$

$$\sup_{0 \le t \le T} |V(t) - mt + M(\Gamma^{J+1})^{1/2} W^{J+1}(mt)| \stackrel{\text{a.s.}}{=} o(T^{1/r'}), \qquad (7.67)$$

$$\sup_{0 \le t \le T} |R^\ell(t) - p^\ell t - (\Gamma^\ell)^{1/2} W^\ell(t)| \stackrel{\text{a.s.}}{=} O(\log T), \qquad (7.68)$$

where p^ℓ is the ℓth column of P', $\ell = 1, \ldots, J$. The covariance matrices Γ^0, Γ^{J+1}, and Γ^ℓ, $\ell = 1, \ldots, J$, follow (7.45)–(7.47). In (7.67), we can choose $r' = r$ when $r < 4$ and any $r' < 4$ when $r \ge 4$, and in (7.68), the approximate bound follows from (5.34) in the remark after Theorem 5.14.

Theorem 7.19 Let Q, Z, and B be the queue-length, the workload, and the busy-time processes. Suppose that the FSAT limits in (7.65)–(7.68) hold with $r > 2$. Then, as $T \to \infty$,

$$\sup_{0 \le t \le T} |Q(t) - \tilde{Q}(t)| \stackrel{\text{a.s.}}{=} o(T^{1/r'}), \qquad (7.69)$$

$$\sup_{0 \le t \le T} |Z(t) - \tilde{Z}(t)| \stackrel{\text{a.s.}}{=} o(T^{1/r'}), \qquad (7.70)$$

$$\sup_{0 \le t \le T} |B(t) - \tilde{B}(t)| \stackrel{\text{a.s.}}{=} o(T^{1/r'}), \qquad (7.71)$$

where

$$\tilde{Q} = \tilde{X} + (I - P')\tilde{Y}, \qquad (7.72)$$

$$\tilde{X}(t) = Q(0) + \theta t + (P - I)\Gamma^{J+1} W^{J+1}((\rho \wedge e)t)$$

$$+ (\Gamma^0)^{1/2} W^0(t) + \sum_{\ell=1}^{J} (\Gamma^\ell)^{1/2} W^\ell((\lambda_\ell \wedge \mu_\ell)t), \qquad (7.73)$$

$$\theta = \alpha - (I - P')\mu, \qquad (7.74)$$

$$\tilde{Y} = \Phi(\tilde{X}),$$

$$\tilde{Z}(t) = M\tilde{Q}(t)$$

$$+ M(\Gamma^{J+1})^{1/2}[W^{J+1}((\rho \wedge e)t) - W^{J+1}(\rho t)], \qquad (7.75)$$

$$\tilde{B}(t) = et - M^{-1}\tilde{Y}(t), \qquad (7.76)$$

Φ is the reflection mapping as defined in Theorem 7.2, and $M = \text{diag}(m)$ is a J-dimensional diagonal matrix whose jth diagonal component is m_j. In (7.69)–(7.71) we can choose $r' = r$ when $r < 4$ and any $r' < 4$ when $r \ge 4$.

Remark 7.20 For $r \geq 4$, the FSAT bounds in (7.69) and (7.70) can be improved as follows:

$$\sup_{0 \leq t \leq T} |Q(t) - \tilde{Q}(t)| = O\left((T \log \log T)^{1/4}(\log T)^{1/2}\right),$$

$$\sup_{0 \leq t \leq T} |Z(t) - \tilde{Z}(t)| = O\left((T \log \log T)^{1/4}(\log T)^{1/2}\right),$$

as $T \to \infty$. A stronger result that (7.69) and (7.70) hold with $r' = r$ for all $r > 2$ may not hold, however. (See Remark 1 after Theorem 6.16.)

Remark 7.21 The process $\tilde{X}$ has the same distribution as $\tilde{X}$ introduced in Section 7.5, namely, it is a Brownian motion starting at $Q(0)$ with drift θ and covariance matrix $\Gamma = (\Gamma_{jk})$ as defined in (7.50). Hence, this theorem provides a theoretical justification for the second-order (Brownian) approximation of the queue-length process proposed in Section 7.5. Specifically, it provides an approximate bound.

Remark 7.22 When $\rho \leq e$, equality (7.75) is Little's law, giving a relation between the limiting queue-length and the limiting workload processes. However, this relation fails when some $\rho_j > 1$, i.e., when there exists at least one strict bottleneck.

Proof (of Theorem 7.19). If (7.65)–(7.67) hold for $r \geq 4$, they clearly hold for $r < 4$. Therefore, it suffices for us to restrict our attention to the case $r < 4$.

Recall the representation of the queue-length process Q and the scaled idle time process Y in Section 7.1 through the reflection mapping $Q = \Phi(X)$ and $Y = \Psi(X)$. If we can show that

$$\sup_{0 \leq t \leq T} |X(t) - \tilde{X}(t)| \overset{\text{a.s.}}{=} o(T^{1/r}), \tag{7.77}$$

then we immediately deduce (7.69) and

$$\sup_{0 \leq t \leq T} |Y(t) - \tilde{Y}(t)| \overset{\text{a.s.}}{=} o(T^{1/r})$$

from the Lipschitz continuity of the mappings Ψ and Φ. The latter implies (7.71).

To prove (7.77), we first rewrite (7.9) as

$$X(t) = Q(0) + [\alpha + (P' - I)\mu]t + \Big[A(t) - \alpha t - (\Gamma^0)^{1/2}W^0(t)\Big]$$

$$+ \sum_{\ell=1}^{J} \Big[R^\ell(S_\ell(B_\ell(t))) - p^\ell S_\ell(B_\ell(t)) - (\Gamma^\ell)^{1/2}W^\ell(S_\ell(B_\ell(t)))\Big]$$

$$+ (P' - I)\left[S(B(t)) - M^{-1}B(t) - \Gamma^{J+1}W^{J+1}(B(t))\right]$$

$$+ \sum_{\ell=1}^{J}(\Gamma^\ell)^{1/2}\left[W^\ell(S_\ell(B_\ell(t))) - W^\ell((\lambda_\ell \wedge \mu_\ell)t)\right]$$

$$+ (P' - I)\left[\Gamma^{J+1}W^{J+1}(B(t)) - \Gamma^{J+1}((\rho \wedge e)t)\right]$$

$$+ (\Gamma^0)^{1/2}W^0(t) + \sum_{\ell=1}^{J}(\Gamma^\ell)^{1/2}W^\ell((\lambda_\ell \wedge \mu_\ell)t)$$

$$+ (P - I)\Gamma^{J+1}W^{J+1}((\rho \wedge e)t).$$

Note that $0 \le B(t) \le et$; the proof of (7.77) follows from the FSAT assumptions (7.65)-(7.68), the FLIL bound (7.58), and Lemma 6.21.

Finally, we prove (7.70). To this end, rewrite (7.3),

$$Z(t) = \Big[V\left(Q(t) + S(B(t))\right) - M(Q(t) + S(B(t)))$$

$$+ M(\Gamma^{J+1})^{1/2}W^{J+1}(M(Q(t) + S(B(t))))\Big]$$

$$+ M\Big[S(B(t)) - M^{-1}B(t) - (\Gamma^{J+1})^{1/2}W^{J+1}(B(t))\Big]$$

$$+ M(\Gamma^{J+1})^{1/2}\left[W^{J+1}(B(t)) - W^{J+1}((\rho \wedge e)t)\right]$$

$$- M(\Gamma^{J+1})^{1/2}\left[W^{J+1}(M(Q(t) + S(B(t)))) - W^{J+1}(\rho t)\right]$$

$$+ M\Big[Q(t) - \tilde{Q}(t)\Big] + \tilde{Z}(t).$$

The proof is then completed based on (7.66)-(7.67) and (7.69), following an argument similar to the one above.   □

7.8 Fluid Limit (FSLLN)

As in the single station case, we allow the initial queue length vector here to depend on n, denoting it by $Q^n(0)$, and assume that as $n \to \infty$,

$$\bar{Q}^n(0) := \frac{1}{n}Q^n(0) \to \bar{Q}(0), \tag{7.78}$$

almost surely. We append a superscript n to all queueing processes that correspond to the initial queue length $Q^n(0)$. For example, Q^n and Z^n

will be used to denote the corresponding queue-length and the workload processes, respectively. It follows from the functional strong law of large numbers (FSLLN, Theorem 5.10) that

$$(\bar{A}^n(t), \bar{S}^n(t), \bar{R}^n(t)) \to (\alpha t, \mu t, P't), \qquad \text{u.o.c.,} \qquad (7.79)$$

as $n \to \infty$, where

$$\bar{A}^n(t) = \frac{1}{n}A(nt), \quad \bar{S}^n(t) = \frac{1}{n}S(nt), \quad \bar{R}^n(t) = \frac{1}{n}R(\lfloor nt \rfloor),$$

and $R(n) = (R^1(n), \ldots, R^J(n))$.

Our objective here is to establish the almost sure convergence of the scaled queueing processes

$$\bar{Q}^n(t) = \frac{1}{n}Q^n(nt), \quad \bar{Z}^n(t) = \frac{1}{n}Z^n(nt), \quad \text{and} \quad \bar{B}^n(t) = \frac{1}{n}B^n(nt),$$

as $n \to \infty$.

Theorem 7.23 Suppose that (7.78)–(7.79) hold. Then, as $n \to \infty$,

$$(\bar{Q}^n, \bar{Z}^n, \bar{B}^n) \to (\bar{Q}, \bar{Z}, \bar{B}) \quad \text{u.o.c.,} \qquad (7.80)$$

where

$$\bar{Q} = \bar{X} + (I - P')\bar{Y},$$
$$\bar{X}(t) = \bar{Q}(0) + \theta t,$$
$$\theta = \alpha - (I - P')\mu,$$
$$\bar{Y} = \Psi(\bar{X}),$$
$$\bar{Z} = M\bar{Q},$$
$$\bar{B}(t) = et - M\bar{Y}(t),$$

where $M = \text{diag}(m)$ is a J-dimensional diagonal matrix whose jth diagonal element equals m_j, $j = 1, \ldots, J$, and Ψ is the reflection mapping in Theorem 7.2 with $R = I - P'$.

Remark 7.24 See the first remark after Theorem 7.13.

When $\bar{Q}(0) = 0$, the limit process takes the following simpler form:

$$\begin{aligned}
\bar{Q}(t) &= (\lambda - \mu)^+ t, \\
\bar{Z}(t) &= (\rho - e)^+ t, \\
\bar{B}(t) &= (\rho \wedge e)t.
\end{aligned}$$

Proof (of Theorem 7.23). Recall the representation of the queue-length process by $Q = \Phi(X)$, with X following (7.9) in Section 7.1. We clearly have a version for the scaled process with $\bar{Q}^n = \Phi(\bar{X}^n)$, where

$$\bar{X}_j^n(t) = \bar{Q}_j^n(0) + \theta_j t + \left[\bar{A}_j^n(t) - \alpha_j t\right] - \left[\bar{S}_j^n(\bar{B}_j^n(t)) - \mu_j \bar{B}_j^n(t)\right]$$
$$+ \sum_{i=1}^{J} \left\{ \left[\bar{R}_j^{i,n}(\bar{S}_i^n(\bar{B}_i^n(t))) - p_{ij}\bar{S}_i^n(\bar{B}_i^n(t))\right] \right.$$
$$\left. + p_{ij}\left[\bar{S}_i^n(\bar{B}_i^n(t)) - \mu_i \bar{B}_i^n(t)\right] \right\}.$$

Since $0 \leq \bar{B}_j^n(t) \leq t$ for all $t \geq 0$ and $j = 1, \ldots, J$, the convergence in (7.78) and (7.79) immediately yields

$$\bar{X}^n \to \bar{X}, \qquad \text{u.o.c.,} \quad \text{as } n \to \infty,$$

with $\bar{X}$ as defined in the theorem. Then the continuity of the reflection mapping (Theorem 7.2) leads to the convergence of $\bar{Q}^n$ and $\bar{Y}^n := \Phi(\bar{X}^n)$. Since $\bar{B}^n = et - M\bar{Y}^n$, the convergence of $\bar{B}^n$ follows.

The proof of the convergence of $\bar{Z}^n$, which is left as an exercise, is similar to the proof for Theorem 6.5.   □

7.9   Diffusion Limit (FCLT)

We first derive the diffusion approximation for the network in Section 7.1. Specifically, we are interested in the weak convergence of the following scaled processes:

$$\hat{Q}^n(t) := \frac{Q(nt) - (\lambda - \mu)^+ nt}{\sqrt{n}},$$
$$\hat{Z}^n(t) := \frac{Z(nt) - (\rho - e)^+ nt}{\sqrt{n}},$$
$$\hat{B}^n(t) := \frac{(\rho \wedge e)nt - B(nt)}{\sqrt{n}},$$

where λ is the effective arrival rate and ρ is the traffic intensity. Let

$$\hat{A}^n(t) := \sqrt{n}\left[\bar{A}^n(t) - \lambda t\right] = \frac{A(nt) - n\lambda t}{\sqrt{n}},$$
$$\hat{S}^n(t) := \sqrt{n}\left[\bar{S}^n(t) - \mu t\right] = \frac{S(nt) - n\mu t}{\sqrt{n}},$$
$$\hat{V}^n(t) := \sqrt{n}\left[\bar{V}^n(t) - mt\right] = \frac{V(\lfloor nt \rfloor) - nmt}{\sqrt{n}},$$
$$\hat{R}^n(t) := \sqrt{n}\left[\bar{R}^n(t) - P't\right] = \frac{R(\lfloor nt \rfloor) - nP't}{\sqrt{n}}.$$

Assume that both the interarrival times of the (exogenous) arrival process A_j $(j \in \mathcal{E})$ and the service time $v_j(1)$ $(j = 1, \ldots, J)$ have finite second moments. Then, it follows from the functional central limit theorem for renewal and i.i.d. summation processes that

$$(\hat{A}^n, \hat{S}^n, \hat{V}^n, \hat{R}^n) \xrightarrow{d} (\hat{A}, \hat{S}, \hat{V}, \hat{R}), \qquad \text{as } n \to \infty, \qquad (7.81)$$

where $\hat{A}$, $\hat{S}$, $\hat{R}^j$ (the jth column of $\hat{R}$), $j = 1, \ldots, J$, are $J + 2$ driftless Brownian motions (each being J-dimensional), and $V(t) = -M\hat{S}(mt)$. The covariance matrices for $\hat{A}$, $\hat{S}$, and $\hat{R}^j$ $(j = 1, \ldots, J)$ follow (7.45)–(7.47), respectively. (Note that $\hat{R}^n$ on the left-hand side of the above convergence is used to denote the scaled routing (matrix) process for the nth network, while $\hat{R}^j$ is used to denote the jth column of $\hat{R}$ (of the right-hand side of the above convergence, the limiting routing process). This might cause some confusion; however, since the former relates to the queueing process and the latter relates to the limiting process, their difference should be clear from the context.)

Let

$$\xi(t) = \hat{A}(t) + \sum_{\ell=1}^{J} \left[\hat{R}^\ell((\lambda_\ell \wedge \mu_\ell)t) + p^\ell \hat{S}_\ell((\rho_\ell \wedge 1)t) \right] - \hat{S}((\rho \wedge e)t). \quad (7.82)$$

It is clear that ξ is a J-dimensional driftless Brownian motion, with the covariance matrix in (7.50). Let $a = \{j : \rho_j < 1\}$, $b = \{j : \rho_j = 1\}$, and $c = \{j : \rho_j > 1\}$ be the sets of nonbottleneck stations, balanced bottleneck stations, and strict bottleneck stations, respectively.

Theorem 7.25 Suppose that the weak convergence in (7.81) holds. Then,

$$(\hat{Q}^n, \hat{Z}^n, \hat{B}^n) \xrightarrow{d} (\hat{Q}, \hat{Z}, \hat{B}), \qquad \text{as } n \to \infty, \qquad (7.83)$$

where

$$\hat{Q}_a = 0,$$
$$\hat{Q}_b = [\xi_b + P'_{ab}(I - P'_a)^{-1}\xi_a] + (I - \hat{P}'_b)\hat{Y}_b, \qquad (7.84)$$
$$\hat{P}_b = P_b + P_{ba}(I - P_a)^{-1}P_{ab}, \qquad (7.85)$$
$$\hat{Y}_b = \Psi_{I - \hat{P}'_b}(\xi_b + P'_{ab}(I - P'_a)^{-1}\xi_a), \qquad (7.86)$$
$$\hat{Q}_c = \xi_c + P'_{ac}(I - P'_a)^{-1}\xi_a - \hat{P}'_{bc}\hat{Y}_b, \qquad (7.87)$$
$$\hat{P}_{bc} = P_{bc} + P_{ba}(I - P_a)^{-1}P_{ac}, \qquad (7.88)$$
$$\hat{Z} = M\hat{Q},$$
$$\hat{B}_a = M_a(I - P'_a)^{-1}[P'_{ba}\hat{Y}_b - \xi_a], \qquad (7.89)$$
$$\hat{B}_b = M_b\hat{Y}_b,$$
$$\hat{B}_c = 0,$$

with $\Psi_{I-\hat{P}_b}$ being the oblique reflection mapping corresponding to the reflection matrix $(I - \hat{P}_b')$.

It should be evident from the proof that other than the weak convergence in (7.81), the theorem does not require the independence among the arrival and service processes and the routing sequence, nor does it require the arrival and service processes to be renewal processes. Without these assumptions, the only difference is in computing the covariance matrix of the Brownian motion ξ.

We take a standard approach by invoking the Skorohod representation theorem, and assume without loss of generality that the probability space is so chosen that the convergence (7.81) holds almost surely under the uniform norm; namely,

$$(\hat{A}^n, \hat{S}^n, \hat{V}^n, \hat{R}^n) \to (\hat{A}, \hat{S}, \hat{V}, \hat{R}), \quad \text{u.o.c.,} \quad \text{as } n \to \infty, \qquad (7.90)$$

almost surely. We note that the convergence (7.90) clearly implies the FS-LLN limit in (7.79) (and clearly (7.78) holds in this case, since $Q^n(0) = Q(0)$ does not vary with n). Therefore, we can invoke Theorem 7.23. In particular, we have

$$\bar{B}^n(t) \to (\rho \wedge e)t, \quad \text{u.o.c.,} \quad \text{as } n \to \infty. \qquad (7.91)$$

We first state three lemmas. The first one is a partial extension of Lemma 6.4, and its proof is left as an exercise.

Lemma 7.26 Let $\{\theta_n, n \geq 1\}$ be a sequence of real numbers, and let $\{u_n, n \geq 1\}$ and $\{x_n, n \geq 0\}$ be two sequences in $\mathcal{D}_0$, where u_n is nondecreasing with $u_n(0) = 0$.

(i) Let

$$z_n(t) = x_n(t) + \theta_n t - u_n(t) + y_n(t),$$
$$y_n(t) = \sup_{0 \leq s \leq t} [-x_n(s) - \theta_n s + u_n(s)]^+.$$

Suppose that

$$x_n \to x \quad \text{u.o.c.,}$$

as $n \to \infty$ with $x \in \mathcal{C}$. If $\theta_n \to -\infty$ as $n \to \infty$. Then,

$$z_n \to 0 \quad \text{u.o.c.} \quad \text{as } n \to \infty.$$

(ii) Let

$$z_n(t) = x_n(t) + \theta_n t + y_n(t),$$
$$y_n(t) = \sup_{0 \leq s \leq t} [-x_n(s) - \theta_n s]^+.$$

Suppose that $-x_n$ is bounded from above by a sequence that converges u.o.c. to a continuous limit. If $\theta_n \to +\infty$ as $n \to \infty$, then

$$y_n \to 0 \qquad \text{u.o.c.} \qquad \text{as } n \to \infty.$$

Lemma 7.27 Suppose that the convergence (7.90) holds. Then almost surely,

$$\hat{Q}_a^n \to 0, \quad \text{u.o.c.}, \quad \text{as } n \to \infty.$$

The proof of this lemma can be done by first applying Lemma 7.26-(i) to each of the subcritical stations, eliminating them from the network to obtain a subnetwork, and repeatedly applying this procedure (to the subnetworks) until all stations in a are eliminated (in view of Lemma 7.7). The details are left as an exercise.

Lemma 7.28 Suppose that the convergence in (7.90) holds. Then, almost surely,

$$\hat{B}_c^n \to 0, \quad \text{u.o.c.}, \quad \text{as } n \to \infty,$$

and equivalently,

$$\hat{Y}_c^n \to 0, \quad \text{u.o.c.}, \quad \text{as } n \to \infty.$$

Proof. Note that $\hat{Y}_j^n = \mu_j \hat{B}_j^n$ for $j \in c$; hence, we need to prove only the second convergence in the lemma. For convenience, let

$$\tilde{Q}^n(t) := \frac{1}{\sqrt{n}} Q(nt) = \hat{Q}^n(t) + (\lambda - \mu)^+ \sqrt{n}t.$$

Note that $\lambda_a < \mu_a$ and $\lambda_b = \mu_b$; hence, $\tilde{Q}_a^n \equiv \hat{Q}_a^n$ and $\tilde{Q}_b^n \equiv \hat{Q}_b^n$. First, we rewrite (7.8) for the scaled process

$$\tilde{Q}^n(t) = \xi^n(t) + \theta\sqrt{n}t + (I - P')\hat{Y}^n(t), \qquad (7.92)$$

where

$$\theta = \alpha - (I - P')\mu,$$

$$\hat{Y}^n(t) = \frac{1}{\sqrt{n}} Y(nt),$$

and the jth component of $\xi^n(t)$ is

$$\xi_j^n(t) = \hat{Q}_j^n(0) + \hat{A}_j^n(t) - \hat{S}_j^n(\bar{B}_j^n(t))$$

$$+ \sum_{i=1}^J \{\hat{R}^{i,n}(\bar{S}_i^n(\bar{B}_i^n(t))) + p_{ij}\hat{S}_i^n(\bar{B}_i^n(t))\}.$$

By the convergence in (7.91) and the random-time-change theorem, (7.90) implies

$$\xi^n \to \xi \qquad \text{u.o.c.,} \qquad \text{as } n \to \infty, \qquad (7.93)$$

where ξ is as in (7.82).

Next, let β be the union of b and c, and rewrite (7.92) in (a, β) block form,

$$\hat{Q}_a^n(t) = \theta_a \sqrt{n}t + \xi_a^n(t) - P_{\beta a}' \hat{Y}_\beta^n(t) + (I - P_a')\hat{Y}_a^n(t), \qquad (7.94)$$
$$\tilde{Q}_\beta^n(t) = \theta_\beta \sqrt{n}t + \xi_\beta^n(t) - P_{a\beta}' \hat{Y}_a^n(t) + (I - P_\beta')\hat{Y}_\beta^n(t). \qquad (7.95)$$

Since the spectral radius of P is less than unity, the inverse of $(I - P_a')$ exists. Solving $\hat{Y}_a^n$ from (7.94) and substituting it into (7.95) yields

$$\tilde{Q}_\beta^n(t) = \chi_\beta^n(t) + (\lambda_\beta - \mu_\beta)\sqrt{n}t + (I - \hat{P}_\beta')\hat{Y}_\beta^n(t), \qquad (7.96)$$

where

$$\chi_\beta^n(t) = \xi_\beta^n(t) + P_{a\beta}'(I - P_a')^{-1}[\xi_\beta^n(t) - \hat{Q}_a^n(t)],$$
$$\hat{P}_\beta = P_\beta + P_{\beta a}(I - P_a)^{-1}P_{a\beta},$$

and we used the following equation:

$$\theta_\beta + P_{a\beta}'(I - P_a')^{-1}\theta_a = \lambda_\beta - \mu_\beta.$$

By Lemma 7.5, $\hat{P}_\beta$ is a substochastic matrix with a spectral radius less than unity. Hence,

$$\hat{Y}_\beta^n = \Psi_{I - \hat{P}_\beta'}\left(\chi_\beta^n + (\lambda_\beta - \mu_\beta)\sqrt{n}\cdot\right),$$

where $\Psi_{I - \hat{P}_\beta'}$ is the oblique reflection matrix with $(I - \hat{P}_\beta')$ being the reflection matrix. Since $\lambda_\beta - \mu_\beta \geq 0$, the least element characterization of the reflection mapping leads to $\hat{Y}_\beta^n \leq \Psi_{I - \hat{P}_\beta'}(\chi_\beta^n)$ for all $n \geq 1$. Since as $n \to \infty$, the sequence χ_β^n converges u.o.c. to a continuous limit (in view of (7.93)), the continuity of the reflection mapping ensures that $\hat{Y}_\beta^n$ is bounded from above by a convergent subsequence.

Finally, from (7.96), we have (noting that $c \subseteq \beta$)

$$\hat{Q}_c^n(t) = \chi_c^n(t) + (\lambda_c - \mu_c)\sqrt{n}t - \left(\hat{P}_\beta'\hat{Y}_\beta^n(t)\right)_c + \hat{Y}_c^n(t).$$

In view of $\lambda_c > \mu_c$, applying Lemma 7.26-(ii) then completes the proof. $\square$

Proof (of Theorem 7.25). Rewrite (7.92) in (a, b, c) block partition:

$$\hat{Q}_a^n(t) = \xi_a^n(t) + \theta_a \sqrt{n}t$$
$$- P'_{ba}\hat{Y}_b^n(t) - P'_{ca}\hat{Y}_c^n(t) + (I - P'_a)\hat{Y}_a^n(t), \qquad (7.97)$$
$$\hat{Q}_b^n(t) = \xi_b^n(t) + \theta_b \sqrt{n}t$$
$$- P'_{ab}\hat{Y}_a^n(t) - P'_{cb}\hat{Y}_c^n(t) + (I - P'_b)\hat{Y}_b^n(t), \qquad (7.98)$$
$$\tilde{Q}_c^n(t) = \xi_c^n(t) + \theta_c \sqrt{n}t$$
$$- P'_{ac}\hat{Y}_a^n(t) - P'_{bc}\hat{Y}_b^n(t) + (I - P'_c)\hat{Y}_c^n(t). \qquad (7.99)$$

Solving $\hat{Y}_a^n$ from (7.97) and substituting the outcome into (7.98) yields

$$\hat{Q}_b^n(t) = \hat{X}_b^n(t) + (I - \hat{P}'_b)\hat{Y}_b^n(t),$$

where

$$\hat{X}_b^n(t) = \xi_b^n(t) + P'_{ab}(I - P'_a)^{-1}[\xi_a^n(t) - \hat{Q}_a^n(t)]$$
$$- \left[P'_{cb} + P'_{ab}(I - P'_a)^{-1}P'_{ca} \right] \hat{Y}_c^n,$$

and $\hat{P}_b$ is as in (7.85). In view of (7.93) and Lemmas 7.27 and 7.28, $\hat{X}^n$ converges u.o.c. to $\xi_b + P'_{ab}(I - P'_a)^{-1}\xi_a$ as $n \to \infty$. Then, from $\hat{Y}_b^n = \Psi_{I-\hat{P}'_b}(\hat{X}_b^n)$ and the continuity of the oblique reflection mapping, we have established the u.o.c. convergence of $\hat{Q}_b^n$ and $\hat{Y}_b^n$ and the equalities in (7.84) and (7.86).

Next, substituting the solution for $\hat{Y}_a^n$ obtained from (7.97) into (7.99) yields

$$\tilde{Q}_c^n(t) = \hat{X}_c^n(t) + (\lambda_c - \mu_c)\sqrt{n}t - \hat{P}'_{bc}\hat{Y}_b^n(t) + (I - \hat{P}'_c)\hat{Y}_c^n(t),$$

where

$$\hat{X}_c^n(t) = \xi_c^n(t) + P'_{ac}(I - P'_a)^{-1} \left[\xi_a^n(t) - \hat{Q}_a^n(t) \right],$$
$$\hat{P}_c = P_c + P_{ca}(I - P_a)^{-1}P_{ac},$$

and $\hat{P}_{bc}$ is as in (7.88). Note that $\hat{Q}_c^n(t) = \tilde{Q}_c^n(t) - (\lambda_c - \mu_c)\sqrt{n}t$; hence, Lemmas 7.27 and 7.28 and the convergence of ξ^n and $\hat{Y}_b^n$ imply the convergence of $\hat{Q}_c^n$ and the equality in (7.87). Thus, we have established the convergence of $\hat{Q}^n$.

Note that

$$\hat{B}^n(t) = M[\hat{Y}^n(t) - (\mu - \lambda)^+ \sqrt{n}t].$$

The u.o.c. convergence of $\hat{B}_b^n$ and $\hat{B}_c^n$ follows immediately from the u.o.c. convergence of $\hat{Y}_b^n$ and $\hat{Y}_c^n$, and the convergence of $\hat{B}_a^n$ then follows from (7.97).

The proof for the convergence of the workload process is left as an exercise. $\qquad\square$

For the rest of this section we consider the diffusion limit for a *sequence of queueing networks* under a *heavy traffic* condition. Most queueing networks that arise from applications usually operate under the condition that the traffic intensity is less than one; that is, none of the stations is a bottleneck. In this case, the diffusion limit suggested by Theorem 7.25 for the queue-length process is zero, which would give us a useless approximation for the queue-length process. To overcome this problem, we can view the network under consideration as one in a sequence of networks whose traffic intensities approach unity in a certain way: a heavy traffic condition. In this case, the queue-length processes corresponding to this sequence will approach a reflected Brownian motion, which provides an approximation for the original queueing network.

We shall index the sequence of generalized Jackson networks by $n = 1, 2, \ldots$. We append a superscript n to all processes and parameters that correspond to the nth network. Specifically, for the nth network, $Q^n(0)$ denotes the initial queue length vector, and A^n and S^n denote the exogenous arrival and service processes. Let V^n be the inverse of S^n. The exogenous arrival rate and the service rate for the nth network are denoted by α^n and μ^n, respectively. For simplicity, we assume that the routing process does not vary with n; let R be the routing process and P be the corresponding routing matrix. We assume that P is a substochastic matrix with a spectral radius less than unity; that is, all the networks in the sequence are open networks.

Now we pause to present a specific construction of the processes A^n and S^n. Let A and S denote the exogenous arrival process and the service process as in Section 7.1, and let α and μ be the corresponding arrival rate and the service rate. We define $A_j^n(t) = A_j((\alpha_j^n/\alpha_j)t)$ for $j \in \mathcal{E}$ and $A_j^n(t) \equiv 0$ for $j \notin \mathcal{E}$, and $S_j^n(t) = S_j((\mu_j^n/\mu_j)t)$, $j = 1, \ldots, J$. We call the network corresponding to the arrival process A and the service process S the limiting network, and call α and μ the exogenous arrival rate and the service rate of the limiting queueing network. We assume that as $n \to \infty$,

$$\hat{Q}^n(0) := \frac{Q^n(0)}{\sqrt{n}} \overset{\mathrm{d}}{\to} \hat{Q}(0), \tag{7.100}$$

$$\sqrt{n}\,(\alpha^n - \alpha) \to c^\alpha, \tag{7.101}$$

$$\sqrt{n}\,(\mu^n - \mu) \to c^\mu, \tag{7.102}$$

where $\hat{Q}(0)$ is a nonnegative J-dimensional random vector, and c^α and c^μ are two J-dimensional vector constants.

Let λ and ρ be, respectively, the effective arrival rate and the traffic intensity for the limiting network. That is, λ is the unique solution to the traffic equation in (7.25), and the components of ρ are $\rho_j = \lambda_j/\mu_j$. The heavy traffic condition is to assume that all stations in the limiting network are balanced bottleneck, namely,

$$\rho = e. \tag{7.103}$$

We note that under the heavy traffic condition (7.103), $\lambda = (I-P')^{-1}\alpha$, and the heavy traffic condition is equivalent to the condition that $(I-P')^{-1}\alpha = \mu$, or $\alpha = (I-P')\mu$.

The diffusion approximation for this sequence of networks is concerned with the weak convergence of the following scaled processes:

$$\hat{Q}^n(t) := \frac{Q^n(nt)}{\sqrt{n}},$$

$$\hat{Z}^n(t) := \frac{Z^n(nt)}{\sqrt{n}},$$

$$\hat{B}^n(t) := \frac{ent - B^n(nt)}{\sqrt{n}},$$

where Q^n, Z^n, and B^n are, respectively, the queue-length process, the workload process and the busy time process for the nth network. Note that the scaling for the queue-length process and the workload process are not centered here, due to the assumption of the heavy traffic condition (7.103). Let

$$\hat{A}^n(t) := \sqrt{n}\left[\bar{A}^n(t) - \lambda^n t\right] \equiv \frac{A^n(nt) - n\lambda^n t}{\sqrt{n}},$$

$$\hat{S}^n(t) := \sqrt{n}\left[\bar{S}^n(t) - \mu^n t\right] \equiv \frac{S^n(nt) - n\mu^n t}{\sqrt{n}},$$

$$\hat{V}^n(t) := \sqrt{n}\left[\bar{V}^n(t) - m^n t\right] \equiv \frac{V^n(\lfloor nt \rfloor) - nm^n t}{\sqrt{n}},$$

$$\hat{R}^n(t) := \sqrt{n}\left[\bar{R}^n(t) - P't\right] \equiv \frac{R(\lfloor nt \rfloor) - nP't}{\sqrt{n}},$$

where $m^n = (m_j^n)$ with $m_j^n = 1/\mu_j^n$. We assume the joint weak convergence of (cf. (7.81))

$$(\hat{A}^n, \hat{S}^n, \hat{V}^n, \hat{R}^n) \overset{d}{\to} (\hat{A}, \hat{S}, \hat{V}, \hat{R}), \qquad \text{as } n \to \infty, \tag{7.104}$$

where $\hat{A}$, $\hat{S}$, $\hat{R}^j$ (the jth column of $\hat{R}$), $j = 1, \ldots, J$, are $J+2$ independent J-dimensional driftless Brownian motions, and $\hat{V}(t) = -M\hat{S}(mt)$. The covariance matrices of $\hat{A}$, $\hat{S}$ and $\hat{R}^\ell$, $\ell = 1, \ldots, J$, are given in (7.45)–(7.47). Under (7.101) and (7.102), the sufficient condition for the convergence in (7.104) is the same as the sufficient condition for the convergence in (7.81).

Theorem 7.29 Suppose that (7.100)–(7.104) hold. Then we have

$$(\hat{Q}^n, \hat{Z}^n, \hat{B}^n) \overset{d}{\to} (\hat{Q}, \hat{Z}, \hat{B}), \qquad \text{as } n \to \infty, \tag{7.105}$$

where

$$\hat{Q} = \hat{X}(t) + (I - P')\hat{Y}(t),$$
$$\hat{X}(t) = \hat{Q}(0) + \xi(t) + \theta t,$$
$$\xi(t) = \hat{A}(t) + \sum_{\ell=1}^{J} \left[\hat{R}^{\ell}(t)(\mu_{\ell}t) + p^{\ell}\hat{S}_{\ell}(t) \right] - \hat{S}(t),$$
$$\theta = c^{\alpha} - (I - P')c^{\mu},$$
$$\hat{Y} = \Psi(\hat{X}),$$
$$\hat{Z} = M\hat{Q},$$
$$\hat{B} = M\hat{Y},$$

with Ψ being the oblique reflection mapping corresponding to a reflection matrix $(I - P')$.

The proof of the theorem is outlined in several exercise problems.

7.10   Closed Networks

We now turn to study the generalized *closed* Jackson network. (The closed network is short for the *irreducible* closed network throughout this section.) The network is the same as described in Section 7.1, but with the exogenous arrival rate $\alpha = 0$ and the routing matrix P being an irreducible stochastic matrix. We first state an oblique reflection mapping for the closed network, and describe a corresponding fluid network and a reflected Brownian motion. Next, in Section 7.10.2 we propose a Brownian approximation through a heuristic argument. We then establish the fluid limit in Section 7.10.3 and the diffusion limit in Section 7.10.4; these provide theoretical support for the approximation proposed in Section 7.10.2. We shall not present a functional law of the iterated logarithm and a functional strong approximation theorem for the closed network, which are much more involved than in the open network. The main difficulty lies in the fact that the oblique reflection mapping corresponding to a closed network is not Lipschitz continuous, a key property used in establishing the FLIL and the FSAT for the open network in Sections 7.6 and 7.7.

7.10.1   Reflection Mapping, Fluid Model, and RBM

Let P be a $J \times J$ irreducible stochastic matrix and $\mathcal{D}_1^J = \{x \in \mathcal{D}_0^J : e'x \equiv 1\}$. Note that the condition $e'x \equiv 1$ in $\mathcal{D}_1^J$ is motivated by the fact that the total number of jobs in a closed network remains invariant over time; hence, with an appropriate scaling the queue-length process will satisfy this condition. We state the following theorem without proof.

Theorem 7.30 (Oblique Reflection Mapping Theorem) Suppose that the $J \times J$ matrix P is an irreducible stochastic matrix. Then, for every $x \in \mathcal{D}_1^J$, there exists a unique pair (y, z) in $\mathcal{D}^{2J}$ satisfying

$$z = x + (I - P')y \geq 0, \tag{7.106}$$
$$dy \geq 0 \text{ and } y(0) = 0, \tag{7.107}$$
$$z_j dy_j = 0, \quad j = 1, \ldots, J. \tag{7.108}$$

We write $y = \Psi(x)$ and $z = \Phi(x)$, and call (Φ, Ψ) the reflection mapping. Therefore, the reflection mapping (Φ, Ψ) is well-defined on $\mathcal{D}_1^J$. In addition, Φ and Ψ are continuous mappings from $\mathcal{D}_1^J$ to $\mathcal{D}^J$ under the uniform norm.

This theorem is almost the same as Theorem 7.2, except that here $R = I - P'$ is a degenerate M-matrix and the reflection mapping is only continuous, instead of Lipschitz continuous. Since P is not convergent in this case, the proof for Theorem 7.2 will not apply here. However, all mathematical properties of the reflected mapping in Section 7.2, such as the least element characterization, nonanticipation, the shift property, and the path continuity, can all be extended to the model here in an obvious way.

Next, analogous to Section 7.3, we describe and characterize a *closed* fluid network. The network consists of J buffers, among which an invariant amount of fluid circulates. The network is described by an initial fluid level vector $z(0) = (z_j(0))$, a *potential outflow rate* (vector) $\mu = (\mu_j)$, and a *flow-transfer matrix* $P = (p_{ij})$. Assume that $z(0)$ is nonnegative, μ is positive, and P is an irreducible stochastic matrix. Also assume, without loss of generality, that $e'z(0) = 1$. The *fluid level process* $z = \{z(t), t \geq 0\}$ and the *cumulative loss process* are jointly determined by

$$z(t) = x(t) + (I - P')y(t) \geq 0,$$
$$dy \geq 0 \text{ and } y(0) = 0,$$
$$z_j(t) dy_j(t) = 0,$$

with

$$x(t) = z(0) - (I - P')\mu t, \qquad t \geq 0.$$

These relations clearly resemble those in (7.21)–(7.24) for the open fluid network. (In fact, the closed model can be obtained from the open model by letting $\alpha = 0$ and requiring P to be an irreducible stochastic matrix.) It follows from Theorem 7.30 that $z = \Phi(x)$ and $y = \Psi(x)$. That is, the fluid level process is a reflected linear process, since x is linear. Note in particular that $e'z(t) = e'z(0) = 1$, i.e., the total fluid level remains at a fixed level of unity.

Let λ be the *effective* inflow rate (vector); it must satisfy the following equation

$$\lambda = P'(\lambda \wedge \mu) \tag{7.109}$$

(cf. (7.25)). Since P is a stochastic matrix, the solution to the above traffic equation is not unique. But it can be shown that there exists a positive maximum solution to the above equation, i.e., there exists a positive solution that dominates all other solutions. (The proof of this fact is left as an exercise.) For convenience, we shall denote this maximum solution by λ, and call it the maximum inflow rate. Call $\rho_j = \lambda_j/\mu_j$ the *traffic intensity* of buffer j. For the corresponding queueing network, call λ the maximum arrival rate and ρ the traffic intensity. Call buffer j a bottleneck if $\rho_j = 1$, and a nonbottleneck if $\rho_j < 1$. Note that there exists at least one bottleneck, and there is no buffer j with $\rho_j > 1$.

Theorem 7.31 Let y and z be respectively the cumulative loss and the fluid level processes of the above closed fluid network with initial fluid level $z(0)$. Let λ and ρ be, respectively, the maximum inflow rate and the traffic intensity. Then, there exists a finite time τ such that

(i) if $\rho_j < 1$ (i.e., buffer j is a nonbottleneck), then

$$z_j(t) = 0, \qquad \text{for } t \geq \tau,$$
$$y_j(t) = y_j(\tau) + (\mu_j - \lambda_j)(t - \tau), \qquad \text{for } t \geq \tau,$$

(ii) if $\rho_j = 1$ (i.e., buffer j is a bottleneck), then

$$z_j(t) = z_j(\tau),$$
$$y_j(t) = 0, \qquad \text{for } t \geq 0.$$

In particular, if $z_j(0) = 0$ for all nonbottleneck buffers, then we can choose $\tau = 0$, and $z(t) = (\lambda - \mu)^+ t$ and $y(t) = (\mu - \lambda)^+ t$.

Note that we can still use the algorithm in Section 7.3 to identify the set of bottlenecks, and hence the set of nonbottlenecks. The proof of the above theorem is left as an exercise.

Finally, we describe a reflected Brownian motion associated with the closed network. Let $\mathcal{S} = \{x \in \Re^J : x \geq 0 \text{ and } e'x = 1\}$ be a J-dimensional unit simplex. Let $X = \{X(t), t \geq 0\}$ be a J-dimensional Brownian motion, starting at $x = X(0) \in \mathcal{S}$, with a drift vector θ and a covariance matrix Γ. We assume that

$$e'\theta = 0 \qquad \text{and} \qquad e'\Gamma = 0. \tag{7.110}$$

Under this condition, the variance of $e'X(t)$ is zero; and hence, with probability one, $e'X(t) \equiv e'X(0) = 1$ for all $t \geq 0$. In other words, $X \in \mathcal{D}_1^J$ with probability one; and the processes Z and Y are well defined, pathwise by $Z = \Phi(X)$ and $Y = \Psi(X)$.

We call Z a reflected Brownian motion (RBM) on the unit simplex $\mathcal{S}$, and Y the corresponding regulator, and write $Z = RBM_x(\theta, \Gamma; I - P')$.

RBMs on an orthant and on a (unit) simplex share similar properties and geometric interpretations. In particular, the RBM on a unit simplex is also a diffusion process (i.e., a strong Markov process with continuous sample path). Since its state space, $\mathcal{S}$, is compact, it always has a stationary distribution (Exercise 36), and the stationary distribution is unique.

With some obvious modifications, the basic adjoint relation (BAR) (in Theorem 7.10) that characterizes the stationary distribution of the RBM on the nonnegative orthant applies to the RBM on the unit simplex as well. Specifically, BAR holds with $\Re_+^J$ replaced by $\mathcal{S}$ and F_j redefined as $F_j = \{x \in \mathcal{S} : x_j = 0\}$.

Similarly, a version of Theorem 7.12 holds as well. Let Λ be a J-dimensional diagonal matrix whose jth component is Γ_{jj}, $j = 1, \ldots, J$. Let η be a fixed solution to

$$2\theta + (I - P')\Lambda\eta = 0. \tag{7.111}$$

[In view of (7.110), the solution to the above equation must exist. In fact, since the rank of $(I - P')$ is $J - 1$, solutions to the above equation form a one-dimensional space.] Suppose that

$$2\Gamma_{jk} = -(p_{jk}\Gamma_{jj} + p_{kj}\Gamma_{kk}) \qquad \text{for all } j \neq k \tag{7.112}$$

holds (cf. (7.51).) Then, the stationary distribution of the RBM on a unit simplex $Z = RBM(\theta, \Gamma; I - P')$ has a product form. Specifically, in this case, the density function takes the form

$$f(z_1, \ldots, z_J) = (N_J)^{-1} \prod_{j=1}^{J} e^{-\eta_j z_j}, \qquad z = (z_j) \in \mathcal{S}, \tag{7.113}$$

where

$$N_J = \int_{\mathcal{S}} \prod_{j=1}^{J} e^{-\eta_j z_j} dz.$$

It is elementary to show that $N_J := N_J(1)$ can be computed by the following iterative formula:

$$N_1(x) = e^{-\eta_1 x}, \tag{7.114}$$

$$N_{k+1}(x) = \int_0^x e^{-\eta_{k+1} y} N_k(x - y) dy, \qquad k = 1, \ldots, J - 1. \tag{7.115}$$

It is left as an exercise to show that the marginal stationary distribution for Z_J in this case is

$$f_J(x) = [N_J(1)]^{-1} N_{J-1}(1 - x) e^{-\eta_J x}, \qquad 0 \leq x \leq 1. \tag{7.116}$$

Finally, note that $\nu_j := \nu_j(F_j)$ gives the push rate of the regulator, namely,

$$\frac{1}{t} E_\pi[Y_j(t)] = \nu_j, \qquad j = 1, \ldots, J.$$

The quantity $\nu = (\nu_j)$ is useful in approximating the average idleness rate for the closed queueing network. When the stationary distribution has a product form (i.e., condition (7.112) holds), it is determined by

$$\nu_J = \frac{1}{2}\Gamma_{JJ}[N_J(1)]^{-1}N_{J-1}(1) \tag{7.117}$$

and

$$\theta + (I - P')\nu = 0. \tag{7.118}$$

[The existence and uniqueness of a solution to the above two equalities are guaranteed by (7.110) and the fact that $(I - P')$ has a rank of $J - 1$.]

7.10.2 Approximating a Generalized Closed Network

We follow the same heuristic argument in Section 7.5 for approximating a generalized open network. The only differences here are that $A \equiv 0$ (no exogenous arrivals) and that the routing matrix P is stochastic.

Consider a closed network with J stations. Assume the service times at station j form an i.i.d. sequence, with a finite mean $m_j = 1/\mu_j$ and a finite coefficient of variation c_j, $j = 1, \ldots, J$. Let $P = (p_{jk})$ be the routing matrix, which is an irreducible stochastic matrix. Let λ be the maximum arrival rate, the maximum solution to the traffic equation (7.109). Let $Q = \{Q(t), t \geq 0\}$ be the queue-length process, and $I = \{I(t), t \geq 0\}$, the idle time process. Let $n = e'Q(0)$ be the total number of jobs in the network. Note that $Q_j(t)/n$ represents the proportion of jobs at station j at time t. In contrast to an open network, an appropriate scaling of an FCLT here should be

$$\hat{Q}^n(t) = \frac{1}{n}Q(n^2t) \quad \text{and} \quad \hat{Y}^n(t) = \frac{1}{n}Y(n^2t),$$

where $Y_j(t) = m_jI_j(t)$. It should be supported by an FCLT in Section 7.10.4 that for large n, $\hat{Q}^n$ and $\hat{Y}^n$ can be approximated, respectively, by the reflected Brownian motion $\tilde{Q} := RBM_{Q(0)/n}(\theta, \Gamma; I - P')$ and its corresponding regulator $\tilde{Y}$, where θ and Γ are given by [cf. (7.39) and (7.50)]

$$\theta = -(I - P')\theta,$$

$$\Gamma_{k\ell} = \sum_{j=1}^{J} \lambda_j \left[p_{jk}(\delta_{k\ell} - p_{j\ell}) + c_j^2(p_{jk} - \delta_{jk})(p_{j\ell} - \delta_{j\ell})\right],$$

$k, \ell = 1, \ldots, J$. Let $Q(\infty)$ and $\tilde{Q}(\infty)$ have the stationary distributions of the queue-length process Q and the reflected Brownian motion $\tilde{Q}$, respectively. Then, we have

$$Q(\infty) \overset{d}{\approx} n\tilde{Q}(\infty).$$

The distribution of $\tilde{Q}(\infty)$, the stationary distribution of the RBM, was characterized in the previous subsection, from which we also know that $E_\pi[\tilde{Y}_j(t)] = \nu_j t$. Hence,

$$E_{Q(\infty)}[I_j(t)] = m_j E_{Q(\infty)}[Y_j(t)] \approx m_j \times n E_{\tilde{Q}(\infty)}[\tilde{Y}_j(t/n^2)] = \frac{m_j \nu_j}{n} t,$$

for $j = 1, \ldots, J$. Let $TH_j(n)$ denote the average throughput rate at station j (with n jobs in the network), which can be approximated by

$$TH_j(n) = \mu_j \left(1 - \frac{E_{Q(\infty)}[I(t)]}{t} \right) \approx \mu_j \left(1 - \frac{m_j \nu_j}{n} \right), \qquad (7.119)$$

for $j = 1, \ldots, J$.

When condition (7.112) holds, the stationary distribution of $\tilde{Q}$ has a product form, with the density function following (7.113), and the parameter $\nu = (\nu_j)$ in the average throughput rate approximation formula (7.119) is determined by (7.117) and (7.118).

7.10.3 Fluid Limit (FSLLN)

Consider the closed network described in the previous subsection. We append a superscript n to all performance measures such as the queue-length process, the workload process and the busy time process, to emphasize their dependence on the total number of jobs in the network. Assume the limit

$$\bar{Q}^n(t) := \frac{1}{n} Q^n(0) \to \bar{Q}(0), \qquad \text{almost surely}, \qquad (7.120)$$

as $n \to \infty$ and necessarily, $e'\bar{Q}(0) = 1$. It follows from the functional strong law of large numbers (FSLLN, Theorem 5.10) that

$$(\bar{S}^n(t), \bar{R}^n(t)) \to (\mu t, P't), \qquad \text{u.o.c.}, \qquad (7.121)$$

as $n \to \infty$, where

$$\bar{S}^n(t) = \frac{1}{n} S(nt), \qquad \bar{R}^n(t) = \frac{1}{n} R(\lfloor nt \rfloor).$$

Theorem 7.32 Let

$$\bar{Q}^n(t) = \frac{1}{n} Q^n(nt), \quad \bar{Z}^n(t) = \frac{1}{n} Z^n(nt), \quad \text{and} \quad \bar{B}^n(t) = \frac{1}{n} B^n(nt).$$

Suppose (7.120) and (7.121) hold. Then, as $n \to \infty$,

$$(\bar{Q}^n, \bar{Z}^n, \bar{B}^n) \to (\bar{Q}, \bar{Z}, \bar{B}), \qquad \text{u.o.c.}, \qquad \text{as } n \to \infty,$$

where

$$\bar{Q} = \bar{X} + (I - P')\bar{Y},$$
$$\bar{X}(t) = \bar{Q}(0) + \theta t,$$
$$\theta = -(I - P')\mu,$$
$$\bar{Y} = \Psi(\bar{X}),$$
$$\bar{Z} = M\bar{Q},$$
$$\bar{B}(t) = et - M\bar{Y}(t),$$

where $M = \mathrm{diag}(m)$ is a J-dimensional diagonal matrix whose jth diagonal element equals m_j, $j = 1, \ldots, J$, and Ψ is the reflection mapping defined in Theorem 7.30.

Remark 7.33 The deterministic process $\bar{Q}$ can be interpreted as the fluid level process of a fluid network with initial state $\bar{Q}(0)$ as described in Section 7.10.1. In particular, Theorem 7.31 provides a characterization for the limit process $\bar{Q}$.

Remark 7.34 When $\bar{Q}_j(0) = 0$ for all nonbottleneck stations j, the limit process takes the following simpler form:

$$\bar{Q}(t) = \bar{Q}(0) + (\lambda - \mu)^+ t,$$
$$\bar{B}(t) = (\rho \wedge e)t.$$

The proof of the above theorem is similar to that of Theorem 7.23, and hence is left as an exercise.

7.10.4 Diffusion Limit (FCLT)

In this subsection, we present a diffusion limit theorem for a sequence of closed networks, under a balanced condition. Let $Q^n(0)$ denote the initial queue length, S^n the service process, and R^n the routing process for the nth network. To construct S^n, let S denote the service process as in Section 7.1, and let μ be the corresponding service rate. We define $S_j^n(t) = S_j((\mu_j^n/\mu_j)t)$, where $\{\mu_j^n, n \geq 1\}$ is a sequence of positive numbers, $j = 1, \ldots, J$. For simplicity, we assume $R^n \equiv R$, i.e., the routing process does not vary with n. Specifically, R is as defined in Section 7.1, and $P = (p_{jk})$ is an irreducible stochastic matrix. Hence, each network in the sequence is a closed network. We assume that as $n \to \infty$,

$$\hat{Q}^n(0) := \frac{Q^n(0)}{n} \xrightarrow{\mathrm{d}} \hat{Q}(0), \qquad (7.122)$$
$$n(\mu^n - \mu) \to c^\mu, \qquad (7.123)$$

where $e'\hat{Q}(0) = 1$ almost surely, and c^μ is a J-dimensional vector.

Let λ and ρ be, respectively, the maximum arrival rate and the traffic intensity for the "limiting" network (the network with service rate μ and the routing matrix P), as defined through the traffic equation in (7.109). The balanced condition is that all stations in the "limiting" network are balanced bottleneck, namely,

$$\rho = e, \qquad \text{or equivalently,} \qquad (I - P')\mu = 0. \tag{7.124}$$

Note that in a closed network, as n, the total number of jobs in the network, increases to infinity, the queue length of at least one station will increase to infinity. In this sense, the network is always under a heavy traffic condition. This is why we use the term "balanced condition" instead of "heavy traffic condition" as was used in the open network case.

The FCLT for this sequence of networks is the weak convergence of the following scaled processes:

$$\hat{Q}^n(t) := \frac{Q^n(n^2 t)}{n},$$

$$\hat{Z}^n(t) := \frac{Z^n(n^2 t)}{n},$$

$$\hat{B}^n(t) := \frac{en^2 t - B^n(n^2 t)}{n},$$

where Q^n, Z^n, and B^n are, respectively, the queue-length process, the workload process and the busy time process for the nth network. Note that the scaling here is different from the one used in the open network. However, they are almost the same, if we replace n by $\sqrt{n}$. Let

$$\hat{S}^n(t) := \frac{S^n(n^2 t) - n^2 \mu^n t}{n},$$

$$\hat{V}^n(t) := \frac{V^n(\lfloor n^2 t \rfloor) - n^2 m^n t}{n},$$

$$\hat{R}^n(t) := \frac{R(\lfloor n^2 t \rfloor) - n^2 P' t}{n},$$

where V^n is the inverse of S^n, $m^n = (m_j^n)$ with $m_j^n = 1/\mu_j^n$. It follows from Theorem 5.11 and (7.123) that

$$(\hat{S}^n, \hat{V}^n, \hat{R}^n) \overset{\text{d}}{\to} (\hat{S}, \hat{V}, \hat{R}), \qquad \text{as } n \to \infty, \tag{7.125}$$

where $\hat{S}$, $\hat{R}^j$ (the jth column of $\hat{R}$), $j = 1, \ldots, J$, are $J + 1$ independent J-dimensional driftless Brownian motions, and $\hat{V}(t) = -M\hat{S}(mt)$. The covariance matrices of $\hat{S}$ and $\hat{R}^\ell$, $\ell = 1, \ldots, J$, are given in (7.46)–(7.47).

Theorem 7.35 Suppose that convergences in (7.122) and (7.125) hold jointly, and that (7.123)–(7.124) hold. Then, we have

$$(\hat{Q}^n, \hat{Z}^n, \hat{B}^n) \overset{\text{d}}{\to} (\hat{Q}, \hat{Z}, \hat{B}), \qquad \text{as } n \to \infty,$$

where

$$\hat{Q} = \hat{X}(t) + (I - P')\hat{Y}(t),$$
$$\hat{X}(t) = \hat{Q}(0) + \xi(t) + \theta t,$$
$$\xi(t) = \sum_{\ell=1}^{J} \left[\hat{R}^\ell(t)(\mu_\ell t) + p^\ell \hat{S}_\ell(t) \right] - \hat{S}(t),$$
$$\theta = -(I - P')c^\mu,$$
$$\hat{Y} = \Psi(\hat{X}),$$
$$\hat{Z} = M\hat{Q},$$
$$\hat{B} = M\hat{Y},$$

with Ψ being the oblique reflection mapping defined in Theorem 7.30.

The proof of the theorem is left as an exercise.

7.11 Notes

Historically, the diffusion limit for generalized Jackson networks was studied before the fluid limit, the FLIL, and the FSAT. Reiman [19] was the first to establish a diffusion approximation for generalized open Jackson networks under a heavy traffic condition, and a foundation for his work is Harrison and Reiman [12] on the oblique reflection mapping and the reflected Brownian motion on a nonnegative orthant. Johnson [16] and Chen and Mandelbaum [4] establish a fluid approximation for a generalized Jackson network; this substantially simplifies the original proof of Reiman [19] and its extensions. In addition, Chen and Mandelbaum establish in [5] the diffusion approximations for generalized Jackson networks, both open and closed models, without assuming the heavy traffic condition. In [3] Chen and Mandelbaum establish the oblique reflection mapping and the reflected Brownian motion on a unit simplex, which is the foundation for studying limit theorems for closed networks.

The homogeneous fluid model is characterized in Chen and Mandelbaum [5]. The stationary distribution of the RBM on a nonnegative orthant is characterized in Harrison and Williams [13], and the stationary distribution of the RBM on a unit simplex is characterized in Harrison et al. [14], along with an outline of how to approximate queueing networks by RBMs. Based on these characterizations, Dai and Harrison [10, 11] propose numerical methods to compute the stationary distributions.

Horváth [15] establishes the strong approximation for a two-station generalized open Jackson network. Chen and Mandelbaum [6] establish the FLIL and the strong approximation for the generalized open Jackson network, and Zhang [20] establishes the strong approximation for the generalized closed Jackson network.

A constructive proof to show that relations (7.1) and (7.2) uniquely determine the queue-length process Q and the busy time process B can be found in Chen and Mandelbaum [4].

The oblique reflection mapping (Theorem 7.2) is proved by Harrison and Reiman [12] for the space $\mathcal{C}_0^J$, but it can be extended easily to the space $\mathcal{D}_0^J$ under the uniform topology (see, for example, Reiman [19], Johnson [16], and Chen and Mandelbaum [3]). The Lipschitz continuity of the reflection mapping in Theorem 7.2 also holds in the Skorohod J_1 and M_1 topologies (Chen and Whitt [8]). In [1] Berman and Lemmons present a long list of equivalent conditions for a matrix to be an M-matrix, including those listed in Lemma 7.1.

Attempts have been made to identify conditions on the matrix R such that the general Skorohod problem in (7.14)–(7.16) has a unique solution. Two special cases are studied in this chapter, one with R being an M-matrix (see Section 7.2) and the other with R being a degenerate M-matrix (see Section 7.10.1). Mandelbaum [18] and Bernard and El Kharroubi [2] have shown that R being completely-S is sufficient for the existence of a pair (y, z) satisfying (7.14)–(7.16) for any given $x \in \mathcal{C}_0^J$; however, uniqueness does not hold. (A squared matrix R is an S-matrix if there exists a vector $x \geq 0$ such that $Rx > 0$, and it is completely-S if all of its principal minors are S-matrices.) This line of research is closely related to the linear complementarity problem; refer to Cottle, Pang, and Stone [9].

The homogeneous fluid network section (Section 7.3) is based on Chen and Mandelbaum [4]. The proofs of all results on the reflected Brownian motion in Section 7.4 can be found in Harrison and Williams [13]. The FLIL (Section 7.6) and the strong approximation (Section 7.7) are based on Chen and Mandelbaum [6]. The fluid approximation (Section 7.8) and the diffusion approximation (Section 7.9) are based on Chen and Mandelbaum [4, 5]. For the closed generalized Jackson network (Section 7.10), its corresponding reflection mapping is characterized in Chen and Mandelbaum [3], and its corresponding reflected Brownian motion is characterized in Harrison et al. [14]. The details of the fluid approximation and the diffusion approximation for the closed network can be found in Chen and Mandelbaum [4, 5]. For an irreducible closed network, Kaspi and Mandelbaum establish in [17] an analytical linkage between the stationary distribution of the queue-length process and the stationary distribution of the approximating RBM; but a similar linkage has not been established for an open queueing network.

7.12   Exercises

1. Suppose that matrix Q is nonnegative with a spectral radius less than unity. Prove that there exists a diagonal matrix D with positive diagonal elements such that matrix $D^{-1}QD$ is a substochastic matrix.

2. (Berman and Plemmons [1]) Prove Lemma 7.1.

3. Let P be a substochastic matrix with a spectral radius less than unity. Then for any J-dimensional vector θ, there exist two J-dimensional vectors $\alpha \geq 0$ and $\mu > 0$ such that $\theta = \alpha - (I - P')\mu$.

4. Consider a J-dimensional linear function $x(t) = x(0) + \theta t$, where $x(0)(\geq 0)$ and θ are J-dimensional vectors. Let Φ and Ψ be as defined in Theorem 7.2. In view of Exercises 1 and 3, give an interpretation of the reflected process $z = \Phi(x)$ and its regulator $y = \Psi(x)$, along a line similar to Section 7.3.

5. Let $x = (x_j) \in \Re^J$ and R be a $J \times J$ matrix. The *linear complementarity problem* $\mathrm{LCP}(x, R)$ is to find $y = (y_j)$ and $z = (z_j)$ (both in $\Re^J$) such that

$$z = x + Ry \geq 0, \qquad\qquad (7.126)$$
$$y \geq 0, \qquad\qquad (7.127)$$
$$z'y = 0. \qquad\qquad (7.128)$$

[cf. relations (7.14)–(7.16)]. The complementarity relation (7.128) says that $y_j > 0$ only when $z_j = 0$, $j = 1, \ldots, J$. Show that if R is an M-matrix, then $\mathrm{LCP}(x, R)$ has a unique solution for all $x \in \Re^J$.

6. Suppose that $J \times J$ matrix R is an M-matrix. Fix $x \in \Re^J_+$ and let y be the unique solution to $\mathrm{LCP}(x, R)$. Show that if $\tilde{y} \in \Re^J$ satisfies (7.126) and (7.127), then $y \leq \tilde{y}$. Conversely, suppose that $y \leq \tilde{y}$ for all $\tilde{y}$ satisfies (7.126) and (7.127); show that y must satisfy (7.128). (This gives a *least element characterization* to the linear complementarity problem.)

7. Prove the least element characterization of the oblique reflection mapping in Section 7.2.

8. An $J \times J$ matrix R is called a P-matrix if all of its principal minors are strictly positive, it is called an S-matrix if there exists a vector $\xi \geq 0$ such that $R\xi > 0$, and it is called a completely-S matrix if all of its principal submatrices are S-matrices. Show that an M matrix must be a P-matrix, and a P-matrix must be a completely-S matrix.

9. Show that R is a completely-S matrix if and only if for every $x \in \Re^J$, $\mathrm{LCP}(x, R)$ has at least one solution.

10. Show that R is a P-matrix if and only if for every $x \in \Re^J$, LCP(x, R) has a unique solution.

11. Show that the Skorohod problem (7.14)–(7.16) has a solution if and only if the reflection matrix R is completely-S.

12. Consider the Skorohod problem (7.14)–(7.16) with

$$R = \begin{pmatrix} 1 & -2 \\ 1 & 1 \end{pmatrix}.$$

It is clear that R is a P-matrix. Show that the solutions to the Skorohod problem are not unique.

13. (Bernard and El Kharroubi [2]) Consider the Skorohod problem (7.14)–(7.16) with

$$x(t) = - \begin{pmatrix} 1 \\ 1 \\ 1 \end{pmatrix} t \quad \text{and} \quad R = \begin{pmatrix} 1 & 3 & 0 \\ 0 & 1 & 3 \\ 3 & 0 & 1 \end{pmatrix}.$$

Show that R is a P-matrix (and hence is completely-S). Fix any $\xi > 0$. Show that using

$$y_t = y_\xi + \begin{pmatrix} 0 \\ t - \xi \\ 0 \end{pmatrix}, \quad z_t = \begin{pmatrix} 2(t - \xi) \\ 0 \\ 2\xi - t \end{pmatrix} \quad \text{on } [\xi, 2\xi],$$

$$y_t = y_{2\xi} + \begin{pmatrix} 0 \\ 0 \\ t - 2\xi \end{pmatrix}, \quad z_t = \begin{pmatrix} 4\xi - t \\ 2t - 4\xi \\ 0 \end{pmatrix} \quad \text{on } [2\xi, 4\xi],$$

$$y_t = y_{4\xi} + \begin{pmatrix} t - 4\xi \\ 0 \\ 0 \end{pmatrix}, \quad z_t = \begin{pmatrix} 0 \\ 8\xi - t \\ 2t - 8\xi \end{pmatrix} \quad \text{on } [4\xi, 8\xi],$$

you may patch together a solution to the Skorohod problem. Thus, the solutions to the Skorohod problem are not unique even when the reflection matrix is a P-matrix.

14. Let R be a $J \times K$ matrix. Matrix R is called an S-matrix if there exists an $\xi \geq 0$ such that $R\xi > 0$. For $x \in \Re^J$, let

$$U(x) = \{y \in \Re^K : x + Ry \geq 0 \text{ and } y \geq 0\}.$$

(a) Show that $U(x)$ is nonempty for all $x \in \Re^J_+$ if and only if R is an S-matrix.

(b) Show that $U(x)$ has a least element for each $x \in \Re^J_+$ (i.e., for each $x \in \Re^J_+$, there exists a $y_0 \in U(x)$ such that $y_0 \leq y$ for all $y \in U(x)$) if and only if R is an S-matrix with exactly one positive element in each row. (This immediately leads to the result in Exercise 6.)

15. Extend the results in Exercise 14 to the functional space, by defining, for each $x \in \mathcal{D}_0^J$, $U(x)$ to be the set of y that satisfies relations (7.14) and (7.15).

16. Use the result of Exercise 5 to prove the reflection mapping theorem for step functions; then extend to all RCLL functions.

17. Refer to Exercises 5 and 6. Formulate the linear complementarity problem for the case where R is a singular M matrix, and establish the existence and uniqueness of its solution. Then provide a least element characterization to this linear complementarity problem.

18. Use the result of Exercise 17 to prove the reflection mapping theorem for step functions; then extend to all RCLL functions.

19. Let z and y be respectively the fluid level process and the cumulative loss process of fluid network (α, μ, P) with initial fluid level $z(0)$. Show that $y(s + t) - y(s) \leq (\mu - \lambda)^+ t$ for all $s, t \geq 0$, and hence $\dot{y}(t) \leq (\mu - \lambda)^+$ whenever $\dot{y}(t)$ exists. [Hint: First prove it for the case $z(0) = 0$; then extend to the general case by using the least element characterization and the shift property of the oblique reflection mapping.]

20. Prove Lemma 7.5. [Hint: Show that
$$\det(I - P) = \det(I - P_a) \cdot \det(I - \hat{P}_b). \,]$$

21. Prove Lemma 7.6.

22. Prove Lemma 7.7.

23. Suppose the condition (7.34) holds. Show that the product form stationary distribution given by Theorem 7.12 satisfies BAR. [Note: If BAR could be proved to be sufficient for π to be a stationary distribution, then this would establish the sufficiency part of Theorem 7.12.] Hint: Use
$$d\nu_j(z) = \frac{1}{2} r_{jj}^{-1} \Gamma_{jj} \eta_j \prod_{k \neq j} \eta_k e^{-\eta_k z_k} \, dz_k,$$

and

$$\frac{\partial f(z)}{\partial z_i}\bigg|_{z_j=0} = \int_0^\infty \left[\frac{\partial f(z)}{\partial z_i} \eta_j - \frac{\partial^2 f(z)}{\partial z_i \partial z_j} \right] e^{-\eta_j z_j} \, dz_j.$$

24. (Harrison and Williams [13]) Suppose that the stationary distribution has a product form, namely (7.35), but do not necessarily assume that

$f_j(z_j)$ has an exponential form. Let

$$\pi_j^*(\lambda_j) = \int_0^\infty e^{-\lambda_j z_j} f_j(z_j) dz_j,$$

$$\nu_j^*(\lambda) = \int_{F_j} e^{-\lambda' z} d\nu_j(z),$$

where $\lambda = (\lambda_1, \ldots, \lambda_J)' \geq 0$ is a dummy (vector) variable.

(a) Show that

$$\left[\frac{1}{2} \lambda' \Gamma \lambda - \theta' \lambda \right] \prod_{j=1}^J \pi_j^*(\lambda_j) - \sum_{j=1}^J r_j' \lambda \nu_j^*(\lambda) = 0.$$

(b) Show that

$$r_{jj} \nu_j^*(\lambda) = \frac{1}{2} \Gamma_{jj} c_j \prod_{k \neq j} \pi_k^*(\lambda_k)$$

for some c_j.

(c) Show that

$$\pi_j^*(\lambda_j) = \frac{\Gamma_{jj} c_j}{\Gamma_{jj} \lambda_j - 2\theta_j - \sum\limits_{k \neq j} r_{jk} r_{kk}^{-1} \Gamma_{kk} c_k},$$

and conclude that the density function f_j must be of the exponential form given by Theorem 7.12 and the condition (7.34) must hold.

[Note: This proves the necessity part of Theorem 7.12.]

25. Prove Lemma 7.26.

26. Use Lemma 7.26 to prove Lemma 7.27.

27. Complete the details of the proof for the equalities (7.57) and (7.58) in Theorem 7.13.

28. Prove the convergence of $\bar{Z}^n$ in Theorem 7.23.

29. Under the condition of Theorem 7.29, establish the fluid approximation for the sequence of the networks. In particular, prove that

$$\bar{B}^n(t) \to et \qquad \text{u.o.c.}, \qquad \text{as } n \to \infty,$$

almost surely, where $\bar{B}^n(t) = B^n(nt)/n$ and B^n is the busy time process for the nth network.

30. Use the result in Exercise 29 to prove Theorem 7.29,

31. Under the condition that P is an irreducible stochastic matrix, the solution to the traffic equation (7.109) is unique up to a multiplicative constant, and the maximum solution is positive. (In other words, let λ^0 be a nonzero solution; then any other solutions must be scalar multiples of λ^0. There exists a (unique) solution λ^*, which dominates all other solutions and all of whose components are positive.) In addition, show that the maximum solution is no greater than μ and one of its components equals the corresponding component of μ.

32. Show that equations (7.111) and (7.109) both have a one-dimensional space of solutions.

33. Prove Theorem 7.31. [Hint: Note that Lemma 7.5 still holds when b is not the empty set.]

34. Show that iterative equalities (7.114) and (7.115) give the appropriate normalizing constant $N_J = N_J(1)$. Show that the density function f for the product-form stationary distribution of the RBM on a unit simplex is unique, even though the choice of η is not.

35. Show that f_J given by (7.116) is the marginal stationary distribution for Z_J.

36. (Harrison et al. [13]) Show that the stationary distribution of an RBM on a unit simplex always exists.

37. Prove Theorem 7.32.

38. Prove Theorem 7.35. [Note that you need first to show that as $n \to \infty$, $B^n(n^2 t)/n^2 \to et$ u.o.c. almost surely.]

39. Use Theorems 7.32 and 7.35 to heuristically derive the approximation proposed in Section 7.10.2.

40. Formulate and prove an exponential rate of convergence for the fluid approximation of the generalized open Jackson network. (Refer to Section 6.8.)

41. Formulate a version of the FLIL Theorem (Theorem 7.13) for a sequence of queueing networks indexed by n, and establish the bound of order $O(\sqrt{n \log \log n})$ as $n \to \infty$. Then use this bound to obtain the fluid limit similar to (7.80).

42. Formulate a version of the strong approximation theorem (Theorem 7.19) for a sequence of queueing networks indexed by n.

(a) Establish the strong approximation for the queue-length process and the workload process with the bound of order $o(n^{1/r})$ as $n \to \infty$ for some $r \in (2, 4)$.

(b) Use the above bound to obtain the diffusion limit of the form (7.105), under the heavy traffic condition.

(c) Repeat the above without the heavy traffic condition.

(d) Use the strong approximation bound to obtain the FLIL bound in Exercise 41.

References

[1] Berman, A. and R.J. Plemmons. (1979). *Nonnegative Matrices in the Mathematical Sciences*, Academic Press.

[2] Bernard, A. and A. El Kharroubi. (1991). Régulation de processus dans le premier orthant de $\Re^n$. *Stochastics and Stochastics Reports*, **34**, 149–167.

[3] Chen, H. and A. Mandelbaum. (1991a). Leontief systems, RBV's and RBM's, in *The Proceedings of the Imperial College Workshop on Applied Stochastic Processes*, ed. by M.H.A. Davis and R.J. Elliotte, Gordon and Breach Science Publishers.

[4] Chen, H. and A. Mandelbaum. (1991b). Discrete flow networks: bottleneck analysis and fluid approximations, *Mathematics of Operations Research*, **16**, 408–446.

[5] Chen, H. and A. Mandelbaum. (1991c). Stochastic discrete flow networks: diffusion approximations and bottlenecks, *Annals of Probability*, **19**, 1463–1519.

[6] Chen, H. and A. Mandelbaum. (1994). Hierarchical modelling of stochastic networks, Part II: strong approximations, in D.D. Yao (ed.), *Stochastic Modeling and Analysis of Manufacturing Systems*, 107–131, Springer-Verlag.

[7] Chen, H. and X. Shen. (2000). Computing the stationary distribution of SRBM in an orthant. In preparation.

214   References

[8] Chen, H. and W. Whitt. (1993). Diffusion approximations for open queueing networks with service interruptions, *Queueing Systems, Theory and Applications*, **13**, 335–359

[9] Cottle, R.W., J.S. Pang and R.E. Stone. (1992). *The Linear Complementarity Problem*, Academic Press.

[10] Dai, J.G. and J.M. Harrison. (1991), Steady-state analysis of reflected Brownian motions: characterization, numerical methods and a queueing application, *Annals of Applied Probability*, **1**, 16–35.

[11] Dai, J.G. and J.M. Harrison. (1992), Reflected Brownian motion in an orthant: numerical methods and a queueing application, *Annals of Applied Probability*, **2**, 65–86.

[12] Harrison, J. M. and M.I. Reiman. (1981). Reflected Brownian motion on an orthant, *Annals of Probability*, **9**, 302–308.

[13] Harrison, J.M. and R.J. Williams. (1987). Brownian models of open queueing networks with homogeneous customer populations, *Stochastics*, **22**, 77–115.

[14] Harrison, J.M., R.J. Williams, and H. Chen. (1990). Brownian models of closed queueing networks with homogeneous customer populations, *Stochastics and Stochastics Reports*, **29**, 37–74.

[15] Horváth, L. (1992). Strong approximations of open queueing networks, *Mathematics of Operations Research*, **17**, 487–508.

[16] Johnson, D. P. (1983). *Diffusion Approximations for Optimal Filtering of Jump Processes and for Queueing Networks*, Ph. D Dissertation, University of Wisconsin.

[17] Kaspi, H. and A. Mandelbaum. (1992). Regenerative closed queueing networks. *Stochastics and Stochastic Reports*, **39**, 239–258.

[18] Mandelbaum, A. (1989). The dynamic complementarity problem (preprint).

[19] Reiman, M.I. (1984). Open queueing networks in heavy traffic, *Mathematics of Operations Research*, **9**, 441–458.

[20] Zhang, H. (1997). Strong approximations of irreducible closed queueing networks, *Advances in Applied Probability*, **29**, 2, 498–522.

8

A Two-Station Multiclass Network

Most of this chapter focuses on a two-station queueing network. We spell out details of the model in the next section, followed by Section 8.2–Section 8.5, where we study, respectively, a corresponding fluid network, the stability of the queueing network via the fluid network, the fluid limit, and the diffusion limit. Finally, in Section 8.6 we present additional network examples that appear to be counterintuitive to what we know from single-class networks. This last section is technically independent of the earlier sections; however, it provides further motivation as to why stability is an important and interesting issue in multiclass networks.

8.1 Model Description and Motivation

Consider the Kumar–Seidman network studied in Section 4.3, which has two stations serving four job classes. The network is as shown in Figure 8.1.

For ease of exposition, we shall make some simplifying assumptions on the arrival and the service processes of the network. Jobs of classes 1 and 3 arrive exogenously to the network following independent Poisson processes at rates α_1 and α_3, respectively. After service completion, a class 1 job turns into a class 2 job, and a class 3 job turns into a class 4 job. Jobs of classes 1 and 4 are served at station 1, and jobs of classes 2 and 3 are served at station 2. Within each class, jobs are served in their order of arrivals, and their service times follow i.i.d. exponential distributions, with

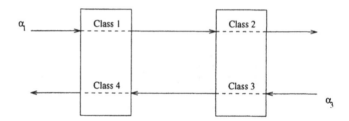

FIGURE 8.1. Kumar–Seidman network

mean m_k for class k, $k = 1, \ldots, 4$. Between the two different classes at the same station, the service follows a preemptive priority rule, with class 4 and class 2 having a higher priority, respectively, at station 1 and station 2. Let $Q_k(t)$ denote the number of class k jobs in the system at time t, $k = 1, \ldots, 4$, and set $Q(t) := (Q_k(t))_{k=1}^4$. Clearly, $Q = \{Q(t), t \geq 0\}$ is a Markov chain, in fact, a 4-dimensional birth–death process.

The first question we ask is under what condition(s) the Markov chain Q is positive recurrent, and hence has a stationary distribution. When Q is positive recurrent, we say the network is stable. Based on what we know from a single-class network, an obvious necessary condition is that the traffic intensity at each station should be less than one, namely,

$$\rho_1 = \alpha_1 m_1 + \alpha_3 m_4 < 1 \quad \text{and} \quad \rho_2 = \alpha_1 m_2 + \alpha_3 m_3 < 1. \quad (8.1)$$

However, the above condition is not sufficient, contrary to the case in single-class networks. Figure 8.2 summarizes the simulation result of the queue-length process, with parameters $\alpha_1 = \alpha_3 = 1$, $m_1 = m_3 = 0.1$, and $m_2 = m_4 = 0.6$. Figure 8.3 shows the total queue length at each station. In this case, the traffic intensities at the two stations are $\rho_1 = \rho_2 = 0.7$, but the total queue length clearly diverges to infinity. The utilization of the server at either station does not appear to converge, but the utilization is always significantly less than the traffic intensity (nominal utilization) of $\rho_1 = \rho_2 = 0.7$, at least at one station. (See Exercise 1 for an explanation.) At the end of the simulation run (at time 50×10^3), the utilizations are 61% and 57%, respectively, for station 1 and station 2.

Next, we simulate the same network example but with deterministic interarrival times and deterministic service times. Figures 8.4 and 8.5 demonstrate that the results are quite similar to the exponential interarrival and service times, and the total queue length again diverges to infinity. (At the end of this simulation run, the utilizations are 60% and 58% at the two stations.) This suggests that as in the single-class case, stability (or instability) is not affected by the variability of the interarrival and service times; but unlike what we see in the single-class case, the usual traffic condition in (8.1) is not enough for stability in the multiclass case. (We note, however, that even for a single-class case, the stability may be affected by the

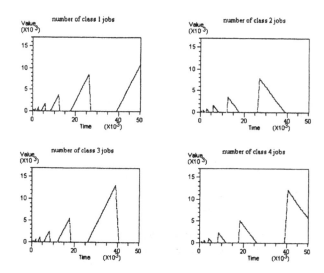

FIGURE 8.2. Queue lengths by class in a Kumar–Seidman network with Poisson arrivals and exponential services

variability if the interarrival times are allowed to be nonrenewal processes.)

Notice that the queue-length processes shown in Figures 8.2–8.5, are very close to a piecewise linear function, with both time and space (number of jobs in the queue) measured in thousands of units (10^3). In other words, $\bar{Q}^n(t) = Q(nt)/n$ with $n = 10^3$ is almost a piecewise linear function, and the total queue length $e'\bar{Q}^n(t)$ increases with time t.

To understand the behavior of the network, the key is to observe that jobs in classes 2 and 4, although served at different stations, can never be served at the same time (with a possible exception for an initial period, if the initial queue length is not zero). For example, if a class 2 job is in service at station 2, then no class 3 job can get through station 2, since it has a lower priority, and hence there will be no class 4 job at station 1. (In our simulation, the initial queue length is zero, and events happen one at a time, particularly so for the deterministic case.) In other words, from time to time there will be forced idling of the station 1 server, waiting for class 3 jobs to get processed at station 2. Analogously, when station 1 is serving jobs from class 4, which has a higher priority over class 1, forced idling could happen to the station 2 server. In short, jobs of class 2 and of class 4 cannot be served at the same time, even though they are served at different service stations. This phenomenon can be viewed as classes 2 and 4 forming a *virtual station*, which creates an additional, "virtual," traffic condition:

$$\alpha_1 m_2 + \alpha_3 m_4 < 1. \tag{8.2}$$

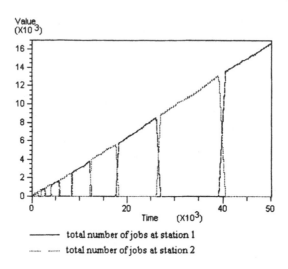

FIGURE 8.3. Queue length by station in a Kumar–Seidman network with Poisson arrivals and exponential services

It is immediate to verify that this condition is violated by the parameters in the above example. Indeed, the above condition, along with the usual traffic condition in (8.1), is a sufficient condition for the network to be stable (i.e., the queue length being positive recurrent). Although we believe it is true, it has not been proved that the network is not stable when (8.2) is violated. On the other hand, when the inequality in (8.2) is reversed (i.e., $\alpha_1 m_2 + \alpha_3 m_4 > 1$), we can indeed prove that the network is not stable.

Note that here and throughout this chapter we adopt the convention that events (such as arrivals and departures of jobs) happen one at a time. Hence, if at some time t_0, a class 1 job finishes service at station 1 and a class 3 job finishes service at station 2, then one of the two jobs is assumed to finish service at time t_0-, and the other has $0+$ unit of service time left. This convention is essential for (8.2) with "<" replaced by "≤" to be a necessary condition for the queue length to be bounded, in the case of deterministic interarrival and service times.

In the next section we consider a fluid network that approximates the queueing network with both state and time scaled by a large n, and show how its fluid level process (which approximates $Q(nt)/n$ as shown in Figures 8.2 and 8.4) can diverge to infinity, even when the traffic condition in (8.1) holds. We then study the stability and the weak stability of this fluid network. In Section 8.3 we relate the stability of a queueing network to the stability of a corresponding fluid network. As a result, we show that the conditions in (8.1) and (8.2) are sufficient for the stability (i.e., positive recurrence) of the queue-length process. However, we have not been

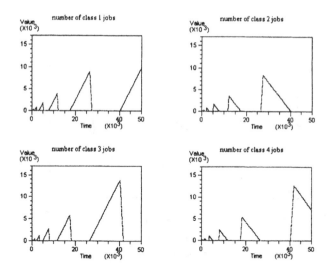

FIGURE 8.4. Kumar–Seidman network with deterministic arrival and service times

able to establish that these conditions are necessary. On the other hand, with the strict inequalities replaced by nonstrict inequalities, these conditions are shown to be both necessary and sufficient for the *weak stability* of the queueing network. The notion of weak stability, when applied to an M/M/1 queue and a Jackson network, is equivalent to the recurrence (including both positive and null recurrences) of the queue-length processes.

Later, in Section 8.4 we shall present the connection between the scaled queueing network and the corresponding fluid network. Specifically, we show that any limit process of $\bar{Q}^n := \{Q(nt)/n, t \geq 0\}$ as $n \to \infty$ must be the fluid level process in the fluid network. In the case when the fluid network is weakly stable, we obtain the fluid approximation for the queue-length process. In Section 8.5 we derive the FCLT for the queue-length process under the heavy traffic condition

$$\rho_1 = \alpha_1 m_1 + \alpha_3 m_4 = 1 \quad \text{and} \quad \rho_2 = \alpha_1 m_2 + \alpha_3 m_3 = 1. \quad (8.3)$$

We show that under the heavy traffic condition, a necessary and sufficient condition for the scaled queue-length process $\hat{Q}^n = \{\hat{Q}^n(t), t \geq 0\}$ to converge weakly as $n \to \infty$ is the condition in (8.2), where

$$\hat{Q}^n(t) := \frac{1}{\sqrt{n}} Q(nt).$$

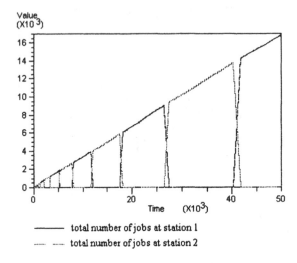

FIGURE 8.5. Kumar–Seidman network with deterministic arrival and service times

8.2   A Fluid Model with Priorities

Here we consider the fluid counterpart of the network shown in Figure 8.1. To recapitulate the model description and notation: There are two stations indexed by $j = 1, 2$, and four classes of fluid indexed by $k = 1, \ldots, 4$. As before, the outflow of classes 1 and 3 turn into fluid of classes 2 and 4, respectively; classes 1 and 4 are processed at station 1, and classes 2 and 3 at station 2. In the fluid model, α_k is the exogenous inflow rate, with $\alpha_2 = \alpha_4 = 0$, whereas $\mu_k = 1/m_k$, $k = 1, \ldots, 4$, is the fluid processing rate, or potential outflow rate. Specifically, for any $t \geq s \geq 0$, $\alpha_k(t - s)$ is the cumulative amount of class k fluid that has flowed into the network over the time interval $[s, t]$; and $\mu_k(t - s)$ is the maximum possible amount of outflow of class k fluid, i.e., if it has been given the full service capacity over $[s, t]$. Let $z_k(0)$ denote the initial fluid level of class k, $k = 1, \ldots, 4$. Set $\alpha = (\alpha_k)_{k=1}^4$, $\mu = (\mu_k)_{k=1}^4$, and $z(0) = (z_k(0))_{k=1}^4$. Assume $\alpha \geq 0$, $\mu > 0$ and $z(0) \geq 0$. For convenience, let $\sigma(k)$, $k = 1, \ldots, 4$, map the fluid class to the station where it is processed, i.e., $\sigma(1) = \sigma(4) = 1$ and $\sigma(2) = \sigma(3) = 2$.

Corresponding to the queue-length process in the queueing network is the fluid level process $z = \{z(t), t \geq 0\}$, where the kth component of $z(t)$, $z_k(t)$, indicates the fluid level of class k at time t. Let $T_k(t)$ denote the total amount of time that station $\sigma(k)$ has devoted to processing class k fluid during the time interval $[0, t]$, $k = 1, \ldots, 4$. Then we have the following flow-balance equations:

$$z_1(t) = z_1(0) + \alpha_1 t - \mu_1 T_1(t),$$
$$z_2(t) = z_2(0) + \mu_1 T_1(t) - \mu_2 T_2(t),$$
$$z_3(t) = z_3(0) + \alpha_3 t - \mu_3 T_3(t),$$
$$z_4(t) = z_4(0) + \mu_3 T_3(t) - \mu_4 T_4(t).$$

These can be succinctly expressed in vector form as follows:

$$z(t) = z(0) + \alpha t - (I - P')DT(t), \qquad (8.4)$$

where

$$\alpha = \begin{pmatrix} \alpha_1 \\ 0 \\ \alpha_3 \\ 0 \end{pmatrix}, \qquad P = \begin{pmatrix} 0 & 1 & 0 & 0 \\ 0 & 0 & 0 & 0 \\ 0 & 0 & 0 & 1 \\ 0 & 0 & 0 & 0 \end{pmatrix},$$

and $D = \operatorname{diag}(\mu)$. Let

$$y_1(t) := t - T_1(t) - T_4(t), \qquad y_3(t) := t - T_2(t) - T_3(t)$$

denote the cumulative idle times (remaining capacity) at the two stations. Then, the work-conserving condition (i.e., no server should be idle if there is fluid to be processed) can be expressed as follows:

$$[z_1(t) + z_4(t)] \, dy_1(t) = 0, \qquad (8.5)$$
$$[z_2(t) + z_3(t)] \, dy_2(t) = 0. \qquad (8.6)$$

That is, at either station, the idle time cannot be increasing while the fluid level is positive. Similarly,

$$y_2(t) := t - T_2(t), \qquad y_4(t) := t - T_4(t)$$

denote the remaining capacity at the two stations after processing the *higher* priority classes (2 and 4). And we have

$$z_2(t)dy_2(t) = 0, \qquad (8.7)$$
$$z_4(t)dy_4(t) = 0. \qquad (8.8)$$

In addition, we also have the following obvious conditions:

$$z(t) \geq 0, \qquad (8.9)$$
$$dT(t) \geq 0 \quad \text{with } T(0) = 0, \qquad (8.10)$$
$$dy(t) \geq 0. \qquad (8.11)$$

In summary, any z and T that satisfy the balance equation in (8.4), the work-conserving condition in (8.5)–(8.6), the priority condition in (8.7)–(8.8), the nonnegative condition in (8.9), and the nondecreasing condition in (8.10)–(8.11), will be called, respectively, a feasible *fluid level* process and a feasible *allocation* process (under the specified priority discipline).

The fluid network is said to be *(strongly) stable* (under the specified priority discipline) if there exists a finite $\delta > 0$ such that for any initial fluid level $z(0)$ with $e'z(0) = 1$ and any corresponding feasible fluid level process z, we have $z(t) = 0$ for all $t \geq \delta$. The fluid network is said to be *weakly stable* if for any feasible fluid level process z with $z(0) = 0$, we have $z(t) = 0$ for all $t \geq 0$.

We first establish some obvious necessary conditions for stability and weak stability.

Lemma 8.1 For the stability of the fluid network specified above, the traffic condition in (8.1) and the virtual traffic condition in (8.2) are both necessary conditions. For weak stability, these are also necessary conditions, with nonstrict inequalities.

Proof   Let

$$C = \begin{pmatrix} 1 & 0 & 0 & 1 \\ 0 & 1 & 1 & 0 \end{pmatrix}.$$

From (8.4), we have

$$CD^{-1}(I - P')^{-1}z(t) = CD^{-1}(I - P')^{-1}z(0) + (\rho - e)t + y_1(t) + y_3(t)$$
$$\geq CD^{-1}(I - P')^{-1}z(0) + (\rho - e)t.$$

The above clearly implies that the traffic condition in (8.1) of the lemma, taking into account that $(I - P')^{-1}$ is a nonnegative matrix (and so are C and D^{-1}). For instance, in the case of stability, suppose $z_j(0) > 0$. Then $\rho_j \geq 1$ implies that the jth component on both sides above is positive for any $t \geq 0$, contradicting the definition of stability.

Next, assume that $\rho \leq e$ and that the condition in (8.2) is violated, with

$$\alpha_1 m_2 + \alpha_3 m_4 > 1. \tag{8.12}$$

(We will consider that the above holds as an equality later.) Then, the above inequality, together with the traffic intensity condition in (8.1), implies

$$\mu_1 > \mu_2, \qquad \mu_3 > \mu_4, \qquad \text{and} \qquad \frac{\alpha_1 \alpha_3}{(\mu_4 - \alpha_3)(\mu_2 - \alpha_1)} > 1. \tag{8.13}$$

First, we construct a pair of feasible (T, z), with $z(0) = (\gamma, 0, 0, 0)'$ (where $\gamma > 0$) such that $z(t) \neq 0$ for all $t \geq 0$. Given $z(0)$, station 1 starts processing class 1 fluid at full capacity. Since $\mu_1 > \mu_2$, this keeps station 2 busy with processing class 2 fluid; hence, no class 3 fluid gets processed, and class 4 stays at zero. This lasts until class 1 reaches zero, at time

$s_1 := \gamma/(\mu_1 - \alpha_1)$. Direct verification shows that for $0 \leq t \leq s_1$,

$$z(t) = \begin{pmatrix} \gamma + (\alpha_1 - \mu_1)t \\ (\mu_1 - \mu_2)t \\ \alpha_3 t \\ 0 \end{pmatrix} \quad \text{and} \quad T(t) = \begin{pmatrix} t \\ t \\ 0 \\ 0 \end{pmatrix},$$

and both are feasible.

Starting from $t = s_1$, station 2 processes class 2 at full capacity, and station 1 is forced to process class 1 at a fraction of its capacity, since $\alpha_1 < \mu_1$ and no fluid of class 4 is available. This lasts until class 2 reaches zero at $s_2 = \gamma/(\mu_2 - \alpha_1)$. Again, it can be directly verified that for $s_1 \leq t \leq s_2$,

$$z(t) = \begin{pmatrix} 0 \\ \frac{\mu_1 - \mu_2}{\mu_1 - \alpha_1}\gamma + (\alpha_1 - \mu_2)(t - s_1) \\ \alpha_3 t \\ 0 \end{pmatrix} \quad \text{and} \quad \dot{T}(t) = \begin{pmatrix} \alpha_1 m_1 \\ 1 \\ 0 \\ 0 \end{pmatrix}$$

jointly satisfy the feasibility conditions. (Note that T is continuous and piecewise linear; hence, it is characterized by its derivative.)

At time $t = s_2$, $z(s_2) = (0, 0, \frac{\alpha_3}{\mu_2 - \alpha_1}\gamma, 0)$. So we can repeat the above argument, reversing the roles of stations 1 and 2, class 1 and class 3, and class 2 and class 4. This leads us to a time

$$t_1 = \frac{\mu_4}{(\mu_4 - \alpha_3)(\mu_2 - \alpha_1)}\gamma$$

when classes 3 and 4 both reach zero, and

$$z(t_1) = \left(\frac{\alpha_1 \alpha_3}{(\mu_4 - \alpha_3)(\mu_2 - \alpha_1)}\gamma, 0, 0, 0 \right)'.$$

Note that $z_k(t_1) = z_k(0) = 0$ for $k = 2, 3, 4$; and $z_1(t_1) > z_1(0)$, following (8.13).

We can repeat the above construction and derive the following general expressions. Let $t_0 = 0$, and

$$t_n = \frac{m_2 \gamma}{\alpha_1 m_2 + \alpha_3 m_4 - 1} \left(\left[\frac{\alpha_1 \alpha_3}{(\mu_2 - \alpha_1)(\mu_4 - \alpha_3)} \right]^n - 1 \right), \quad n \geq 1,$$

$$\gamma_n = \gamma \left[\frac{\alpha_1 \alpha_3}{(\mu_2 - \alpha_1)(\mu_4 - \alpha_3)} \right]^n, \quad n \geq 0. \tag{8.14}$$

Then, for $n = 0, 1, 2, \ldots$, we have

$$z(t) = \begin{pmatrix} \gamma_n + (\alpha_1 - \mu_1)(t - t_n) \\ (\mu_1 - \mu_2)(t - t_n) \\ \alpha_3(t - t_n) \\ 0 \end{pmatrix}, \quad \dot{T}(t) = \begin{pmatrix} 1 \\ 1 \\ 0 \\ 0 \end{pmatrix}, \tag{8.15}$$

for $t_n \le t < t_n + \frac{\gamma_n}{\mu_1 - \alpha_1}$;

$$z(t) = \begin{pmatrix} 0 \\ \frac{\mu_1 - \mu_2}{\mu_1 - \alpha_1}\gamma_n + (\alpha_1 - \mu_2)(t - [t_n + \frac{\gamma_n}{\mu_1 - \alpha_1}]) \\ \alpha_3(t - t_n) \\ 0 \end{pmatrix},$$

$$\dot{T}(t) = \begin{pmatrix} \alpha_1 m_1 \\ 1 \\ 0 \\ 0 \end{pmatrix},$$

(8.16)

for $t_n + \frac{\gamma_n}{\mu_1 - \alpha_1} \le t < t_n + \frac{\gamma_n}{\mu_2 - \alpha_1}$;

$$z(t) = \begin{pmatrix} \alpha_1(t - [t_n + \frac{\gamma_n}{\mu_2 - \alpha_1}]) \\ 0 \\ \frac{\alpha_3 \gamma_n}{\mu_2 - \alpha_1} + (\alpha_3 - \mu_3)(t - [t_n + \frac{\gamma_n}{\mu_2 - \alpha_1}]) \\ (\mu_3 - \mu_4)(t - [t_n + \frac{\gamma_n}{\mu_2 - \alpha_1}]) \end{pmatrix}, \quad \dot{T}(t) = \begin{pmatrix} 0 \\ 0 \\ 1 \\ 1 \end{pmatrix}, (8.17)$$

for $t_n + \frac{\gamma_n}{\mu_2 - \alpha_1} \le t < t_n + \frac{\mu_3 \gamma_n}{(\mu_3 - \alpha_3)(\mu_2 - \alpha_1)}$;

$$z(t) = \begin{pmatrix} \alpha_1(t - [t_n + \frac{\gamma_n}{\mu_2 - \alpha_1}]) \\ 0 \\ 0 \\ \frac{\alpha_3(\mu_3 - \mu_4)}{(\mu_2 - \alpha_1)(\mu_3 - \alpha_3)}\gamma_n + (\alpha_3 - \mu_4)(t - [t_n + \frac{\mu_3 \gamma_n}{(\mu_3 - \alpha_3)(\mu_2 - \alpha_1)}]), \end{pmatrix}$$

(8.18)

$$\dot{T}(t) = \begin{pmatrix} 0 \\ 0 \\ \alpha_3 m_3 \\ 1 \end{pmatrix},$$

for $t_n + \frac{\mu_3 \gamma_n}{(\mu_3 - \alpha_3)(\mu_2 - \alpha_1)} \le t < t_{n+1}$.

The feasibility of (z, T) constructed above is directly verified. Note that under the condition in (8.12),

$$z_1(t_n) = \gamma_n = \left[\frac{\alpha_1 \alpha_3}{(\mu_2 - \alpha_1)(\mu_4 - \alpha_3)} \right]^n \gamma \to +\infty, \qquad \text{as } n \to \infty,$$

implying that the network cannot be stable.

Finally, when (8.12) holds as an equality,

$$\alpha_1 m_2 + \alpha_3 m_4 = 1,$$

the network is still not stable, because in this case the processes z and T can still be constructed through (8.15)–(8.18) with

$$t_n = \frac{n\mu_4 \gamma}{(\mu_2 - \alpha_1)(\mu_4 - \alpha_3)}.$$

We then have $z_1(t_n) = \gamma$ for all $n \geq 1$.

To show that under the condition in (8.12) the network is not weakly stable, we need to construct a feasible pair, z and T, such that z can increase from a zero initial fluid level to a nonzero level. From (8.12), we have

$$t_n \to t_{-\infty} := -\frac{m_2\gamma}{\alpha_1 m_2 + \alpha_3 m_4 - 1} \quad \text{and} \quad \gamma_n \to 0, \quad \text{as } n \to -\infty.$$

We can obtain such a pair as follows: Let $z(0) = T(0) = 0$, let $t_n := t_n - t_{-\infty}$, and for $t > 0$, $z(t)$ and $T(t)$ are defined through (8.15)–(8.18) for all integers n between $-\infty$ and $+\infty$. $\qquad\square$

Also note that whenever $\rho \leq e$,

$$z(t) \equiv 0 \quad \text{and} \quad T(t) = (\alpha_1 m_1, \alpha_1 m_2, \alpha_3 m_3, \alpha_3 m_4)' t$$

give another feasible pair. In other words, the feasible fluid level process and the feasible unused capacity process are not unique in this case.

Theorem 8.2 The two-station fluid network specified earlier is stable if and only if both the traffic condition in (8.1) and the virtual-station condition in (8.2) hold. The network is weakly stable if and only if those conditions hold as nonstrict inequalities.

Proof. That the conditions are necessary has already been established through the above construction. Here we prove the sufficiency by constructing a linear Lyapunov function. Specifically, we want to show that there exists a positive $h = (h_k)_{k=1}^4$ such that $f(t) := h'z(t)$ will reach zero in a finite time for the case of stability, and will stay at zero for the case of weak stability. We prove the case of stability, and leave the case of weak stability as an exercise.

Note that both T and z are Lipschitz continuous functions, in view of (8.10)–(8.11) and (8.4); hence, they are absolutely continuous, and in particular differentiable almost everywhere. Therefore, it suffices to show that there exists an $\epsilon > 0$ such that $\dot{f}(t) = h'\dot{z}(t) < -\epsilon$, whenever $f(t) > 0$ (equivalently $z(t) \neq 0$) and $\dot{T}(t)$ exists.

Pick a fixed $t > 0$ where $\dot{T}(t)$ exists and hence $\dot{z}(t)$ exists. Set $x := \dot{T}(t)$. In view of (8.4), the condition $\dot{f}(t) < -\epsilon$ can be written as

$$h_1(\alpha_1 - \mu_1 x_1) + h_2(\mu_1 x_1 - \mu_2 x_2)$$
$$+ h_3(\alpha_3 - \mu_3 x_3) + h_4(\mu_3 x_3 - \mu_4 x_4) < -\epsilon. \qquad (8.19)$$

Next, we derive conditions that x must satisfy. From (8.10)–(8.11), we have

$$x \geq 0, \qquad (8.20)$$

$$x_1 + x_4 \leq 1, \quad \text{and} \quad x_2 + x_3 \leq 1. \qquad (8.21)$$

Since z is nonnegative, if $z_k(t) = 0$, then its left derivative must be nonpositive and its right derivative must be nonnegative. Hence, the existence of $\dot{z}(t)$ implies $\dot{z}_k(t) = 0$ whenever $z_k(t) = 0$ for $k = 1, \ldots, 4$. Therefore, in view of (8.4), the conditions in (8.5)–(8.6) and (8.7)–(8.8) can be rewritten as

$$(\alpha_1 - \mu_1 x_1 + \mu_3 x_3 - \mu_4 x_4)(1 - x_1 - x_4) = 0, \tag{8.22}$$

$$(\alpha_3 + \mu_1 x_1 - \mu_2 x_2 - \mu_3 x_3)(1 - x_2 - x_3) = 0, \tag{8.23}$$

$$(\mu_3 x_3 - \mu_4 x_4)(1 - x_4) = 0, \tag{8.24}$$

$$(\mu_1 x_1 - \mu_2 x_2)(1 - x_2) = 0. \tag{8.25}$$

To recapitulate, it suffices to find $h > 0$ such that the inequality in (8.19) holds for all x satisfying (8.20)–(8.25), with one exception, which we now explain. When $x_1 + x_4 < 1$ and $x_2 + x_3 < 1$, we have $z(t) = 0$. In this case, it is not necessary to have $\dot{f}(t) < -\epsilon$, or equivalently (8.19); and in fact, they do not hold. This case, in view of (8.22)–(8.25), corresponds to $x = D^{-1}(I - P')^{-1}\alpha := \beta$ with $\beta_1 = \alpha_1 m_1$, $\beta_2 = \alpha_1 m_2$, $\beta_3 = \alpha_3 m_3$, and $\beta_4 = \alpha_3 m_4$.

What remains is purely algebraic manipulation. For each $x \neq \beta$ satisfying (8.20)–(8.25), we obtain an inequality from (8.19). Since there is only a finite number of inequalities involved, we can replace "$< -\epsilon$" in (8.19) by "< 0." To identify all x satisfying (8.20)–(8.25), we consider all possible cases based on whether $x_2 < 1$ or $x_2 = 1$, $x_4 < 1$ or $x_4 = 1$, $x_1 + x_4 < 1$ or $x_1 + x_4 = 1$, and $x_2 + x_3 < 1$ or $x_2 + x_3 = 1$. These lead to 8 possible cases:

Case 1: $x = (0, 1, 0, 1)'$;

Case 2: $x = (1, 1, 0, 0)'$;

Case 3: $x = (0, 0, 1, 1)'$;

Case 4: $x = (0, 0, \alpha_3 m_3, 1)'$

Case 5: $x = (\alpha_1 m_1, 1, 0, 0)'$;

Case 6: $x = (\alpha_1 m_1, \alpha_1 m_2, 1 - \alpha_1 m_2, m_4(1 - \alpha_1 m_2)/m_3)'$;

Case 7: $x = (1 - \alpha_3 m_4, m_2(1 - \alpha_3 m_4)/m_1, \alpha_3 m_3, \alpha_3 m_4)'$;

Case 8: $0 \leq x_2, x_4 < 1$, $x_2 + x_3 = 1$ and $x_1 + x_4 = 1$.

In each of cases 1–5, the inequality in (8.19) becomes

$$h_1 \alpha_1 - h_2 \mu_2 + h_3 \alpha_3 - h_4 \mu_4 < 0, \tag{8.26}$$

$$h_1(\alpha_1 - \mu_1) + h_2(\mu_1 - \mu_2) + h_3 \alpha_3 < 0. \tag{8.27}$$

$$h_1 \alpha_1 + h_3(\alpha_3 - \mu_3) + h_4(\mu_3 - \mu_4) < 0, \tag{8.28}$$

$$h_1 \alpha_1 + h_4(\alpha_3 - \mu_4) < 0, \tag{8.29}$$

$$h_2(\alpha_1 - \mu_2) + h_3 \alpha_3 < 0. \tag{8.30}$$

It is easy to verify that in cases 6 and 7, the inequality (8.19) always holds for $h > 0$ and $\rho < e$.

For case 8, in view of (8.24) and (8.25), $x_2 < 1$ and $x_4 < 1$ imply

$$\mu_1 x_1 = \mu_2 x_2 \qquad \text{and} \qquad \mu_3 x_3 = \mu_4 x_4.$$

If $\mu_1 \mu_3 = \mu_2 \mu_4$, then there is a solution to (8.20)–(8.25) if and only if $\mu_1 = \mu_2$ and $\mu_3 = \mu_4$. In this case, for any $0 \le u \le 1$, $x_1 = x_2 = u$ and $x_3 = x_4 = 1 - u$ satisfy (8.20)–(8.25); the inequality in (8.19) is equivalent to

$$h_1(\alpha_1 - \mu_1) + h_3 \alpha_3 < 0 \quad \text{and} \quad h_1 \alpha_1 + h_3(\alpha_3 - \mu_3) < 0,$$

which are the same as (8.27) and (8.28). When $\mu_1 \mu_3 \ne \mu_2 \mu_4$, (8.20)–(8.25) has a solution under case 8 if and only if

$$0 \le \frac{\mu_1(\mu_4 - \mu_3)}{\mu_2 \mu_4 - \mu_1 \mu_3} < 1 \quad \text{and} \quad 0 \le \frac{\mu_3(\mu_2 - \mu_1)}{\mu_2 \mu_4 - \mu_1 \mu_3} < 1. \tag{8.31}$$

When the solution does exist, it is unique:

$$x_1 = \frac{\mu_2(\mu_4 - \mu_3)}{\mu_2 \mu_4 - \mu_1 \mu_3}, \qquad x_2 = \frac{\mu_1(\mu_4 - \mu_3)}{\mu_2 \mu_4 - \mu_1 \mu_3},$$

$$x_3 = \frac{\mu_4(\mu_2 - \mu_1)}{\mu_2 \mu_4 - \mu_1 \mu_3} < 1, \qquad x_4 = \frac{\mu_3(\mu_2 - \mu_1)}{\mu_2 \mu_4 - \mu_1 \mu_3} < 1;$$

and the inequality in (8.19) is equivalent to

$$h_1 \left[\alpha_1 - \frac{\mu_1 \mu_2(\mu_4 - \mu_3)}{\mu_2 \mu_4 - \mu_1 \mu_3} \right] + h_3 \left[\alpha_3 - \frac{\mu_3 \mu_4(\mu_2 - \mu_1)}{\mu_2 \mu_4 - \mu_1 \mu_3} \right] < 0. \tag{8.32}$$

In summary, if $\mu_1 \mu_3 \ne \mu_2 \mu_4$ and (8.31) holds, the sufficient condition for the existence of a linear Lyapunov function is that the set of inequalities in (8.26)–(8.30) and (8.32) has a positive solution $h = (h_k)_{k=1}^4$; otherwise, the sufficient condition is that the set of inequalities in (8.26)–(8.30) has a positive solution h. It is left as an exercise to show that the solution does exist under the condition of the theorem. □

8.3 Stability

The following theorem relates the stability of the queueing network to the stability of its corresponding fluid network.

Theorem 8.3 The queueing network (as described in Section 8.1) is stable, i.e., the queue-length process Q is positive recurrent, if the corresponding fluid network (as described in Section 8.2) is stable.

Remark 8.4 The stability of the fluid network guarantees the convergence in (8.38), which is in fact sufficient for the stability of the queueing network. The proof (provided toward the end of this section) is based on the convergence in (8.38) and is general enough for a general multiclass queueing network with any number of classes and stations under a priority service discipline.

In view of Theorem 8.2, we have the following theorem.

Theorem 8.5 The Kumar–Seidman queueing network is stable if both the traffic condition in (8.1) and the virtual-station condition in (8.2) hold.

The converse of the above may also hold, although we do not have a proof at this point. We can, however, prove a weaker converse. Specifically, if any of the inequalities in (8.1) and (8.2) is violated by a *strict* inequality in the reversed direction, then, the Kumar–Seidman network is not stable; in fact, in this case, the total queue length $e'Q(t)$ will diverge to infinity. To formally state this result, we introduce a notion of weak stability.

The Kumar–Seidman queueing network is said to be *weakly stable* if for each class k, $Q_k(t)/t$ converges to zero as $t \to \infty$. Let $A_k(t)$ and $D_k(t)$, respectively, denote the number of arrivals and the number of departures of class k jobs during $[0, t]$. It is clear that

$$Q_k(t) = Q_k(0) + A_k(t) - D_k(t).$$

Hence, if the queueing network is weakly stable and if at least one of the following limits below exists, then they are equal,

$$\lim_{t\to\infty} \frac{A_k(t)}{t} = \lim_{t\to\infty} \frac{D_k(t)}{t}. \tag{8.33}$$

This equality simply states that the rate of (total) arrivals equals the rate of departures. It follows from Lemma 5.8 in Section 5.4 that the equality (8.33) is actually equivalent to

$$\lim_{n\to\infty} \frac{1}{n} A_k(nt) = \lim_{n\to\infty} \frac{1}{n} D_k(nt) \qquad \text{u.o.c.} \qquad t \geq 0. \tag{8.34}$$

The above equality implies that

$$\lim_{n\to\infty} \frac{1}{n} Q(nt) = 0, \qquad \text{u.o.c.;} \tag{8.35}$$

i.e., the fluid limit of the queue-length process Q is zero. When applying the above definition to the generalized Jackson network, it follows from its fluid approximation that the generalized Jackson network is weakly stable if and only if its traffic intensity ρ is less than or equal to e. It should be clear that if the queueing network is not weakly stable, then it must not be stable.

Theorem 8.6 The Kumar–Seidman queueing network is weakly stable if and only if $\rho \leq e$ and

$$\alpha_1 m_2 + \alpha_3 m_4 \leq 1 \tag{8.36}$$

hold. In particular, if at least one of the inequalities $\rho_j \leq 1$, $j = 1, 2$, and (8.36) fails, then the Kumar–Seidman queueing network is not weakly stable, and hence not stable.

This theorem follows from Theorem 8.10 in the next section, if one notes the connection between the weak stability and the fluid limit of the queue-length process being zero.

Finally, we return to the proof of Theorem 8.5.

Proof (of Theorem 8.5). We note that this proof is actually general enough for a multiclass queueing network with any number of stations and classes under a priority service discipline. Within this proof only, we let $\mathcal{S} = \mathcal{Z}_+^K$ denote the set of K-dimensional vectors of nonnegative integers, the state space of Q. (In this case $K = 4$.) It is clear that Q is a continuous-time Markov chain (CTMC), actually a (multidimensional) birth–death process. Let $X = \{X_n, n \geq 0\}$ be its embedded discrete-time Markov chain (DTMC). Note that Q is irreducible, and that the transition rate of Q at every state $x \in \mathcal{S}$ is bounded below by $q_{\min} := \sum_{k=1}^K \alpha_k > 0$. Hence, it is elementary to show that CTMC Q is positive recurrent if its embedded DTMC X is positive recurrent (see Exercise 9). Hence, the following two lemmas complete the proof. □

Lemma 8.7 Suppose that the fluid network is stable. Then there exist a $\delta > 0$ and a finite set $B \subset \mathcal{S}$ such that

$$\mathsf{E}_x[\tau_B(\delta)] < \infty \qquad \text{for all } x \in \mathcal{S}, \tag{8.37}$$

where $\tau_B(\delta) = \inf\{t \geq \delta : Q(t) \in B\}$.

Lemma 8.8 Under the conclusion of Lemma 8.7, we have
(i)

$$\mathsf{E}_x[\sigma_B] < \infty \qquad \text{for all } x \in \mathcal{S},$$

where $\sigma_B = \inf\{n \geq 1 : X_n \in B\}$.
(ii) the embedded DTMC X is positive recurrent.

In the following proofs of these two lemmas, we denote the norm of a vector $x \in \Re^K$ by $|x|$ (instead of $\|x\|$) for convenience.

Proof (of Lemma 8.7). It follows from the stability of the fluid network that there exists a $\delta > 0$ such that

$$\lim_{|x| \to \infty} \frac{1}{|x|} \mathsf{E}_x |Q(|x|\delta)| = 0. \tag{8.38}$$

(See Exercise 13.) Hence, for $0 < \epsilon < 1$, there exists an $N \geq 1$ such that

$$\mathsf{E}_x|Q(|x|\delta)| \leq (1 - \epsilon)|x| \qquad \text{for all } |x| \geq N. \qquad (8.39)$$

Let $B = \{x \in \mathcal{S} : |x| \leq N\}$. It is easy to show that there exists a finite $b > \epsilon$ such that

$$\mathsf{E}_x|Q(\delta)| \leq b - \epsilon \qquad \text{for all } x \in B. \qquad (8.40)$$

Let

$$n(x) = \left\{ \begin{array}{ll} |x|\delta, & x \notin B, \\ \delta, & x \in B. \end{array} \right.$$

Combining the inequalities (8.39) and (8.40) yields

$$\mathsf{E}_x|Q(n(x))| \leq |x| - \frac{\epsilon}{\delta}n(x) + b\mathbf{1}_B(x) \qquad \text{for all } x \in \mathcal{S}. \qquad (8.41)$$

Next, we construct an "embedded" DTMC $Y = \{Y_\ell, \ell \geq 0\}$ of Q sampled at time $s(\ell)$ $(\ell = 1, 2, \ldots)$ as follows. Let

$$s(0) = 0, \qquad Y_0 = Q(s(0)) = Q(0),$$
$$s(\ell) = s(\ell - 1) + n(Y_{\ell-1}),$$
$$Y_\ell = Q(s(\ell)), \qquad \ell \geq 1.$$

Then Y is a DTMC with the transition probability

$$\bar{\mathsf{P}}(x, A) = \mathsf{P}^{n(x)}(x, A), \qquad x \in \mathcal{S}, A \subset \mathcal{S},$$

where $\mathsf{P}^t(x, A) = \mathsf{P}(Q(t) \in A | Q(0) = x)$. (Note that $n(x)$ grows linearly with x; hence, Y samples Q in big steps when $|Q(t)|$ is big.)

Note that $\mathsf{E}_x|Q(n(x))| = \mathsf{E}_x|Y_1|$. Let $\bar{\tau}_B = \inf\{\ell \geq 1 : Y_\ell \in B\}$. It follows from their definitions that

$$\tau_B(\delta) \leq s(\bar{\tau}_B) = \sum_{k=0}^{\bar{\tau}_B - 1} n(Y_k). \qquad (8.42)$$

In view of $Y_{k+1} = Q(s(k) + n(Y_k))$ and (8.41), we have (using the strong Markov property of Q)

$$\mathsf{E}\big[|Y_{k+1}|\big|Y_k\big] \leq |Y_k| - \frac{\epsilon}{\delta}n(Y_k) + b\mathbf{1}_B(Y_k),$$

which implies

$$\frac{\epsilon}{\delta}n(Y_k) \leq \mathsf{E}\big[|Y_k| - |Y_{k+1}|\big|Y_k\big] + b\mathbf{1}_B(Y_k).$$

Using the above in (8.42) yields

$$\frac{\epsilon}{\delta} \tau_B(\delta) \le \sum_{k=0}^{\bar{\tau}_B-1} \mathsf{E}\left[|Y_k| - |Y_{k+1}| \big| Y_k\right] + b \sum_{k=0}^{\bar{\tau}_B-1} \mathbf{1}_B(Y_k)$$

$$\le \sum_{k=0}^{\bar{\tau}_B-1} \mathsf{E}\left[|Y_k| - |Y_{k+1}| \big| Y_k\right] + b.$$

Hence,

$$\frac{\epsilon}{\delta} \mathsf{E}_x \tau_B(\delta)$$

$$\le \mathsf{E}_x \sum_{k=0}^{\bar{\tau}_B-1} \mathsf{E}\left[|Y_k| - |Y_{k+1}| \big| Y_k\right] + b$$

$$= \sum_{k=0}^{\infty} \mathsf{E}_x \left[\mathbf{1}_{\{\bar{\tau}_B \ge k+1\}} \mathsf{E}\left[|Y_k| - |Y_{k+1}| \big| Y_1, \ldots, Y_k\right]\right] + b$$

$$= \sum_{k=0}^{\infty} \mathsf{E}_x \left[\mathsf{E}\left[(|Y_k| - |Y_{k+1}|) \mathbf{1}_{\{\bar{\tau}_B \ge k+1\}} \big| Y_1, \ldots, Y_k\right]\right] + b$$

$$= \sum_{k=0}^{\infty} \mathsf{E}_x \left[[|Y_k| - |Y_{k+1}|] \mathbf{1}_{\{\bar{\tau}_B \ge k+1\}} \right] + b$$

$$= \mathsf{E}_x \sum_{k=0}^{\bar{\tau}_B-1} [|Y_k| - |Y_{k+1}|] + b$$

$$\le \mathsf{E}_x |Y_0| + b \le |x| + b,$$

where the exchange of the expectation and the summation in the first equality is justified by the fact that

$$\mathbf{1}_{\{\bar{\tau}_B \ge k+1\}} \left[\mathsf{E}\left[|Y_k| - |Y_{k+1}| \big| Y_1, \ldots, Y_k\right] + b\mathbf{1}_B(Y_k)\right]$$

is nonnegative and the summation of $\mathbf{1}_{\{\bar{\tau}_B \ge k+1\}} \mathbf{1}_B(Y_k)$ over k from 1 to ∞ is bounded by one. □

Proof (of Lemma 8.8). Let S_k be the kth jump time of Q. If we have

$$\mathsf{E}_x S_{\sigma_B} < \infty, \tag{8.43}$$

then we have $\mathsf{E}_x[\sigma_B] < \infty$, since

$$
\begin{aligned}
\mathsf{E}_x S_{\sigma_B} &= \mathsf{E}_x \sum_{k=1}^{\sigma_B} (S_k - S_{k-1}) \\
&= \sum_{k=1}^{\infty} \mathsf{E}_x[(S_k - S_{k-1})\mathbf{1}_{\{\sigma_B \geq k\}}] \\
&= \sum_{k=1}^{\infty} \mathsf{E}\Big\{\mathsf{E}_x\Big[(S_k - S_{k-1})\mathbf{1}_{\{\sigma_B \geq k\}}\Big|Y_{k-1}\Big]\Big\} \\
&= \sum_{k=1}^{\infty} \mathsf{E}\Big\{\mathsf{E}_x\Big[(S_k - S_{k-1})\Big|Y_{k-1}\Big] \cdot \mathsf{E}_x\Big[\mathbf{1}_{\{\sigma_B \geq k\}}\Big|Y_{k-1}\Big]\Big\} \\
&\geq \frac{1}{q_{\min}} \sum_{k=1}^{\infty} \mathsf{E}\Big\{\mathsf{E}_x\Big[\mathbf{1}_{\{\sigma_B \geq k\}}\Big|Y_{k-1}\Big]\Big\} \\
&= \frac{1}{q_{\min}} \mathsf{E}_x \sigma_B,
\end{aligned}
$$

where the fourth equality holds due to the fact that given Y_{k-1}, $S_k - S_{k-1}$ and $\mathbf{1}_{\{\sigma_B \geq k\}}$ are conditionally independent, and the inequality holds if we recall that $q_{\min}$ gives a lower bound on the transition rate of Q. (The conditional independence follows from the strong Markov property of Q and the fact that $S_k - S_{k-1}$ is adapted to $\sigma(Q(t) : t \geq S_{k-1})$, and $\{\sigma_B \geq k\} = \{\sigma_B \leq k-1\}^c$ is adapted to $\sigma(Q(t) : t \leq S_{k-1})$ or $\sigma(Y_i : i \leq k-1)$.)

Now we return to the proof of (8.43). Let M be any given positive constant; we have

$$
\begin{aligned}
\mathsf{E}_x(S_{\sigma_B} \wedge M) \\
&= \mathsf{E}_x\left[S_{\sigma_B} \wedge M \big| S_1 \leq \delta\right] \mathsf{P}_x(S_1 \leq \delta) + \mathsf{E}_x\left[S_{\sigma_B} \wedge M \big| S_1 > \delta\right] \mathsf{P}_x(S_1 > \delta) \\
&\leq \mathsf{E}_x\left[\tau_B(\delta) \big| S_1 \leq \delta\right] \mathsf{P}_x(S_1 \leq \delta) \\
&\quad + \mathsf{E}_x\left[\delta + [(S_1 - \delta) + (S_{\sigma_B} - S_1)] \wedge M \big| S_1 > \delta\right] \mathsf{P}_x(S_1 > \delta) \\
&\leq \mathsf{E}_x\left[\tau_B(\delta)\right] + [\delta + \mathsf{E}_x(S_{\sigma_B} \wedge M)] \mathsf{P}_x(S_1 > \delta),
\end{aligned}
$$

where the memoryless property of the exponential distribution for S_1 is used for the last inequality. Hence,

$$
\mathsf{E}_x(S_{\sigma_B} \wedge M) \leq \frac{[\mathsf{E}_x \tau_B(\delta) + \delta \mathsf{P}_x(S_1 > \delta)]}{\mathsf{P}_x(S_1 \leq \delta)}.
$$

Letting $M \to \infty$ in the above proves (8.43).

Finally, we show that X is positive recurrent. Let $\tilde{P}$ denote the transition matrix of X. Note that X is irreducible and recurrent; hence, there exists a $\pi \neq 0$ such that $\pi = \tilde{P}'\pi$ (see Exercise 11). Let

$$\pi^0(x) = \sum_{y \in B} \pi(y) \mathsf{E}_y \left(\sum_{\ell=1}^{\sigma_B} 1_{\{X_\ell = x\}} \right)$$

$$= \sum_{y \in B} \pi(y) \sum_{\ell=1}^{\infty} \mathsf{P}_y(Y_\ell = x, \sigma_B \geq \ell).$$

Then it is left as an exercise (see Exercise 12) to show that

$$\pi^0 = \tilde{P}'\pi^0 \qquad \text{and} \qquad \pi^0(\mathcal{S}) < \infty. \tag{8.44}$$

Hence, $\pi^0/\pi^0(\mathcal{S})$ is a stationary distribution of X; this establishes that X is positive recurrent. □

8.4 Fluid Limit

In this section and the next one we will allow more general arrival and service processes in the two-station network studied earlier. Let $A_1 = \{A_1(t), t \geq 0\}$ and $A_3 = \{A_3(t), t \geq 0\}$ denote the exogenous arrival processes of class 1 and class 3 jobs, and let $S_k = \{S_k(t), t \geq 0\}$ denote the (potential) service process for class k, $k = 1, \ldots, 4$. Assume that these are all renewal (counting) processes, with the underlying renewal intervals having finite means. Set $Q(0) = (Q_k(0))$, with $Q_k(0)$ being the initial queue length of class k jobs. We assume that A_i, $Q_k(0)$ and S_k, $i = 1, 3$ and $k = 1, \ldots, 4$, are defined on the same probability space; and for convenience, we also assume that they are mutually independent. In keeping with the notation used in the first two sections, let $\alpha = (\alpha_i)$ (with $\alpha_2 = \alpha_4 = 0$) denote the arrival rates, $\mu = (\mu_k)$, the service rates, and $m = (m_k) = (1/\mu_k)$, the mean service times.

As before, $T = \{T(t), t \geq 0\}$ denotes the allocation process; in particular, $T_k(t)$ denotes the cumulative amount of time that class k jobs have been served during $(0, t]$. The queue-length process Q and the allocation process T must jointly satisfy

$$Q_1(t) = Q_1(0) + A_1(t) - S_1(T_1(t)), \tag{8.45}$$
$$Q_2(t) = Q_2(0) + S_1(T_1(t)) - S_2(T_2(t)), \tag{8.46}$$
$$Q_3(t) = Q_3(0) + A_3(t) - S_3(T_3(t)), \tag{8.47}$$
$$Q_4(t) = Q_4(0) + S_3(T_3(t)) - S_4(T_4(t)), \tag{8.48}$$

and

$$T_1(t) = \int_0^t 1\{Q_4(s) = 0, Q_1(s) > 0\}ds, \tag{8.49}$$

$$T_2(t) = \int_0^t 1\{Q_2(s) > 0\}ds, \tag{8.50}$$

$$T_3(t) = \int_0^t 1\{Q_2(s) = 0, Q_3(s) > 0\}ds, \tag{8.51}$$

$$T_4(t) = \int_0^t 1\{Q_4(s) > 0\}ds. \tag{8.52}$$

The relations in (8.45)–(8.48) are the flow-balance equations, whereas (8.49)-(8.52) specify the priority discipline. From (8.45)–(8.52), one can constructively generate the queue length Q and the allocation process T.

Similar to $y_k(t)$, the idle-time process here can be expressed as follows:

$$Y_1(t) = t - T_1(t) - T_4(t) = \int_0^t 1\{Q_1(s) = 0, Q_4(s) = 0\}ds, \tag{8.53}$$

$$Y_2(t) = t - T_2(t) = \int_0^t 1\{Q_2(s) = 0\}ds, \tag{8.54}$$

$$Y_3(t) = t - T_2(t) - T_3(t) = \int_0^t 1\{Q_2(s) = 0, Q_3(s) = 0\}ds, \tag{8.55}$$

$$Y_4(t) = t - T_4(t) = \int_0^t 1\{Q_4(s) = 0\}ds. \tag{8.56}$$

The above directly implies the following work-conserving relations:

$$\int_0^\infty [Q_1(t) + Q_4(t)]dY_1(t) = 0, \tag{8.57}$$

$$\int_0^\infty Q_2(t)dY_2(t) = 0, \tag{8.58}$$

$$\int_0^\infty [Q_2(t) + Q_3(t)]dY_3(t) = 0, \tag{8.59}$$

$$\int_0^\infty Q_4(t)dY_4(t) = 0. \tag{8.60}$$

Next, in view of the relations in (8.53)–(8.56), we apply centering to (8.45)–(8.48):

$$Q_1(t) = X_1(t) + \alpha_1 t + \mu_1 Y_1(t) - \mu_1 Y_4(t), \tag{8.61}$$

$$Q_2(t) = X_2(t) - \mu_2 t - \mu_1 Y_1(t) + \mu_2 Y_2(t) + \mu_1 Y_4(t), \tag{8.62}$$

$$Q_3(t) = X_3(t) + \alpha_3 t - \mu_3 Y_2(t) + \mu_3 Y_3(t), \tag{8.63}$$

$$Q_4(t) = X_4(t) - \mu_4 t + \mu_3 Y_2(t) - \mu_3 Y_3(t) + \mu_4 Y_4(t), \tag{8.64}$$

where

$$X_1(t) = Q_1(0) + [A_1(t) - \alpha_1 t] - [S_1(T_1(t)) - \mu_1 T_1(t)], \tag{8.65}$$

$$X_2(t) = Q_2(0) + [S_1(T_1(t)) - \mu_1 T_1(t)]$$
$$- [S_2(T_2(t)) - \mu_2 T_2(t)], \tag{8.66}$$

$$X_3(t) = Q_3(0) + [A_3(t) - \alpha_3 t] - [S_3(T_3(t)) - \mu_3 T_3(t)], \tag{8.67}$$

$$X_4(t) = Q_4(0) + [S_3(T_3(t)) - \mu_3 T_3(t)]$$
$$- [S_4(T_4(t)) - \mu_4 T_4(t)]. \tag{8.68}$$

Set

$$\bar{Q}^n(t) := \frac{1}{n} Q(nt), \qquad \bar{T}^n(t) := \frac{1}{n} T(nt);$$

and $\beta := (\beta_k)_{k=1}^4$ with $\beta_k := \alpha_1 m_k$, $k = 1, 2$, and $\beta_k =: \alpha_3 m_k$, $k = 3, 4$.

Proposition 8.9 Any subsequence of $\{(\bar{Q}^n, \bar{T}^n), n \geq 1\}$ has a convergent subsequence. Any limit point of $\{(\bar{Q}^n, \bar{T}^n), n \geq 1\}$ must give a pair of feasible fluid level process and feasible allocation process of the fluid network as described in Section 8.2 with zero initial fluid level.

Proof. It follows from the FSLLN for renewal processes that

$$\bar{A}^n(t) := \frac{1}{n} A(nt) \to \alpha t \quad \text{and} \quad \bar{S}^n(nt) := \frac{1}{n} S(nt) \to \mu t, \quad \text{u.o.c.}, \tag{8.69}$$

as $n \to \infty$. Note that $0 \leq \bar{T}_k^n(t) - \bar{T}_k^n(s) \leq t - s$ for all $t \geq s \geq 0$, $k = 1, \ldots, 4$. Hence, the sequence $\{\bar{T}^n, n \geq 1\}$ is equicontinuous, which implies that any of its subsequence has a convergent subsequence. In view of (8.69) and (8.45)–(8.48), for any convergent subsequence of $\{\bar{T}^n, n \geq 1\}$, $\{\bar{Q}^n, n \geq 1\}$ converges along the same subsequence.

Let $\bar{Q}$ and $\bar{T}$ be a pair of limit points of $\{(\bar{Q}^n, \bar{T}^n), n \geq 1\}$. In view of (8.45)–(8.64) and making use of the (one-dimensional) reflection mapping theorem, it must be that $\bar{Q}$ and $\bar{T}$ jointly satisfy relations (8.4)–(8.11) with $z(0) = 0$, and $\bar{Q}$ and $\bar{T}$ replacing z and T. □

Theorem 8.10 For the Kumar–Seidman queueing network under the given priority discipline, the limit,

$$(\bar{Q}^n, \bar{T}^n) \to (\bar{Q}, \bar{T}) \qquad \text{u.o.c.} \qquad \text{as } n \to \infty \tag{8.70}$$

exists with $\bar{Q} = 0$ and $\bar{T}(t) = \beta t$ if and only if $\rho \leq e$ and the condition in (8.36) holds.

Proof. By Proposition 8.9, to prove the sufficiency, we only need to argue that the limit point of $\bar{T}^n$ does not depend on the subsequence we choose. By Theorem 8.2, taking into account $\bar{Q} = 0$, we have $\bar{Q}(t) \equiv 0$, and hence $\bar{T}(t) = \beta t$ for $t \geq 0$, regardless the choice of the subsequence.

The necessity of the condition $\rho \leq e$ follows from the same argument as in the proof of Lemma 8.1. For the necessity of the condition in (8.36), set $\tau := \inf\{t \geq 0 : Q_2(t) = 0 \text{ or } Q_4(t) = 0\}$. Since classes 2 and 4 have higher priority, we have

$$\tau \leq \max\{V_2(Q_2(0)), V_4(Q_4(0))\} < \infty,$$

where $V_k(\ell)$ denotes the ℓth renewal epoch of S_k (with $V_k(0) = 0$), $k = 2, 4$. Furthermore, after τ, whenever a class 2 job is in service, no class 4 job will enter station 1; similarly, whenever a class 4 job is in service, no class 2 job will enter station 2. In other words, after τ, the two stations cannot serve a class 2 job (at station 2) and a class 4 job (at station 1) at the same time. This implies that for $t \geq s \geq \tau$,

$$[T_2(t) + T_4(t)] - [T_2(s) + T_4(s)] \leq t - s.$$

Taking limits, we have

$$[\bar{T}_2(t) + \bar{T}_4(t)] - [\bar{T}_2(s) + \bar{T}_4(s)] \leq t - s$$

for all $t \geq s \geq 0$. This, together with $\bar{T}(t) = \beta t$, implies $\beta_2 + \beta_4 \leq 1$, which is the same as (8.36).   □

8.5 Diffusion Limit

In this section we further assume that the renewal intervals underlying the processes A_i and S_k, $i = 1, 3$ and $k = 1, \ldots, 4$, have finite variances. Let $c_{0,i}$ denote the coefficient of variation of the interarrival times, $i = 1, 3$, and c_k the coefficient of variation of the service times, $k = 1, \ldots, 4$.

Theorem 8.11 Under the heavy traffic condition in (8.3), the scaled queue-length process $\hat{Q}^n$ converges weakly as $n \to \infty$ if and only if the virtual-station condition in (8.2) holds. When $\hat{Q}^n \overset{d}{\to} \hat{Q}$ as $n \to \infty$, we have $\hat{Q}_2 = \hat{Q}_4 = 0$, and

$$\begin{pmatrix} \hat{Q}_1 \\ \hat{Q}_3 \end{pmatrix} = \begin{pmatrix} \tilde{X}_1 \\ \tilde{X}_3 \end{pmatrix} + H \begin{pmatrix} \hat{Y}_1 \\ \hat{Y}_3 \end{pmatrix} \geq 0, \qquad (8.71)$$

$$d\hat{Y}_k(\cdot) \geq 0, \quad \text{with } \hat{Y}_k(0) = 0, \ k = 1, 3, \qquad (8.72)$$

$$\int_0^\infty \hat{Q}_k(t) d\hat{Y}_k(t) = 0, \ k = 1, 3, \qquad (8.73)$$

where

$$\tilde{X}_1 = \frac{m_2 m_4 \hat{S}_2(\beta_2 t) - m_1 m_3 \hat{S}_1(\beta_1 t) + m_3 m_4[\hat{S}_3(\beta_3 t) - \hat{S}_4(\beta_4 t)]}{m_1 m_3 - m_2 m_4}$$
$$+ \hat{A}_1(t), \tag{8.74}$$

$$\tilde{X}_3 = \frac{m_2 m_4 \hat{S}_4(\beta_4 t) - m_1 m_3 \hat{S}_3(\beta_3 t) + m_1 m_2[\hat{S}_1(\beta_1 t) - \hat{S}_2(\beta_2 t)]}{m_1 m_3 - m_2 m_4}$$
$$+ \hat{A}_3(t), \tag{8.75}$$

$$H = \frac{1}{m_1 m_3 - m_2 m_4} \begin{pmatrix} m_3 & -m_4 \\ -m_2 & m_1 \end{pmatrix}, \tag{8.76}$$

with $\hat{A}_i$ ($i = 1, 3$) and $\hat{S}_k$ ($k = 1, \dots, 4$) being six independent driftless Brownian motions with variances $\alpha_i c_{0,i}^2$ ($i = 1, 3$) and $\mu_k c_k^2$ ($k = 1, \dots, 4$), respectively.

Some clarification is in order.

1. Here for simplicity, we do not introduce a sequence of queueing networks with arrival and service processes indexed by n; but the extension to that setup is straightforward. The theorem can also be generalized to allow the initial queue length to vary with n, as long as $\hat{Q}^n(0) := Q^n(0)/\sqrt{n}$ converges weakly to 0 as $n \to \infty$. However, if as $n \to \infty$, $\hat{Q}^n(0)$ converges weakly to $\hat{Q}(0) \neq 0$, then the extension is not obvious, particularly, the weak convergence of $(\hat{Q}_2^n, \hat{Q}_4^n)$ to 0 may fail. In this case, one may have to consider the unconventional diffusion approximation in Harrison and Williams [25] for a related closed network. (To mimic the formulation in [25], one can consider $\bar{Q}^n(0) := Q^n(0)/n$ converging weakly to $\bar{Q}(0)$ as $n \to \infty$ with $\bar{Q}_1(0) \neq 0$ and $\bar{Q}_k(0) = 0$, $k = 2, 3, 4$.)

2. It is easy to check that under the heavy traffic condition in (8.3), the virtual-station condition in (8.2) holds if and only if $m_1 m_3 - m_2 m_4 > 0$, which is equivalent to H being an M-matrix. Hence, under the condition of the theorem, the limiting process $(\hat{Q}_1, \hat{Q}_3)$, together with the process $(\hat{Y}_1, \hat{Y}_3)$, is uniquely determined by the relations in (8.71)–(8.73) (referring to Theorem 7.2); and the limiting process is the reflected Brownian motion on the nonnegative orthant with H being the reflection matrix. (See Theorems 7.2 and 7.8.)

3. Let $\hat{Y}_k^n(t) = Y(nt)/\sqrt{n}$, $k = 1, 3$. In the proof below we will actually establish the joint weak convergence of $(\hat{Q}_1^n, \hat{Q}_3^n, \hat{Y}_1^n, \hat{Y}_3^n)$ to $(\hat{Q}_1, \hat{Q}_3, \hat{Y}_1, \hat{Y}_3)$ as $n \to \infty$.

We prove the main theorem by establishing the following three propositions.

Proposition 8.12 With $\rho \leq e$ and the condition in (8.36), we have

$$(\hat{Q}_2^n, \hat{Q}_4^n) \xrightarrow{d} 0, \qquad \text{as } n \to \infty. \tag{8.77}$$

Proposition 8.13 Under the heavy traffic condition in (8.3), if (8.2) is replaced by an *equality*, i.e.,

$$\alpha_1 m_2 + \alpha_3 m_4 = 1,$$

then neither $\hat{Q}_1^n$ nor $\hat{Q}_3^n$ converges weakly as $n \to \infty$.

Proposition 8.14 Under the heavy traffic condition in (8.3), if (8.2) holds, then

$$(\hat{Q}_1^n, \hat{Q}_3^n) \xrightarrow{d} (\hat{Q}_1, \hat{Q}_3), \qquad \text{as } n \to \infty, \tag{8.78}$$

where $(\hat{Q}_1, \hat{Q}_3)$ satisfies (8.71)–(8.73).

Let

$$\hat{A}_k^n(t) := \frac{A_k(nt) - \alpha_k nt}{\sqrt{n}}, \qquad k = 1, 3,$$

$$\hat{S}_k^n(t) := \frac{S_k(nt) - \mu_k nt}{\sqrt{n}}, \qquad k = 1, \ldots, 4.$$

Then by the functional central limit theorem for renewal processes we have

$$(\hat{A}_1^n, \hat{A}_3^n, \hat{S}_1^n, \hat{S}_2^n, \hat{S}_3^n, \hat{S}_4^n) \xrightarrow{d} (\hat{A}_1, \hat{A}_3, \hat{S}_1, \hat{S}_2, \hat{S}_3, \hat{S}_4) \tag{8.79}$$

as $n \to \infty$, where $\hat{A}_k$, $k = 1, 3$, $\hat{S}_k$, $k = 1, \ldots, 4$, are independent driftless Brownian motions. The variance for $\hat{A}_k$ is $\alpha_k c_{0,k}^2$, $k = 1, 3$, and the variance for $\hat{S}_k$ is $\mu_k c_k^2$, $k = 1, \ldots, 4$.

Following the Skorohod representation theorem, we assume that the probability space is so chosen that for each sample path, the convergence in (8.79) holds under the uniform topology, i.e., as $n \to \infty$,

$$(\hat{A}_1^n, \hat{A}_3^n, \hat{S}_1^n, \hat{S}_2^n, \hat{S}_3^n, \hat{S}_4^n) \to (\hat{A}_1, \hat{A}_3, \hat{S}_1, \hat{S}_2, \hat{S}_3, \hat{S}_4) \qquad \text{u.o.c.} \tag{8.80}$$

Since the assumptions in Propositions 8.12-8.14 either coincide with or imply the assumptions in Theorem 8.10, we shall assume the fluid convergence in (8.70) when we prove these propositions. Also note that the convergence in (8.80) clearly implies the FSLLN limit

$$(\bar{A}_1^n, \bar{A}_3^n, \bar{S}_1^n, \bar{S}_2^n, \bar{S}_3^n, \bar{S}_4^n) \to (\bar{A}_1, \bar{A}_3, \bar{S}_1, \bar{S}_2, \bar{S}_3, \bar{S}_4) \qquad \text{u.o.c.} \tag{8.81}$$

as $n \to \infty$, where $\bar{A}_k(t) = \alpha_k t$, $k = 1, 3$, and $\bar{S}_k(t) = \mu_k t$, $k = 1, \ldots, 4$.

Next, we rewrite the relations in (8.61)–(8.64) for the scaled queue-length process:

$$\hat{Q}_1^n(t) = \hat{X}_1^n(t) + \alpha_1\sqrt{n}t + \mu_1\hat{Y}_1^n(t) - \mu_1\hat{Y}_4^n(t), \tag{8.82}$$

$$\hat{Q}_2^n(t) = \hat{X}_2^n(t) - \mu_2\sqrt{n}t - \mu_1\hat{Y}_1^n(t) + \mu_2\hat{Y}_2^n(t) + \mu_1\hat{Y}_4^n(t), \tag{8.83}$$

$$\hat{Q}_3^n(t) = \hat{X}_3^n(t) + \alpha_3\sqrt{n}t - \mu_3\hat{Y}_2^n(t) + \mu_3\hat{Y}_3^n(t), \tag{8.84}$$

$$\hat{Q}_4^n(t) = \hat{X}_4^n(t) - \mu_4\sqrt{n}t + \mu_3\hat{Y}_2^n(t) - \mu_3\hat{Y}_3^n(t) + \mu_4\hat{Y}_4^n(t), \tag{8.85}$$

where

$$\hat{X}_1^n(t) = \hat{Q}_1^n(0) + \hat{A}_1^n(t) - \hat{S}_1^n(\bar{T}_1^n(t)),$$

$$\hat{X}_2^n(t) = \hat{Q}_2^n(0) + \hat{S}_1^n(\bar{T}_1^n(t)) - \hat{S}_2^n(\bar{T}_2^n(t)),$$

$$\hat{X}_3^n(t) = \hat{Q}_3^n(0) + \hat{A}_3^n(t) - \hat{S}_3^n(\bar{T}_3^n(t)),$$

$$\hat{X}_4^n(t) = \hat{Q}_4^n(0) + \hat{S}_3^n(\bar{T}_3^n(t)) - \hat{S}_4^n(\bar{T}_4^n(t)),$$

and $\hat{Y}_k^n(t) = Y_k(nt)/\sqrt{n}$. It follows from Theorem 8.10 that $\bar{T}^n(t)$ converges to βt u.o.c. Hence, the convergence in (8.80), along with the random-time change theorem, implies

$$(\hat{X}_1^n, \hat{X}_2^n, \hat{X}_3^n, \hat{X}_4^n) \to (\hat{X}_1, \hat{X}_2, \hat{X}_3, \hat{X}_4) \qquad \text{u.o.c.} \qquad \text{as } n \to \infty, \tag{8.86}$$

where

$$\hat{X}_1(t) = \hat{A}_1(t) - \hat{S}_1(\beta_1 t),$$

$$\hat{X}_2(t) = \hat{S}_1(\beta_1 t) - \hat{S}_2(\beta_2 t),$$

$$\hat{X}_3(t) = \hat{A}_3(t) - \hat{S}_3(\beta_3 t),$$

$$\hat{X}_4(t) = \hat{S}_3(\beta_3 t) - \hat{S}_4(\beta_4 t).$$

As a further preparation for the proof of Proposition 8.12, we need the following two lemmas.

Lemma 8.15 Assume $\rho \le e$ and (8.36) holds. Consider $k = 2$ or 4. For any sequence $\{n_\ell, \ell \ge 1\}$ with $n_\ell \to \infty$ as $\ell \to \infty$, and any bounded sequence $\{(s_{n_\ell}, t_{n_\ell}), \ell \ge 1\}$ with $t_{n_\ell} > s_{n_\ell}$, if $\hat{Q}_k^{n_\ell}(t) > 0$ for $t \in [s_{n_\ell}, t_{n_\ell}]$, then

$$t_{n_\ell} - s_{n_\ell} \to 0, \qquad \text{as } \ell \to \infty.$$

Proof. Consider $k = 2$; the proof for $k = 4$ is completely analogous. Use contradiction, and assume without loss of generality that $t_{n_\ell} \to t_0'$ and $s_{n_\ell} \to s_0'$ with $t_0' > s_0'$. Then, we can choose t_0 and s_0 with $t_0' \ge t_0 > s_0 \ge s_0'$ such that $t_{n_\ell} > t_0 > s_0 > s_{n_\ell}$ for ℓ large enough, and hence $\hat{Q}_2^{n_\ell}(t) > 0$ for $t \in [s_0, t_0]$. This, together with (8.51), implies

$$T_3(n_\ell t_0) - T_3(n_\ell s_0) = 0,$$

which, in view of (8.47), leads to

$$\bar{Q}_3^{n_\ell}(t_0) - \bar{Q}_3^{n_\ell}(s_0) = \bar{A}_3^{n_\ell}(t_0) - \bar{A}_3^{n_\ell}(s_0) \to \alpha_3(t_0 - s_0),$$

as $\ell \to \infty$, where the convergence in (8.81) is used. Hence, we have

$$\limsup_{n \to \infty} \bar{Q}_3^n(t_0) = \limsup_{n \to \infty}(\bar{Q}_3^n(s_0) + [\bar{Q}_3^n(t_0) - \bar{Q}_3^n(s_0)])$$
$$\geq \lim_{n \to \infty}[\bar{Q}_3^n(t_0) - \bar{Q}_3^n(s_0)] = \alpha_3(t_0 - s_0) > 0,$$

contradicting Theorem 8.10.   □

Lemma 8.16 Assume $\rho \leq e$ and that (8.36) holds. Then, the sequences $\{\hat{Q}_2^n, n \geq 1\}$ and $\{\hat{Q}_4^n, n \geq 1\}$ are C-tight (for each sample path on the chosen probability space). (In this case, the sequence $\{\hat{Q}_k^n, n \geq 1\}$ is said to be C-tight if given any fixed $T > 0$, for each $\epsilon > 0$, there exist a $\delta > 0$ and an $n_0 \geq 1$ such that

$$\sup_{\substack{0 \leq s, t \leq T \\ |s-t| \leq \delta}} |\hat{Q}_k^n(s) - \hat{Q}_k^n(t)| < \epsilon,$$

for all $n \geq n_0$.)

Proof.   Suppose that $\{\hat{Q}_2^n, n \geq 1\}$ is not C-tight. Then, for some $\epsilon > 0$, there exist sequences $\{n_k, k \geq 1\}$ and $\{(s_{n_k}, t_{n_k}), k \geq 1\}$ such that

$$0 < t_{n_k} - s_{n_k} \to 0, \qquad \text{as } k \to \infty,$$
$$|\hat{Q}_2^{n_k}(t_{n_k}) - \hat{Q}_2^{n_k}(s_{n_k})| > \epsilon,$$
$$\hat{Q}_2^{n_k}(t) > 0 \qquad \text{for } t \in [s_{n_k}, t_{n_k}].$$

Following Lemma 8.15, we can assume that

$$\hat{Q}_2^{n_k}(t_{n_k}) - \hat{Q}_2^{n_k}(s_{n_k}) > \epsilon.$$

(Otherwise, we set $t_{n_k} = s_{n_k}$ and then $s_{n_k} = \sup\{t < t_{n_k} : \hat{Q}_2^{n_k}(t) \leq \hat{Q}_2^{n_k}(0)\}$, and note that $\hat{Q}^{n_k}(0) \to 0$ as $n_k \to \infty$.) From (8.83) and (8.54), we have

$$\epsilon < \hat{Q}_2^{n_k}(t_{n_k}) - \hat{Q}_2^{n_k}(s_{n_k})$$
$$= [\hat{X}^{n_k}(t_{n_k}) - \hat{X}^{n_k}(s_{n_k})] - \mu_2\sqrt{n_k}(t_{n_k} - s_{n_k}) - \mu_1[\hat{Y}_1^{n_k}(t_{n_k}) - \hat{Y}_1^{n_k}(s_{n_k})]$$
$$+ \mu_1[\hat{Y}_4^{n_k}(t_{n_k}) - \hat{Y}_4^{n_k}(s_{n_k})]$$
$$\leq [\hat{X}^{n_k}(t_{n_k}) - \hat{X}^{n_k}(s_{n_k})] - \mu_2\sqrt{n_k}(t_{n_k} - s_{n_k}) + \mu_1[\hat{Y}_4^{n_k}(t_{n_k}) - \hat{Y}_4^{n_k}(s_{n_k})].$$

Note that

$$[\hat{Y}_4^{n_k}(t_{n_k}) - \hat{Y}_4^{n_k}(s_{n_k})] \leq \sqrt{n_k}(t_{n_k} - s_{n_k}),$$

which follows from (8.56), and that $\mu_1 \leq \mu_2$, which follows from $\rho \leq e$ and (8.36). Hence, we have

$$\epsilon < \hat{Q}_2^{n_k}(t_{n_k}) - \hat{Q}_2^{n_k}(s_{n_k}) \leq \hat{X}^{n_k}(t_{n_k}) - \hat{X}^{n_k}(s_{n_k}).$$

In view of (8.86) and the fact that $t_{n_k} - s_{n_k} \to 0$ as $k \to \infty$, we know that the right-hand side of the above inequality converges to 0 as $k \to \infty$, and hence a contradiction. The proof for the C-tightness of $\{\hat{Q}_4^n, n \geq 1\}$ follows a similar argument. $\square$

We are now ready to prove the propositions.

Proof (of Proposition 8.12).   We prove only the case for $\hat{Q}_2^n$, since the proof for $\hat{Q}_4^n$ is analogous. From Lemma 8.16, it suffices to show that any convergent subsequence of $\{\hat{Q}_2^n, n \geq 1\}$ converges to zero. Suppose

$$\hat{Q}_2^{n_k} \to \hat{Q}_2 \qquad \text{u.o.c.} \qquad \text{as } n \to \infty,$$

where $n_k \to \infty$ as $k \to \infty$. Suppose $\hat{Q}_2(t_0) > 0$ for some finite $t_0 > 0$; we will reach a contradiction. Since the sequence $\{\hat{Q}_2^{n_k}, k \geq 1\}$ is C-tight, it follows from Theorem 10.2 in Chapter 3 of Ethier and Kurtz [8] that $\hat{Q}_2$ is continuous. Hence, there exist $\delta > 0$ and $\epsilon > 0$ such that

$$\hat{Q}_2(t) > \epsilon \qquad \text{for } t \in [t_0 - \delta, t_0 + \delta].$$

Consequently, for k large enough,

$$\hat{Q}_2^{n_k}(t) > 0 \qquad \text{for } t \in [t_0 - \delta, t_0 + \delta],$$

contradicting Lemma 8.15. $\square$

Proof (of Proposition 8.13).   In this case, we have $m_1 = m_2$ and $m_3 = m_4$. From (8.83) and (8.85), we have

$$\hat{Y}_1^n(t) - \hat{Y}_3^n(t) = [m_2 \hat{X}_2^n(t) - m_4 \hat{X}_4^n(t)] - [m_2 \hat{Q}_2^n(t) - m_4 \hat{Q}_4^n(t)].$$

By (8.86) and Proposition 8.12, the following weak convergence holds as $n \to \infty$:

$$\hat{Y}_1^n(t) - \hat{Y}_3^n(t) \overset{d}{\to} m_2 \hat{X}_2(t) - m_4 \hat{X}_4(t)$$
$$= m_2[\hat{S}_1(\beta_1 t) - \hat{S}_2(\beta_2 t)] + m_4[\hat{S}_3(\beta_3 t) - \hat{S}_4(\beta_4 t)],$$

where the limit is clearly a nondegenerate Brownian motion. Note that both $\hat{Y}_1^n$ and $\hat{Y}_3^n$ are nondecreasing processes and that a (nondegenerate) Brownian motion has infinite variation. Therefore, neither $\hat{Y}_1^n$ nor $\hat{Y}_3^n$ converges weakly.

Now suppose $\hat{Q}_1^n$ converges weakly. We then claim that $[\hat{Y}_2^n - \beta_3 \sqrt{n} t]$ converges weakly. To see this, we obtain from (8.82) and (8.83),

$$\hat{Q}_1^n(t) + \hat{Q}_2^n(t) = \hat{X}_1^n(t) + \hat{X}_2^n(t) + \mu_2[\hat{Y}_2^n(t) - \beta_3 \sqrt{n} t]$$

(where we used the fact that $\alpha_1 m_2 + \alpha_3 m_3 = 1$ and $\beta_3 = \alpha_3 m_3$), and then apply (8.86) and Proposition 8.12. Next, rewrite (8.84) as

$$\hat{Q}_3^n(t) = \hat{X}_3^n(t) - \mu_3[\hat{Y}_2^n(t) - \beta_3\sqrt{n}t] + \mu_3\hat{Y}_3^n(t).$$

In view of (8.59), we have

$$\int_0^\infty \hat{Q}_3^n(t)d\hat{Y}_3^n(t) = 0.$$

Since $\hat{X}_3^n(t) - \mu_3[\hat{Y}_2^n(t) - \beta_3\sqrt{n}t]$ converges weakly, it follows from the continuity of the one-dimensional reflection mapping (Theorem 6.1) that $\hat{Y}_3^n$ converges weakly, and we reach a contradiction. Hence, $\hat{Q}_1^n$ does not converge weakly as $n \to \infty$. The proof for the nonconvergence of $\hat{Q}_3^n$ follows a similar argument. □

Proof (of Proposition 8.14).   Set $\Delta := \mu_2\mu_4 - \mu_1\mu_3$. Note that $\Delta > 0$ under the assumed conditions in the proposition. Solving $\hat{Y}_2^n$ and $\hat{Y}_4^n$ from (8.83) and (8.85) yields

$$\hat{Y}_2^n(t) = \frac{1}{\Delta} \times \left\{ \mu_4[\hat{Q}_2^n(t) - \hat{X}_2^n(t) + \mu_1\hat{Y}_1^n(t) + \mu_2\sqrt{n}t] \right.$$
$$\left. - \mu_1[\hat{Q}_4^n(t) - \hat{X}_4^n(t) + \mu_3\hat{Y}_3^n(t) + \mu_4\sqrt{n}t] \right\}$$

$$\hat{Y}_4^n(t) = \frac{1}{\Delta} \times \left\{ \mu_2[\hat{Q}_4^n(t) - \hat{X}_4^n(t) + \mu_3\hat{Y}_3^n(t) + \mu_4\sqrt{n}t] \right.$$
$$\left. - \mu_3[\hat{Q}_2^n(t) - \hat{X}_2^n(t) + \mu_1\hat{Y}_1^n(t) + \mu_2\sqrt{n}t] \right\}.$$

Next substituting the above into (8.82) and (8.84) yields

$$\begin{pmatrix} \hat{Q}_1^n(t) \\ \hat{Q}_3^n(t) \end{pmatrix} = \begin{pmatrix} \tilde{X}_1^n(t) \\ \tilde{X}_3^n(t) \end{pmatrix} + H \begin{pmatrix} \hat{Y}_1^n(t) \\ \hat{Y}_3^n(t) \end{pmatrix}, \qquad (8.87)$$

where

$$\tilde{X}_1^n(t) = \hat{X}_1^n(t) - \frac{\mu_1}{\Delta}\{\mu_2[\hat{Q}_4^n(t) - \hat{X}_4^n(t)] - \mu_3[\hat{Q}_2^n(t) - \hat{X}_2^n(t)]\},$$
$$\tilde{X}_3^n(t) = \hat{X}_3^n(t) - \frac{\mu_3}{\Delta}\{\mu_4[\hat{Q}_2^n(t) - \hat{X}_2^n(t)] - \mu_1[\hat{Q}_4^n(t) - \hat{X}_4^n(t)]\}.$$

From (8.86), we have

$$(\tilde{X}_1^n, \tilde{X}_3^n) \to (\tilde{X}_1, \tilde{X}_3) \qquad \text{u.o.c.} \qquad \text{as } n \to \infty \qquad (8.88)$$

with $\tilde{X}_1$ and $\tilde{X}_3$ following (8.74) and (8.75). The relations in (8.57) and (8.59) imply that

$$\int_0^\infty \hat{Q}_k^n(t)d\hat{Y}_k^n(t) = 0, \qquad k = 1, 3.$$

In view of (8.87) and (8.88), the continuity of the oblique reflection mapping (see Remark 1 after the proposition) implies the convergence in (8.78) with the limit as identified in the proposition. □

8.6 More Examples

Our first example is a variation of the Kumar–Seidman network with one additional station, as shown in Figure 8.6. The difference here is that a class 1 job after service at station 1 turns into a class 5 job at station 3, before turning into a class 2 job at station 2. Let m_k denote the mean service

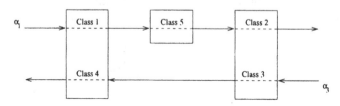

FIGURE 8.6. A three-station variation of a Kumar–Seidman network

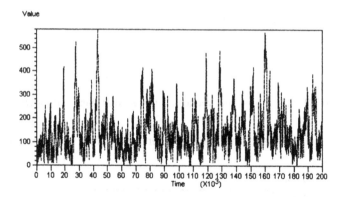

FIGURE 8.7. Total queue length of the three station variation of Kumar–Seidman network with Poisson arrivals and exponential services

time of class k, $k = 1, \ldots, 5$. We note that when $m_5 = 0$, the network reduces to the Kumar–Seidman network. Now choose $\alpha_1 = \alpha_3 = 1$, $m_1 = m_3 = 0.1$, $m_2 = m_4 = 0.6$, and $m_5 = 0.7$. Then, simulation indicates that the network is stable: The plot of the queue-length processes indicates that they will reach the steady state, under both exponential and deterministic interarrival and service times. Figure 8.7 shows the total queue length of the exponential case with a simulation run of 200×10^3 units of time. Indeed, we can show through the corresponding fluid network that the queueing network is indeed stable. (Proving that the corresponding fluid network is stable is left as an exercise.) Note, however, that if m_5 is reduced to zero while all other parameters remain the same, then the network becomes the Kumar–Seidman network, and we know that it is not stable. In other words, by reducing the mean service time, m_5 from 0.7 to 0, we can turn a stable network into an unstable network. This example illustrates that a multiclass queueing network may become unstable when we reduce the mean service time (or equivalently, increase the service rate). In particular, speeding the service may in fact increase the total queue length (work in process) in a multiclass network. This is in sharp contrast to the monotonicity results in single-class networks (refer to Chapter 3).

Next, we consider another variation of the Kumar–Seidman network as shown in Figure 8.8. The network has two stations and six job classes. The

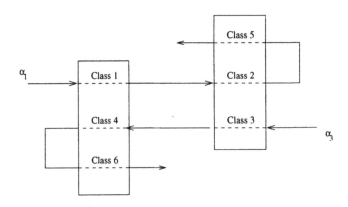

FIGURE 8.8. A six-class version of the Kumar–Seidman network

transition of classes and the station where each class is served are both

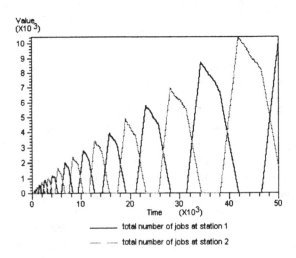

FIGURE 8.9. Queue length by station in a six-class variation of a Kumar–Seidman network with $m = (0.35, 0.25, 0.35, 0.25, 0.3, 0.3)$, Poisson arrivals, and exponential services

indicated in the figure. The following priorities are assumed at the two stations: At station 1, class 4 has the highest priority, class 6 the next, and class 1 the lowest; at station 2, class 2 has the highest priority, class 5 the next, and class 3 the lowest. Note that if we combine classes 2 and 5 into one class and combine classes 4 and 6 into another class, and make the service times of each of the newly formed classes equal to the sum of the original service times, then the network is reduced to the Kumar–Seidman network. Let m_k denote the mean service time of class k jobs, $k = 1, \ldots, 6$, and let $m = (m_k)_{k=1}^{6}$. We fix $\alpha_1 = \alpha_3 = 1$ in the following numeric examples. By choosing $m = (0.35, 0.25, 0.35, 0.25, 0.3, 0.3)$, the simulation result in Figure 8.9 indicates that this network is not stable. Note that with the given parameters, the specified service discipline is the well-known rule of shortest (expected) processing (service) time first (SPT). That is, this network is not stable under the SPT discipline. We also note that both the simulation and the analysis in the introduction and Section 8.2-Section 8.3 indicate that the Kumar–Seidman network may not be stable under the scheduling rule of shortest (expected) remaining processing time first (SRP). Note that the SPT and the SRP rules are known to be optimal in minimizing the total number of jobs in the system for certain single-station multiclass queues (e.g., refer to Example 11.17 in Chapter 11), and they have been among the most widely used scheduling rules, as heuristics, in other more general settings.

This six-class network is also not stable if we choose $m_k = 0.3$, $k = 1, \ldots, 6$, as shown in Figure 8.10. Under the FIFO service discipline, this

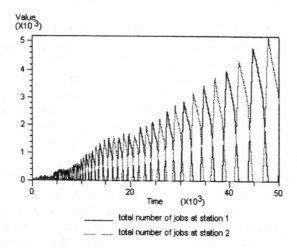

FIGURE 8.10. Queue length by station in a six-class variation of a Kumar–Seidman network with $m = (0.3, 0.0.3, 0.3, 0.3, 0.3, 0.3)$, Poisson arrivals and exponential services

network belongs to the class of Kelly networks (Chapter 4, in particular the deterministic routing model at the end of Section 4.3); it is known to have a product-form stationary distribution when the usual traffic condition is in force (i.e., $\rho < e$), and is hence stable.

Consider the linkage between this six-class network with the Kumar–Seidman network. It seems that in addition to the usual traffic condition, the stability condition for this network should be

$$\alpha_1(m_2 + m_5) + \alpha_3(m_4 + m_6) < 1.$$

When the above becomes a strict inequality in reverse, the network is unstable. We leave it as an exercise to find the stability and the weak stability conditions in this case.

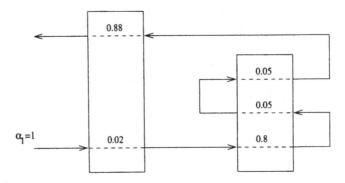

FIGURE 8.11. A variation of the Bramson network

Finally, we turn to the FIFO service discipline. We show by examples that the counterintuitive phenomenon that appeared in a network with priority service discipline may happen under a FIFO service discipline as well. Consider a two-station network as shown in Figure 8.11. Jobs arrive from outside with a unit rate. Each job first visits station 1, then visits station 2 three consecutive times, and finally visits station 1 for the second time and then leaves the network. The numbers in the figure indicate the mean service times for each visit. The service discipline is FIFO. The traffic intensities at both stations are 0.9 (at station 1, $0.02 + 0.88 = 0.9$, and at station 2, $0.8 + 0.05 + 0.05 = 0.9$). Figures 8.12 and 8.13 illustrate the total queue length at each station in a simulation run of 50×10^3 time units; and we consider both exponential and deterministic interarrival and service times. The plots clearly indicate that the network is unstable under both distributions. However, we do not have a proof. Here is an open problem to finish off this example. Suppose that the mean service times are fixed. It is clear that if we reduce the arrival rate to a substantially lower value, then the network will become stable. The question is, What is the critical value such that stability is guaranteed if the arrival rate is kept below this value, whereas exceeding it will result in an unstable network?

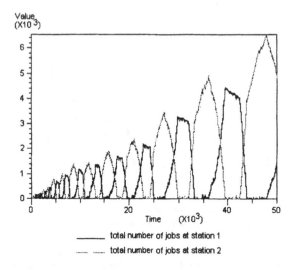

FIGURE 8.12. Queue length by station in a variation of Bramson network with Poisson arrivals and exponential services

A variation of the above network, with one additional service station, is depicted in Figure 8.14. Decreasing the mean service time at this added station to zero reduces the network to the previous two-station variation of the Bramson network. A simulation run (with Poisson arrivals and exponential service times) indicates that this network is stable; see Figure 8.15 (whereas the previous model is not). This example shows that in a multiclass queueing network, the network may become unstable when we speed up the service, even under FIFO.

8.7 Notes

The Kumar–Seidman network was first studied by Kumar and Seidman [27] for a network with deterministic arrival and service times. Whitt [33] showed that the usual heavy traffic limit theorem does not exist for this network under a FIFO service discipline. Rybko and Stolyar [31] proved that this network may not be stable under the traffic condition for the case of Poisson arrivals and exponential service times. Botvich and Zamyatin [4] first observed the virtual station phenomenon and established the sufficient condition for the stability, and Dai and Weiss [21] extended this to the nonexponential case using the fluid network approach.

  The heterogeneous fluid model first appeared in Chen [9] and Chen and Mandelbaum [11]. The presentation of Section 8.2 follows Chen and Zhang [15]. Theorem 8.3 was first introduced in Rybko and Stolyar [31], and was later generalized by Dai [16], who showed that for a general multiclass

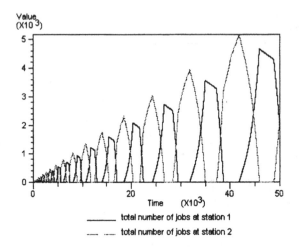

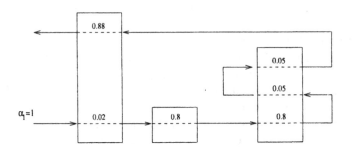

FIGURE 8.13. Queue length by station in a variation of Bramson network with deterministic arrivals and services

FIGURE 8.14. A Three-Station Variation of Bramson Network

queueing network under various service disciplines such as FIFO, priority and head-of-line processor-sharing, the queueing network is stable (i.e., the queue-length processes are positive Harris recurrent) if the corresponding fluid network is stable. This provides a very powerful approach to analyzing the stability of a queueing network. The proof of Theorem 8.5 is due to Hengqing Ye.

The weak stability and the fluid approximation for a multiclass queueing network are topics studied in Chen [10], and the diffusion approximation for the Kumar–Seidman network in Chen and Zhang [14]. The notion of C-tightness used in the diffusion section is different from the definition of the tightness in Billingsley [3]. Since the limiting processes under consideration are all continuous, it is sufficient here to characterize the C-tightness; see Pollard [30] for more details.

Other than the counterexamples brought forth by Kumar–Seidman [27] and Rybko and Stolyar [31] for queueing networks under priority service

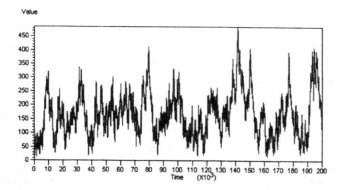

FIGURE 8.15. Total queue length in a three-station variation of the Bramson network with Poisson arrivals and exponential service times

disciplines, Seidman [32] and Bramson [5] first demonstrated the instability of the FIFO service discipline. Bramson [5] considered a network similar to the one in Figure 8.11, but after visiting station 1, jobs make thousands of repeated visits to station 2 (instead of station 3 as in our example here) before visiting station 1 again. It was proved that the network is unstable under a set of parameters with the traffic intensity strictly less than one. (Note that through simulation we have only observed, but not proved, that the network in Figure 8.11 is not stable.) Later, Bramson [6] provided another example that shows that for any given $\epsilon > 0$, there exists a queueing network with traffic intensity less than ϵ at each station but it is not stable under a FIFO service discipline. Dumas [23] provided a three-station network example where the stability region of the network under a priority service discipline is not monotone in service rates. However, it was proved in Chen [10] that the stability region of a queueing network is monotone in the exogenous arrival rate. Banks and Dai [1] also provided a number of counterexamples through simulation.

Those examples have in recent years inspired extensive studies on the issue: whether a multiclass queueing network is stable under a given service discipline, or whether a queueing network is stable under all work-conserving disciplines (in which case the stability is termed *global stability*). Rybko and Stolyar [31] establish the stability of the Kumar–Seidman network under FIFO service discipline. Using a quadratic Lyapunov function, Kumar and Meyn [26] and Chen [10] establish some sufficient conditions for the global stability. Botvich and Zamyatin [4] introduce the usage of a piecewise Lyapunov function to the study of the stability in a Markovian setting. Dai and Weiss [21] and Down and Meyn [22] use piecewise Lyapunov functions to the study of the stability of fluid networks, with more general network structures. In particular, Dai and Weiss [21] establishes the stability of a reentrant line network under a first-buffer-first-served and a last-buffer-first-served discipline. With the piecewise linear Lyapunov approach, Dai and Vande Vate [19, 20] provide a necessary and sufficient condition for the global stability of a two-station fluid network with deterministic routing. (See also Bertsimas et al. [2].) By using a linear Lyapunov function, Chen and Zhang [15] obtains a sufficient condition for the stability of a fluid network under a priority service discipline; when applied to the three-station network, this sufficient condition recovers and extends the nonmonotone and nonconvex stability region identified by Dumas [23]. In studying a three-station network, Dai et al. [18] find that the global stability region is not monotone, and also identify cases in which the network is stable but does not satisfy the sufficient condition in Chen and Zhang [15]. By using a piecewise linear Lyapunov function, Chen and Ye [12] weaken the sufficient condition for the stability, which can identify in particular all the stability region found in Dai et al. [18]. There have been relatively few studies on the stability of a queueing network under FIFO; some of these include Bramson [7] and Chen and Zhang [13]. In establishing the stability of a fluid network, almost all of the above approaches use a Lyapunov function; in particular, Ye and Chen [34] show that a fluid network is stable if and only if there exists a Lyapunov function. On the other hand, Meyn [28] and Dai [17] provide a partial converse to the sufficient condition given in Dai [16], but a recent work of Bramson [8] shows that the converse may not be true in general, by providing an example network where the queueing network is stable but its corresponding fluid network is not stable.

Refer to Section 10.7 for the literature on the diffusion approximation of multiclass queueing networks.

8.8   Exercises

1. Consider the pair of feasible allocation process T and fluid level process z for the Kumar–Seidman network.

   (a) Show that $T(t)/t$ does not converge as $t \to \infty$.

   (b) Find

   $$\liminf_{t \to \infty} \frac{T_1(t) + T_4(t)}{t} \quad \text{and} \quad \limsup_{t \to \infty} \frac{T_1(t) + T_4(t)}{t}.$$

   (c) Find a subsequence $\{t_n, n \geq 1\}$ such that the lim inf in the above is achieved along this subsequence, i.e.,

   $$\liminf_{t \to \infty} \frac{T_1(t) + T_4(t)}{t} = \lim_{n \to \infty} \frac{T_1(t_n) + T_4(t_n)}{t_n}.$$

   Then does the limit

   $$\lim_{n \to \infty} \frac{T_2(t_n) + T_3(t_n)}{t_n}$$

exist? Does this limit (or its lim sup) coincide with

$$\limsup_{t\to\infty} \frac{T_2(t) + T_3(t)}{t}.$$

(d) Compare various limits in the above with ρ (the traffic intensity), and then interpret.

2. In Section 8.2, a pair of feasible allocation process and fluid level process is constructed under the condition (8.12). Can you construct some other feasible processes? Can you identify all of them?

3. Suppose that the traffic condition in (8.1) and the virtual traffic condition in (8.2) hold. Show that if $\mu_1\mu_3 \neq \mu_2\mu_4$ and (8.31) holds, the set of inequalities in (8.26)–(8.30) and (8.32) has a positive solution $h = (h_k)_{k=1}^4$; otherwise, the set of inequalities in (8.26)–(8.30) has a positive solution h.

4. Prove the weak stability part of Theorem 8.2.

5. Write down the corresponding fluid network of the network as shown in Figure 8.6, and show that it is stable under the parameters given in Section 8.6. In general, what is the sufficient condition on the set of network parameters for this network to be stable?

6. Formulate a version of Proposition 8.9 with $\bar{Q}(0) \neq 0$.

7. Consider the six-class two-station network in Section 8.6, as shown in Figure 8.8. Find a necessary and sufficient condition for the stability of its corresponding fluid network.

8. Formulate a general multiclass fluid network under a work-conserving discipline. Show that if the fluid network is stable, then the traffic intensity ρ is strictly less than e.

9. Consider an irreducible continuous-time Markov chain $Z = \{Z(t), t \geq 0\}$. Assume that the transition rates of Z at all of its states are bound below by a positive constant. Show that Z is positive recurrent if and only if its embedded discrete-time Markov chain is.

10. Prove the convergence (8.38) under the condition that the corresponding fluid network is stable.

11. (Meyn and Tweedie [29]) Let $X = \{X_n, n \geq 1\}$ be an irreducible and recurrent discrete time Markov chain with state-space S. Let $\tau_x = \min\{n \geq 1 : X_n = x\}$ for $x \in S$ and define the n-step taboo probability

$$P^{x,n}(y, A) = P_y(X_n \in A, \tau_x \geq n), \qquad x, y \in S, A \subseteq S.$$

Let

$$\pi_x(A) = \sum_{n=1}^{\infty} P^{x,n}(x, A).$$

(a) Show that $\pi_x(x) := \pi_x(\{x\}) = 1$.

(b) Let $P(x, A)$ be the one-step transition probability of X. Show that for $x \in S$ and $A \subseteq S$,

$$P(x, A) = P^{x,1}(x, A),$$

$$\sum_{y \in S \setminus \{x\}} P^{x,n}(x, y) P(y, A) = P^{x,n+1}(x, A).$$

Then show that for any fixed $x \in S$, $\pi_x(\cdot)$ is invariant, i.e.,

$$\pi_x(A) = \sum_{y \in S} \pi_x(y) P(y, A).$$

12. Establish (8.44) by proving the following.

(a) Show by induction that

$$\sum_{y \in B} \left[\pi(y) \sum_{\ell=1}^{n} P_y(X_\ell = x, \sigma_B \geq \ell) \right] \leq \pi(x), \qquad x \in S,$$

holds for $n = 0, 1, \ldots$, to conclude that $\pi^0(x) \leq \pi(x)$ for all $x \in S$.

(b) Show that $\pi^0(B) = \pi(B)$ to conclude that $\pi^0(x) = \pi(x)$ for all $x \in B$.

(c) Show that π^0 is invariant, i.e., $\pi^0 = \tilde{P}' \pi^0$.

(d) Show that

$$\pi^0(S) = \sum_{y \in B} \pi(y) E_y \sigma_B$$

to conclude that $\pi^0(S) < \infty$.

13. Consider the multiclass queueing network described in Section 8.3. Let Q^x denote its queue-length process with initial state $Q^x(0) = x$, and let

$$\bar{Q}^x(t) = \frac{1}{|x|} Q^x(|x|t).$$

Let S be the state space of Q.

(a) Show that the sequence $\{\bar{Q}^x(\cdot) : x \in S\}$ is equicontinuous. Hence, every subsequence of the sequence has a further u.o.c. convergent subsequence.

(b) Show that every limit process of the above sequence must be a feasible fluid level process of the corresponding fluid network as described in Section 8.3.

(c) Show that if the corresponding fluid level process is stable, then there exists a finite $\delta > 0$ such that

$$\limsup_{|x| \to \infty} \frac{1}{|x|} |Q^x(|x|\delta)| = 0.$$

(d) Show the above convergence implies the convergence (8.38).

References

[1] Banks, J. and J.G. Dai. (1996). Simulation studies of multiclass queue-ing networks. *IIE Transactions*, **29**, 213–219.

[2] Bertsimas, D., D. Gamarnik, and J. Tsitsiklis. (1996). Stability con-ditions for multiclass fluid queueing networks. *IEEE Transactions on Automatic Control*, **41**, 1618–1631.

[3] Billingsley, P. (1968). *Convergence of Probability Measures*, Wiley, New York.

[4] Botvich, D.D. and A.A. Zamyatin. Ergodicity of conservative com-munication networks. *Rapport de recherche* 1772, **INRIA**, October, 1992.

[5] Bramson, M. (1994a). Instability of FIFO queueing networks. *Annals of Applied Probability*, **4**, 414–431.

[6] Bramson, M. (1994b). Instability of FIFO queueing networks with quick service times. *Annals of Applied Probability*, **4**, 693–718.

[7] Bramson, M. (1996). Convergence to equilibria for fluid models of FIFO queueing networks. *Queueing Systems, Theory and Applications*, **22**, 5–45.

[8] Bramson, M. (1999). A stable queueing network with unstable fluid model. *Annals of Applied Probability*, **9**, 818–853.

[9] Chen, H. (1988). A heterogeneous fluid model and its applications in multiclass queueing networks. Submitted to George Nicholson Student Paper Competition, Operations Research Society of America (Honorable Mention).

[10] Chen, H. (1995). Fluid approximations and stability of multiclass queueing networks: Work-conserving discipline. *Annals of Applied Probability*, **5**, 637–655.

[11] Chen, H. and A. Mandelbaum. (1991). Open heterogeneous fluid networks, with applications to multiclass queues. Preprint.

[12] Chen, H. and H. Ye. (1999). Piecewise linear Lyapunov function for the stability of priority multiclass queueing networks. *Proceedings of the 28th IEEE Conference on Decision and Control*, 931–936.

[13] Chen, H. and H. Zhang. (1997). Stability of multiclass queueing networks under FIFO service discipline. *Mathematics of Operations Research.* **22**, 691–725.

[14] Chen, H. and H. Zhang. (1998). Diffusion approximations for Kumar–Seidman network under a priority service discipline. *Operations Research Letters*, **23**, 171–181.

[15] Chen, H. and H. Zhang. (2000). Stability of multiclass queueing networks under priority service disciplines. *Operations Research*, **48**, 26–37.

[16] Dai, J.G. (1995). On positive Harris recurrence of multiclass queueing networks: a unified approach via fluid limit models. *Annals of Applied Probability*, **5**, 49–77.

[17] Dai, J.G. (1996). A fluid-limit model criterion for instability of multiclass queueing networks. *Annals of Applied Probability.* **6**, 751–757.

[18] Dai, J.G., J.J. Hasenbein, and J.H. Vande Vate. (1999) Stability of a three-station fluid network, *Queueing Systems, Theory and Applications*, **33**, 293–325.

[19] Dai, J.G. and J.H. Vande Vate. (1996). Virtual stations and the capacity of two-station queueing networks (preprint).

[20] Dai, J.G. and J.H. Vande Vate. (2000). The stability of two-station multi-type fluid networks. *Operations Research* (to appear).

[21] Dai, J.G. and G. Weiss. (1996). Stability and instability of fluid models for certain reentrant lines. *Mathematics of Operations Research*, **21**, 115–134.

[22] Down, D. and S. Meyn. (1994). Piecewise linear test functions for stability of queueing networks. *Proceeding of the 33rd Conference on Decision and Control*, 2069–2074.

[23] Dumas,V. (1997). A multiclass network with nonlinear, nonconvex, nonmonotonic stability conditions. *Queueing Systems, Theory and Applications*, **25**, 610–623.

[24] Ethier, S.N. and T.G. Kurtz. (1986). *Markov Processes*, Wiley, New York.

[25] Harrison, J.M. and R.J. Williams. (1996). A multiclass closed network with unconventional heavy traffic behavior. *Annals of Applied Probability*. **6**, 1–47.

[26] Kumar, P.R. and S. Meyn. (1995). Stability of queueing networks and scheduling polities. *IEEE Transactions on Automatic Control*, **40**, 251–260.

[27] Kumar, P.R. and T.I. Seidman. (1990). Dynamic instabilities and stabilization methods in distributed real-time scheduling of manufacturing systems. *IEEE Transactions on Automatic Control*, **35**, 289–298.

[28] Meyn, S.P. (1995). Transience of multiclass queueing networks via fluid limit models. *Annals of Applied Probability*, **5**, 946–957.

[29] Meyn, S. and R.L. Tweedie. (1993). *Markov Chains and Stochastic Stability*. Springer-Verlag, London.

[30] Pollard, D. (1984). *Convergence of Stochastic Processes*, Springer-Verlag.

[31] Rybko, A.N. and A.L. Stolyar. (1992). Ergodicity of stochastic processes describing the operations of open queueing networks. *Problemy Peredachi Informatsii*, **28**, 2–26.

[32] Seidman, T.I. (1993). First come first served can be unstable. *IEEE Transactions on Automatic Control*, **39**, 2166–2171.

[33] Whitt, W. (1993). Large fluctuations in a deterministic multiclass network of queues, *Management Science*, **39**, 1020–1028.

[34] Ye, H. and H. Chen. (2000). Lyapunov method for the stability of fluid networks. Preprint.

9
Multiclass Feedforward Networks

In a feedforward network, jobs after service completion at a station can be routed only to a downstream station. Other than this restriction on routing, the network has more features than the generalized Jackson network studied in Chapter 7. In particular, jobs at each station come from different classes, with different service time distributions, and upon service completion will follow different routing sequences. In addition, job classes are partitioned into priority groups. Within the same group, jobs are served in the order of arrivals, i.e., following a first-in-first-out (FIFO) discipline. Among different groups, jobs are served under a preassigned preemptive priority discipline. Therefore, the model includes a network with a pure FIFO service discipline or a pure priority service discipline as a special case.

In the feedforward network, arrivals to each station either come from an external source or from an upstream station. This makes it possible to analyze the network in a recursive manner, from the upstream stations to the downstream stations. Specifically, we first focus on a single station multiclass queue, and show that if the arrival process satisfies a certain approximation such as the functional law of the iterated logarithm approximation or the strong approximation, then the departure process also satisfies a similar approximation. This analysis can then be extended to the feedforward network, where the arrival process to a downstream station is simply the departure process from an upstream station. This way, we obtain the FLIL and the functional strong approximation for the whole network. In addition to the limit theorems, we propose approximations for performance measures such as the queue lengths and the sojourn times, and use numerical examples to illustrate the accuracy of the approximations.

The chapter is organized as follows. The single station model is studied first, with the model introduced in Section 9.1, and the FLIL and the strong approximations established in Section 9.2 and Section 9.3, respectively. The network model is then presented in Section 9.4, and its FLIL and strong approximations are presented in Sections 9.5 and 9.6. In Section 9.7 approximations are proposed based on the limit theorems, and numerical examples are presented in Section 9.8.

9.1   The Single Station Model

Consider a single-server station serving K classes of jobs. Let $\mathcal{K} = \{1, \ldots, K\}$ denote the set of job class indices. Jobs of all classes arrive exogenously, wait for service and after service completion leave the system. To specify the service discipline, we partition $\mathcal{K}$ into L groups, $g_1, \ldots, g_L$. For any $\ell < \ell'$, a job of a class in g_ℓ has a higher preemptive-resume priority over any job of any class in $g_{\ell'}$; as a result, the presence of the latter job has no impact on the former job. In this sense, for $\ell < \ell'$, a job of a class in g_ℓ *does not see* any job of any class in $g_{\ell'}$. Within each group, jobs are served under FIFO discipline. Let π be a mapping from $\mathcal{K}$ (job class index set) to $\mathcal{L} := \{1, \ldots, L\}$ (job group index set); specifically, $\pi(k) = \ell$ if and only if $k \in g_\ell$, i.e., class k is in group $g_{\pi(k)}$. A job of class k is referred to as a (group) g_ℓ job or it is in g_ℓ, if $k \in g_\ell$. The station is assumed to have an infinite waiting room. We note that with $L = 1$, the station models a multiclass queue under a (pure) FIFO service discipline, and with $L = K$ (implying that each group g_ℓ contains a single class) the station models a multiclass queue under a (pure) priority service discipline. Figure 9.1 is an example of a single station queue with $K = 4$ classes. Assume that classes 1 and 2 have equal priority (hence, all of their jobs are served under a

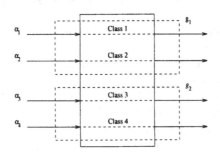

FIGURE 9.1. A multiclass single-station queue

FIFO discipline), and classes 3 and 4 have equal priority. Further, assume

that jobs of classes 1 and 2 have higher priority over jobs of classes 3 and 4. With the above notation, we have $L = 2$ groups, with $g_1 = \{1, 2\}$ and $g_2 = \{3, 4\}$ and with $\pi(1) = \pi(2) = 1$ and $\pi(3) = \pi(4) = 2$.

The queue is characterized by the following primitive data: K counting processes $A_k = \{A_k(t), t \geq 0\}$ ($k \in \mathcal{K}$), and K sequences of nonnegative random variables $v_k = \{v_k(n), n \geq 1\}$ ($k \in \mathcal{K}$), all defined on the same probability space. The (integer-valued) quantity $A_k(t)$ indicates the number of class k jobs that have arrived (exogenously) to the system during $[0, t]$. We call $A = (A_k)$ an exogenous arrival process. The random variable $v_k(n)$ is the service time required for the nth class k job. We call $v = (v_k)$ the service time sequence. We assume that initially there is no job in the system.

Let $u_k(1)$ be the arrival time of the first class k job and $u_k(n)$, $n > 1$, the interarrival time between the $(n-1)$st and the nth class k jobs (corresponding to jump points of A_k). We call $u = (u_k)$ with $u_k = \{u_k(n), n \geq 1\}$ the interarrival time sequence. We introduce the summation,

$$V_k(0) = 0, \quad V_k(n) = \sum_{m=1}^{n} v_k(m), \quad n \geq 1, \quad k \in \mathcal{K},$$

and let $V = (V_k)$.

The performance measures of interest are the L-dimensional aggregated workload process $Z = (Z_\ell)$ with $Z_\ell = \{Z_\ell(t), t \geq 0\}$ ($\ell \in \mathcal{L}$), the K-dimensional queue length process $Q = (Q_k)$ with $Q_k = \{Q_k(t), t \geq 0\}$ ($k \in \mathcal{K}$), and the L-dimensional cumulative idle time process $Y = (Y_\ell)$ with $Y_\ell = \{Y_\ell(t), t \geq 0\}$ ($\ell \in \mathcal{L}$). All of these processes are nonnegative processes. The quantity $Z_\ell(t)$ indicates the total amount of immediate work for the station embodied in jobs that are in groups 1 to ℓ and that are queued or in service at the station at time t. The quantity $Q_k(t)$ is integer-valued and indicates the number of class k jobs at the station at time t. The quantity $Y_\ell(t)$ indicates the cumulative amount of time that the server does not serve jobs in groups 1 to ℓ during $[0, t]$. It is clear that Y must be nondecreasing with $Y(0) = 0$.

To describe the dynamics of the queue, we need some additional notation.

- $D_k(t)$ counts the number of departures of class k jobs from the station during $[0, t]$ after their service completions;

- $W_k(t)$ is the workload process of class k jobs;

- $\tau_\ell(t)$ is the arrival time of the g_ℓ job that has most recently completed service ($\tau_\ell(t) \equiv 0$ if there have been no service completions for group ℓ);

- $\nu_k(t)$ is the partial service time (if any) that has been performed on a class k job during $(\tau_\ell(t), t]$, where $k \in g_\ell$;

- $S_k(t)$ is the sojourn time of class k jobs at time t, denoting the time that will be spent at the station by the first class k job to arrive at time greater than or equal to t;

- $\eta_k(t)$ is the time at which the first class k job arrives during $[t, \infty)$;

- $\mathcal{T}_k(t)$ is the time that a class k job would spend at the station if it arrived at time t.

Making use of the quantities defined above, we can express the dynamics of the queue as follows:

$$Q_k(t) = A_k(t) - D_k(t), \tag{9.1}$$

$$W_k(t) = V_k(A_k(t)) - V_k(D_k(t)) - \nu_k(t), \tag{9.2}$$

$$Y_\ell(t) = t - \sum_{i=1}^{\ell} \sum_{k \in g_i} T_k(t), \tag{9.3}$$

$$Z_\ell(t) = \sum_{i=1}^{\ell} \sum_{k \in g_i} W_k(t) = \sum_{i=1}^{\ell} \sum_{k \in g_i} V_k(A_k(t)) - t + Y_\ell(t), \tag{9.4}$$

$$D_k(t) = S_k(T_k(t)) = A_k(\tau_\ell(t)), \quad \text{where} \quad \ell = \pi(k), \tag{9.5}$$

$$0 \le \nu_k(t) \le \max_{1 \le n \le A_k(t)} v_k(n), \tag{9.6}$$

$$0 \le \eta_k(t) - t \le u_k(A_k(t) + 1), \tag{9.7}$$

$$S_k(t) = T_k(\eta_k(t)). \tag{9.8}$$

The equations in (9.1) and (9.2) are flow-balance relations in terms of jobs and work, respectively. The second equality in (9.5) follows from the FIFO service discipline among jobs in group g_ℓ; specifically, any job in group g_ℓ that has arrived before $\tau_\ell(t)$ must have finished service by time t.

The dynamics of $\mathcal{T}_k$ can be described by a recursive relationship:

$$T_k(t) = Z_{\pi(k)}(t-) + \sum_{\ell=1}^{\pi(k)-1} \sum_{i \in g_\ell} \left[V_i(A_i(\mathcal{T}_i(t) + t)) - V_i(A_i(t)) \right]$$
$$+ V_i(A_i(t)) - V_i(A_i(t) - 1), \tag{9.9}$$

where $Z_{\pi(k)}(t)$ is the current workload at time t (contributed by those jobs having priority no less than class k) just before this class k job arrives, $V_j(A_j(T_k(t) + t)) - V_j(A_j(t))$ is the workload of class j jobs that arrive after time t and before the completion of this class k job, and $V_k(A_k(t)) - V_k(A_k(t) - 1)$ is the service time required for this job. The summation term in (9.9) is the total amount of work carried by those higher priority jobs that have arrived during the sojourn time of the job in question (the class k job).

Note the workload process in (9.4); under the work-conserving (i.e., non-idling) condition, $Y_\ell(\cdot)$ can increase at time t only when $Z_\ell(t) = 0$. Hence, the pair (Z_ℓ, Y_ℓ) jointly satisfies the one-dimensional reflection mapping theorem, which yields

$$Y_\ell(t) = \sup_{0 \leq s \leq t} \left[s - \sum_{i=1}^{\ell} \sum_{k \in g_i} V_k(A_k(t)) \right]. \tag{9.10}$$

We assume that there exist a long-run average arrival rate and an average service time, namely,

$$\frac{A(t)}{t} \to \lambda \qquad \text{as } t \to \infty,$$

$$\frac{V(n)}{n} \to m \qquad \text{as } n \to \infty.$$

We shall call λ_k, the kth coordinate of λ, the (exogenous) arrival rate of a class k job, and call m_k, the kth coordinate of m, the average service time of a class k job (alternatively the mean service time of class k job when $v_k(n)$ has the same finite mean for all $n \geq 0$). Define

$$\beta_\ell = \sum_{k \in g_\ell} \lambda_k m_k, \qquad \rho_\ell = \sum_{i=1}^{\ell} \beta_i, \qquad \text{and} \qquad \rho \equiv \beta_L, \tag{9.11}$$

Note that β_ℓ is the aggregated traffic intensity for job classes in g_ℓ, ρ_ℓ is the aggregated traffic intensity for all classes in $g_1 \cup \cdots \cup g_\ell$ ($\rho_0 \equiv 0$), and ρ is the traffic intensity of the service station.

We shall assume that the traffic intensity ρ is less than or equal to 1 throughout this chapter. To avoid dealing with some trivial cases, we assume that $\lambda > 0$ and $m > 0$. (Otherwise, $\rho \leq 1$ may not imply $\rho_\ell < 1$ for $\ell < L$.)

9.2   FLIL: The Single Station Case

For any two stochastic processes $X, \bar{X} \in \mathcal{D}^K$, we write

$$X(t) \overset{\ell i\ell}{\approx} \bar{X}(t), \qquad \text{or equivalently} \qquad X \overset{\ell i\ell}{\approx} \bar{X}$$

if as $T \to \infty$,

$$\|X - \bar{X}\|_T \overset{a.s.}{=} O(\sqrt{T \log \log T}).$$

A stochastic process X is said to have an FLIL approximation $\bar{X}(t) = at$ for some (deterministic) constant $a \in \Re^K$, if

$$X(t) \overset{\ell i\ell}{\approx} at.$$

The key result of this section is to show that if the primitive data (the input process) have FLIL approximations, then the departure process (the output process) and the key performance measures of the queue also have FLIL approximations. To this end, assume that all the primitive data, the exogenous arrival process and the service process have FLIL approximations: For $k \in \mathcal{K}$,

$$A_k(t) \overset{\ell i \ell}{\approx} \lambda_k t, \tag{9.12}$$

$$V_k(t) \overset{\ell i \ell}{\approx} m_k t, \tag{9.13}$$

By Theorem 5.13, a sufficient condition for the above approximations is that for each $k \in \mathcal{K}$, the (exogenous) interarrival time sequence u_k and the service time sequence v_k are i.i.d. sequences with finite variances. The main results follow.

Theorem 9.1 Suppose that the FLIL assumptions (9.12)–(9.13) hold. Assume that the traffic intensity ρ is less than or equal to 1. Then for all $\ell \in \mathcal{L}$ and $k \in \mathcal{K}$,

$$(Z_\ell(t), W_k(t), Q_k(t), S_k(t)) \overset{\ell i \ell}{\approx} (0, 0, 0, 0), \tag{9.14}$$

$$(D_k(t), Y_\ell(t), \tau_\ell(t)) \overset{\ell i \ell}{\approx} (\lambda_k t, (1 - \rho_\ell)t, t). \tag{9.15}$$

The following two lemmas are useful in proving the above theorem.

Lemma 9.2   (i) If $Y(t) \overset{\ell i \ell}{\approx} at$, then there exist positive constants b and T_0 such that $|Y(t)| \leq bt$ for $t \geq T_0$ with probability one.

(ii) Suppose that $X(t) \overset{\ell i \ell}{\approx} \bar{X}(t)$, and suppose that there exist positive constants b and T_0 such that $Y(t) \leq bt$ for $t \geq T_0$ with probability one. Then

$$X(Y(t)) \overset{\ell i \ell}{\approx} \bar{X}(Y(t)).$$

(iii) Let $X \in \mathcal{D}_0^K$, and

$$Y(t) = \sup_{0 \leq u \leq t} [-X(u)]^+ \quad \text{and} \quad Z(t) = X(t) + Y(t).$$

Suppose that

$$X(t) \overset{\ell i \ell}{\approx} \bar{X}(t).$$

Then

$$Y(t) \overset{\ell i \ell}{\approx} \bar{Y}(t) \quad \text{and} \quad Z(t) \overset{\ell i \ell}{\approx} \bar{Z}(t),$$

where

$$\bar{Y}(t) = \sup_{0 \leq u \leq t} [-\bar{X}(u)]^+ \quad \text{and} \quad \bar{Z}(t) = \bar{X}(t) + \bar{Y}(t).$$

In particular, if $\bar{X}(t) = at$ for some $a \in \Re^K$, then $\bar{Y}(t) = [-a]^+t$ and $\bar{Z}(t) = a^+t$.

Proof. The lemma follows clearly from the definition of $\overset{\ell i \ell}{\approx}$ and the Lipschitz continuity of the one dimensional reflection mapping. □

Lemma 9.3 Suppose that the FLIL assumptions (9.12)–(9.13) hold. Then for $k \in \mathcal{K}$,

$$\nu_k(t) \overset{\ell i \ell}{\approx} 0,$$

$$\eta_k(t) \overset{\ell i \ell}{\approx} t.$$

If we further assume that $\rho \leq 1$, then there exist positive constants b and T_0 such that $\mathcal{T}_k(t) \leq bt$ for $t \geq T_0$.

Proof. From (9.6), we have for $k \in \mathcal{K}$,

$$
\begin{aligned}
\|\nu_k(t)\|_T &\leq \sup_{0 \leq t \leq T} \left\{ \max_{1 \leq n \leq A_k(t)} v_k(n) \right\} \\
&= \sup_{0 \leq t \leq T} v_k(A_k(t)) \\
&\leq \sup_{0 \leq t \leq T} \{V_k(A_k(t)) - V_k(A_k(t) - 1)\} \\
&\leq \sup_{0 \leq t \leq T} |V_k(A_k(t)) - m_k A_k(t)| \\
&\quad + \sup_{0 \leq t \leq T} |V_k(A_k(t) - 1) - m_k(A_k(t) - 1)| + m_k \\
&\overset{\ell i \ell}{\approx} O(\sqrt{T \log \log T}),
\end{aligned}
$$

where Lemma 9.2 (ii) is used to obtain the last inequality. The above proves $\nu_k(t) \overset{\ell i \ell}{\approx} 0$. In view of (9.7), a similar argument would prove $\eta_k(t) \overset{\ell i \ell}{\approx} t$.

Finally, we establish $\mathcal{T}_k(t) \leq bt$ for some $t \geq T_0$. It follows from the assumptions (9.12) and (9.13) and Lemma 9.2 (ii) that $V_k(A_k(t)) \overset{\ell i \ell}{\approx} \lambda_k m_k t$. Hence, there exist positive constants a and T_1 such that with probability one,

$$V_k(A_k(t)) \leq \lambda_k m_k t + a\sqrt{t \log \log t} \qquad \text{for } t \geq T_1.$$

From (9.9), we deduce that with probability one,

$$
\begin{aligned}
\mathcal{T}_k(t) &\leq Z_{\pi(k)}(t-) + \rho_{\pi(k)-1}\mathcal{T}_k(t) + \rho_{\pi(k)-1}t \\
&\quad + a \sum_{\ell=1}^{\pi(k)-1} \sum_{j \in g_\ell} \sqrt{(\mathcal{T}_k(t) + t) \log \log(\mathcal{T}_k(t) + t)} \\
&\quad + V_k(A_k(t)) - V_k(A_k(t) - 1) - \sum_{\ell=1}^{\pi(k)-1} \sum_{k \in g_\ell} V_k(A_k(t)),
\end{aligned}
$$

for $t \geq T_1$. Then for any $\epsilon > 0$, there exists a $T_1' \geq T_1$ such that the above inequality implies that with probability one

$$(1 - \rho_{\pi(k)-1} - \epsilon)\mathcal{T}_k(t) \leq Z_{\pi(k)}(t) + \rho_{\pi(k)-1}t$$
$$+ a \sum_{i=1}^{\pi(k)-1} \sum_{j \in g_i} \sqrt{t \log \log t} + V_k(A_k(t))$$

for $t \geq T_1'$. (See Exercise 2.) Since we assume that $\rho \leq 1$, we know that $\rho_{\pi(k)-1} < 1$ and hence $1 - \rho_{\pi(k)-1} - \epsilon > 0$ for $\epsilon > 0$ small enough. The above inequality, together with Lemma 9.2 (i) and the FLIL approximation for Z_ℓ in Theorem 9.1, yield the desired result. Even though the proof of Theorem 9.1 (which is to be provided below) makes use of this lemma, the proof of the FLIL approximation for Z_ℓ in Theorem 9.1 does not depend on this lemma. Therefore, the proof is complete.   □

Proof (of Theorem 9.1). By Lemma 9.2 (i)–(ii), we have

$$\sum_{i=1}^{\ell} \sum_{k \in g_i} V_k(A_k(t)) - t \stackrel{\ell i \ell}{\approx} \sum_{i=1}^{\ell} \sum_{k \in g_i} \lambda_k m_k t - t = (\rho_\ell - 1)t.$$

Then by Lemma 9.2 (iii) and in view of (9.4) and (9.10), we establish the FLIL approximations for Z_ℓ in (9.14) and for Y_ℓ in (9.15).

By Lemma 9.3, in view of (9.2) and (9.5), we have

$$\sum_{i \in g_\ell} W_i(t) = \sum_{i \in g_\ell} [V_i(A_i(t)) - V_i(D_i(t)) - \nu_i(t)]$$
$$\stackrel{\ell i \ell}{\approx} \sum_{i \in g_\ell} [m_i \lambda_i t - m_i \lambda_i \tau_\ell(t)] = \beta_\ell(t - \tau_\ell(t)),$$

where we also used Lemma 9.2 and the fact that $0 \leq \tau(t) \leq t$. The above, in view of (9.4), can be rewritten as

$$Z_\ell(t) - Z_{\ell-1}(t) \stackrel{\ell i \ell}{\approx} \beta_\ell(t - \tau_\ell(t)),$$

which combined with the FLIL approximation for Z implies the FLIL approximation for τ_ℓ in (9.15).

For $k \in \mathcal{K}$, let $\ell = \pi(k)$; the FLIL approximation for the departure process D_k in (9.15) follows from

$$D_k(t) = A_k(\tau_\ell(t)) \stackrel{\ell i \ell}{\approx} \lambda_k \tau_\ell(t) \stackrel{\ell i \ell}{\approx} \lambda_k t.$$

Similarly, the FLIL approximation for the queue length process Q_k in (9.14) is proved by observing that

$$Q_k(t) = A_k(t) - D_k(t) \stackrel{\ell i \ell}{\approx} \lambda_k t - \lambda_k t = 0,$$

and the FLIL approximation for the workload process W_k in (9.14) is proved by observing that

$$W_k(t) = V_k(A_k(t)) - V_k(D_k(t)) - \nu_k(t)$$
$$\overset{\ell i \ell}{\approx} m_k[A_k(t) - D_k(t)] \overset{\ell i \ell}{\approx} 0.$$

Finally, we establish the FLIL approximation for the sojourn time process S_k in (9.14). Noting (9.9) and Lemmas 9.2 and 9.3, we have

$$\mathcal{T}_k(t) \overset{\ell i \ell}{\approx} \sum_{i=1}^{\pi(k)-1} \sum_{j \in g_i} [\lambda_j m_j (\mathcal{T}_k(t) + t) - \lambda_j m_j t]$$
$$= \rho_{\pi(k)-1} \mathcal{T}_k(t),$$

implying $\mathcal{T}_k(t) \overset{\ell i \ell}{\approx} 0$ since $\rho_{\pi(k)-1} < 1$. Since $S_k(t) = \mathcal{T}_k(\eta_k(t))$, the above combined with Lemma 9.3 yields the FLIL approximation in (9.14).   □

9.3 Strong Approximation: The Single Station Case

The key result of this section is to show that if the primitive data of the queue have strong approximations, then the performance measures (such as the workload process, the queue length process and the sojourn time process) and the output process (namely the departure process) also have strong approximations.

To describe the main result, we first formally define what we mean by a process having a strong approximation. A function $x \in \mathcal{D}^K$ is said to be *strong continuous* with degree r, or r-strong continuous, for some $r \in (2, 4)$, if

$$\sup_{\substack{0 \le u,v \le T \\ |u-v| \le h(T)}} \|x(u) - x(v)\| = o(T^{1/r}), \quad \text{as } T \to \infty, \tag{9.16}$$

where $h(T) \equiv \sqrt{T \log \log T}$, and it is simply said to be strong continuous if it is r-strong continuous for all $r \in (2, 4)$. (We note that an r-strong continuous function may not be continuous.) It follows from Lemma 6.21 that with probability one, the sample path of a standard Brownian motion is strong continuous. A stochastic process $X = \{X(t), t \ge 0\}$ in $\mathcal{D}^K$ is said to be an r-strong continuous process for some $r \in (2, 4)$, if with probability one, the sample path of this version is r-strong continuous on a probability space. For simplicity, if a version of a process is r-strong continuous, then we assume throughout this chapter that the process under consideration is this version. A stochastic process is simply said to be strong continuous if it is r-strong continuous for all $r \in (2,4)$. We say a stochastic process X has *a strong approximation* if for some $r \in (2,4)$, there exists a probability space on which a version of X (for simplicity we still write it as X) and an r-strong continuous stochastic process $\hat{X}$ are defined such that

$$\sup_{0 \le t \le T} |X(t) - mt - \hat{X}(t)| \overset{\text{a.s.}}{=} o(T^{1/r}),$$

where m is a (deterministic) constant. When the above equality holds, we also say that X has a strong approximation $\tilde{X} = \{\tilde{X}(t), t \ge 0\}$ with $\tilde{X}(t) = mt + \hat{X}(t)$. For any two processes $X, \tilde{X} \in \mathcal{D}^K$, if for some $r \in (2,4)$,

$$\sup_{0 \le t \le T} |X(t) - \tilde{X}(t)| \overset{\text{a.s.}}{=} o(T^{1/r}),$$

then we write

$$X(t) \overset{r}{\approx} \tilde{X}(t), \qquad \text{or equivalently,} \qquad X \overset{r}{\approx} \tilde{X}.$$

The strong approximation assumptions for the primitive data are that processes A_k, S_k and V_k are defined on an appropriate probability space such that for some $r \in (2,4)$ and for $k \in \mathcal{K}$,

$$A_k(t) \overset{r}{\approx} \tilde{A}_k(t) \equiv \lambda_k t + \hat{A}_k(t), \tag{9.17}$$

$$V_k(t) \overset{r}{\approx} \tilde{V}_k(t) \equiv m_k t + \hat{V}_k(t), \tag{9.18}$$

where $\lambda_k \ge 0$, $m_k \ge 0$, and $\hat{A}_k$ and $\hat{V}_k$ are r-strong continuous. Note in particular that we need not assume that $\hat{A}_k$ and $\hat{V}_k$ are Brownian motions. On the other hand, if we assume that the sequence of interarrival times u_k and the sequence of service times v_k are mutually independent non-negative i.i.d. sequences having finite rth moment with $r \in (2,4)$, then by Theorem 5.14 we can have (9.17)–(9.18), with

$$\hat{A}_k(t) = \lambda_k^{1/2} c_{0,k} B_{0,k}(t), \tag{9.19}$$

$$\hat{V}_k(t) = m_k c_k B_{1,k}(t), \tag{9.20}$$

where $\lambda_k = 1/E(u_k(n))$, $m_k = E(v_k(n))$, $c_{0,k}$ and c_k are respectively the coefficients of variations of $u_k(n)$ and $v_k(n)$, and $B_{0,k}(t)$ and $B_{1,k}(t)$, $k \in \mathcal{K}$, are mutually independent standard Brownian motions.

Throughout the chapter we shall assume that $r \in (2,4)$.

Theorem 9.4 Suppose that the strong approximation assumptions (9.17)–(9.18) hold with $\hat{A}_k$ and $\hat{S}_k$ being r-strong continuous for some $r \in (2,4)$. Assume that the traffic intensity satisfies $\rho \le 1$. Then for $\ell \in \mathcal{L}$ and $k \in \mathcal{K}$,

$$(Z_\ell, S_k, D_k, Q_k, W_k) \overset{r}{\approx} (\tilde{Z}_\ell, \tilde{S}_k, \tilde{D}_k, \tilde{Q}_k, \tilde{W}_k), \tag{9.21}$$

where

$$\tilde{Z}_\ell(t) = \tilde{N}_\ell(t) + \tilde{Y}_\ell(t), \tag{9.22}$$

$$\tilde{N}_\ell(t) = (\rho_\ell - 1)t + \sum_{i=1}^{\ell} \sum_{k \in g_i} [m_k \hat{A}_k(t) + \hat{V}_k(\lambda_k t)], \tag{9.23}$$

$$\tilde{Y}_\ell(t) = \sup_{0 \le s \le t} [-\tilde{N}_\ell(s)]^+, \tag{9.24}$$

$$\tilde{S}_k(t) = \frac{\tilde{Z}_{\pi(k)}(t)}{1 - \rho_{\pi(k)-1}}. \tag{9.25}$$

$$\tilde{D}_k(t) = \lambda_k t + \hat{A}_k(t) - \frac{\lambda_k}{\beta_{\pi(k)}}[\tilde{Z}_{\pi(k)}(t) - \tilde{Z}_{\pi(k)-1}(t)], \tag{9.26}$$

$$\tilde{Q}_k(t) = \frac{\lambda_k}{\beta_{\pi(k)}}[\tilde{Z}_{\pi(k)}(t) - \tilde{Z}_{\pi(k)-1}(t)], \tag{9.27}$$

$$\tilde{W}_k(t) = \frac{\lambda_k m_k}{\beta_{\pi(k)}}[\tilde{Z}_{\pi(k)}(t) - \tilde{Z}_{\pi(k)-1}(t)] = m_k \tilde{Q}_k(t), \tag{9.28}$$

and $\tilde{Z}_\ell$ is r-strong continuous.

Remark 9.5 Since $\tilde{Z}_\ell$ is an r-strong continuous process, $\tilde{S}_k$, $\tilde{Q}_k$, and $\tilde{W}_k$ are r-strong continuous. In particular, let

$$\hat{D}_k(t) = \hat{A}_k(t) - \frac{\lambda_k}{\beta_{\pi(k)}}[\tilde{Z}_{\pi(k)}(t) - \tilde{Z}_{\pi(k)-1}(t)];$$

then $\hat{D}_k$ is r-strong continuous, and the departure process D_k has the strong approximation

$$D_k(t) \stackrel{r}{\approx} \lambda_k t + \hat{D}_k(t).$$

Note that in a feedforward network, the departure process (or a probabilistic split of the departure process) may be the arrival process to a downstream station; hence, the above property is essential in extending the strong approximation of the single station queue to the network case.

Remark 9.6 The second equality in (9.28) is Little's law for the strong approximation limits of the workload and the queue-length process.

Remark 9.7 The strong approximation of the sojourn time in (9.25) should be better interpreted as the strong approximation to the *sojourn queue time* which is the time between the arrival of the job and the time that it just begins service. As we can see from the proof of the sojourn time approximation, we approximate $V_k(A_k(t)) - V_k(A_k(t) - 1)$, the service time, by zero. If we replace $V_k(A_k(t)) - V_k(A_k(t) - 1)$ by zero in (9.9), then the new T_k is exactly the sojourn queue time. For a single-class single-station queue, the sojourn time of a job equals its sojourn queue time plus its service time. For preemptive priority queueing networks, even if a job is in service, it may well be interrupted by another arriving job with higher priority. Hence, in general, the sojourn time of a job should be longer than or equal to its sojourn queue time plus its service time. Our numeric examples suggest that approximating the service time by its mean would yield an improved strong approximation for the sojourn time (9.25),

$$\tilde{S}_k(t) = \frac{\tilde{Z}_{\pi(k)}(t) + m_k}{1 - \rho_{\pi(k)-1}}. \tag{9.29}$$

(In fact, our numeric experiments involve estimating only the mean sojourn time. To approximate the sojourn time *distribution*, it may be better to replace m_k by a random variable that has the same distribution as the service time distribution of a class k job and is independent of $\tilde{Z}_{\pi(k)}$.) For an M/G/1 preemptive priority queue, the approximated steady-state mean sojourn time is

$$\mathsf{E}\mathcal{S}_k \quad = \quad \frac{1}{1 - \rho_{\pi(k)-1}}\Big(\frac{\sum_{i=1}^{\pi(k)} m_i^2(1 + b_i^2)}{2(1 - \rho_{\pi(k)})} + m_k\Big), \tag{9.30}$$

which is the same as the exact mean sojourn time, see (3.39) in Kleinrock [4].

In order to prove the theorem, we establish some properties that relate to the FLIL approximation, strong continuity, and strong approximation.

Lemma 9.8   (i) A Wiener process (i.e., a standard Brownian motion) is a strong continuous process.

(ii) If a process has a strong approximation, then it must have a FLIL approximation. Specifically, if

$$X(t) \overset{r}{\approx} mt + \hat{X}(t)$$

with $\hat{X} = \{\hat{X}(t), t \geq 0\}$ being r-strong continuous, then we must have

$$X(t) \overset{\ell i \ell}{\approx} mt.$$

(iii) A linear combination of r-strong continuous functions is r-strong continuous, and a (deterministic) linear combination of r-strong continuous processes is r-strong continuous.

(iv) Suppose that $\tau = \{\tau(t), t \geq 0\}$ is a process with $\tau(t) \in \Re_+$ for all $t \geq 0$ having an FLIL approximation:

$$\tau(t) \overset{\ell i \ell}{\approx} \alpha t.$$

If $X = \{X(t), t \geq 0\}$ is an r-strong continuous process, then

$$X(\tau(t)) \overset{\ell i \ell}{\approx} X(\alpha t).$$

If $X = \{X(t), t \geq 0\}$ has a strong approximation,

$$X(t) \overset{r}{\approx} at + \hat{X}(t)$$

(with $\hat{X}$ being r-strong continuous), then

$$X(\tau(t)) \overset{r}{\approx} a\tau(t) + \hat{X}(\tau(t)) \overset{r}{\approx} a\tau(t) + \hat{X}(\alpha t).$$

(v) Assume that both processes X and Y have the strong approximations:

$$X(t) \overset{r}{\approx} at + \hat{X}(t),$$
$$Y(t) \overset{r}{\approx} bt + \hat{Y}(t),$$

with the range of $Y(t)$ in $\Re_+$ for all $t \geq 0$. Then,

$$X(Y(t)) \overset{r}{\approx} abt + [a\hat{Y}(t) + \hat{X}(bt)].$$

(vi) Let $X \in \mathcal{D}_0^K$,

$$Y(t) = \sup_{0 \leq u \leq t} [-X(u)]^+, \qquad \text{and} \qquad Z(t) = X(t) + Y(t).$$

Suppose that

$$X(t) \overset{r}{\approx} \tilde{X}(t),$$

with $\tilde{X}(0) \geq 0$. Then

$$Y(t) \overset{r}{\approx} \tilde{Y}(t) \qquad \text{and} \qquad Z(t) \overset{r}{\approx} \tilde{Z}(t),$$

where

$$\tilde{Y}(t) = \sup_{0 \leq u \leq t} [-\tilde{X}(u)]^+ \qquad \text{and} \qquad \tilde{Z}(t) = \tilde{X}(t) + \tilde{Y}(t).$$

Proof. Part (i) follows from Lemma 6.21, and the parts (ii)–(iv) clearly follow from the definitions of the strong continuity, the FLIL approximations, and the strong approximations. For (v), we have

$$X(Y(t)) \overset{r}{\approx} aY(t) + \hat{X}(Y(t))$$
$$\overset{r}{\approx} a[bt + \hat{Y}(t)] + \hat{X}(bt),$$

where we used (ii) and (iv) of this lemma. The last part (vi) follows from the Lipschitz continuity of the one-dimensional reflection mapping.   □

Lemma 9.9 Suppose that the strong approximations (9.17) and (9.18) hold. Then for $k = 1, \ldots, K$,

$$\nu_k(t) \overset{r}{\approx} 0,$$
$$\eta_k(t) \overset{r}{\approx} t.$$

The proof of this lemma is similar to the proof of Lemma 9.3, and is left as an exercise.

Proof (of Theorem 9.4). First, by Lemma 9.8 (ii), the strong approximation assumptions (9.17)–(9.18) imply the FLIL assumptions (9.12)–(9.13); hence, Theorem 9.1 prevails. In the rest of the proof we shall repeatedly use Lemma 9.8 and the results of Theorem 9.1 without explicitly referring to them.

We rewrite the net-put process as

$$N_\ell(t) = \sum_{i=1}^\ell \sum_{j \in g_i} V_j(A_j(t)) - t$$
$$\overset{r}{\approx} \sum_{i=1}^\ell \sum_{j \in g_i} [\lambda_j m_j t + m_j \hat{A}_j(t) + \hat{V}_j(\lambda_j t)] - t = \tilde{N}_\ell(t).$$

Then the strong approximation for the workload process Z_ℓ in (9.21) follows from Lemma 9.8 (vi). That the process $\tilde{Z}_\ell$ is r-continuous is stated as Lemma 9.10 below (with the proof left as an exercise).

In view of (9.2) and (9.5), we have

$$\sum_{i \in g_\ell} W_i(t) = \sum_{i \in g_\ell} \left\{ [V_i(A_i(t)) - V_i(A_i(\tau(t)))] - \nu_i(t) \right\}$$
$$\overset{r}{\approx} \sum_{i \in g_\ell} \left\{ [\lambda m_i t + m_i \hat{A}_i(t) + \hat{V}_i(\lambda_i t)] \right.$$
$$\left. - [\lambda_i m_i \tau_\ell(t) + m_i \hat{A}_i(t) + \hat{V}_i(\lambda_i t)] \right\}$$
$$= \beta_\ell [t - \tau_\ell(t)],$$

where we also used Lemma 9.9. In view of the strong approximation for Z_ℓ, the above implies

$$\beta_\ell [t - \tau_\ell(t)] \overset{r}{\approx} \tilde{Z}_\ell(t) - \tilde{Z}_{\ell-1}(t).$$

Now fix $k \in \mathcal{K}$ and $\ell = \pi(k)$. Using the above, we have

$$
\begin{aligned}
D_k(t) &= A_k(\tau_\ell(t)) \\
&\overset{r}{\approx} \lambda_k \tau_\ell(t) + \hat{A}_k(t) \\
&\overset{r}{\approx} \lambda_k t + \hat{A}_k(t) - \frac{\lambda_k}{\beta_\ell}[\tilde{Z}_\ell(t) - \tilde{Z}_{\ell-1}(t)];
\end{aligned}
$$

this proves the strong approximation for the departure process D_k.

Similarly, we prove the strong approximation for the queue-length process Q_k by observing that

$$
\begin{aligned}
Q_k(t) &= A_k(t) - D_k(t) \\
&\overset{r}{\approx} [\lambda_k t + \hat{A}_k(t)] - \tilde{D}_k(t) = \tilde{Q}_k(t),
\end{aligned}
$$

and the strong approximation for the workload process W_k by observing that

$$
\begin{aligned}
W_k(t) &= V_k(A_k(t)) - V_k(D_k(t)) - \nu_k(t) \\
&\overset{r}{\approx} [\lambda_k m_k t + m_k \hat{A}_k(t) + \hat{V}_k(\lambda_k(t))] - [m_k \tilde{D}_k(t) + \hat{V}_k(\lambda_k t)] \\
&= m_k \tilde{Q}_k(t).
\end{aligned}
$$

Finally, we establish the strong approximation for the sojourn time process $\mathcal{S}_k$. Note that $\mathcal{T}_k(t) \overset{\ell i \ell}{\approx} 0$, as shown in the proof of Theorem 9.1; we have from (9.9),

$$
\begin{aligned}
\mathcal{T}_k(t) \overset{r}{\approx} \tilde{Z}_{\pi(k)}(t) + \sum_{i=1}^{\pi(k)-1} \sum_{j \in g_i} & \Big\{ [\lambda_j m_j(\mathcal{T}_k(t) + t) + m_j \hat{A}_j(t) + \hat{V}_j(\lambda_j t)] \\
& - [\lambda_j m_j t + m_j \hat{A}_j(t) + \hat{V}_j(\lambda_j t)] \Big\} \\
= \tilde{Z}_{\pi(k)}(t) + \rho_{\pi(k)-1} \mathcal{T}_k(t);
\end{aligned}
$$

hence,

$$
\mathcal{T}_k(t) \overset{r}{\approx} \frac{1}{1 - \rho_{\pi(k)-1}} \tilde{Z}_{\pi(k)}(t).
$$

Recall that $\mathcal{S}_k(t) = \mathcal{T}_k(\eta_k(t))$; the above, together with Lemmas 9.8 and 9.9, implies the strong approximation for the sojourn time process with the limit given by (9.25). $\qquad \square$

Lemma 9.10 Suppose that $x \in \mathcal{D}_0$ is an r-strong continuous function $(2 < r < 4)$. Let

$$
f(t) \equiv \sup_{0 \leq s \leq t} [-\theta s - x(s)]^+ - [-\theta]^+ t, \tag{9.31}
$$

Then f is also an r-strong continuous function.

9.4 The Feedforward Network Model

9.4.1 Primitive Data and Assumptions

The queueing network consists of a set of J service stations, indexed by $j = 1, \ldots, J$, serving K classes of jobs, indexed by $k = 1, \ldots, K$. There are L priority groups, denoted by g_ℓ, $\ell = 1, \ldots, L$. Let $\mathcal{K} = \{1, \ldots, K\}$, $\mathcal{J} = \{1, \ldots, J\}$ and $\mathcal{L} = \{1, \ldots, L\}$. Let π be a many-to-one mapping from $\mathcal{K}$ to $\mathcal{L}$, and let σ be a many-to-one mapping from $\mathcal{L}$ to $\mathcal{J}$. Specifically, job class k belongs to the priority group $g_{\pi(k)}$, and jobs from group g_ℓ are served exclusively at station $\sigma(\ell)$. (For simplicity, we define $\pi(0) = 0$ and $\sigma(0) = 0$.) That is, jobs of different classes in the same group are served at the same service station. The composition $\sigma \circ \pi$ is a many-to-one mapping from $\mathcal{K}$ to $\mathcal{J}$. By a possible renumbering, we assume without loss of generality that if $\ell < m$, then jobs in group g_ℓ have a preemptive priority over jobs in group g_m ($\ell, m = 1, \ldots, L$). Jobs within a group are served in the order of arrival. The network is a feedforward queueing network in the sense that any job at station i can turn into another class at station j only if $j > i$ ($i, j = 1, \ldots, J$).

To facilitate our analysis, we use the following convention on indexing priority groups and job classes:

- Every station has at least one priority group, and every priority group has at least one job class; hence, necessarily, $K \geq L \geq J$.

- For $k, j = 1, \ldots, K$, $\pi(k) \leq \pi(j)$ if $k < j$. Therefore, jobs of class 1 must be in group 1, and jobs of class K must be in group L.

- For $\ell, m = 1, \ldots, L$, $\sigma(\ell) \leq \sigma(m)$ if $\ell < m$. Thus, jobs from group 1 must be served at station 1, and jobs from group L must be served at station J.

For any $k \in \mathcal{K}$, let $h(k) = \pi(k) - 1$ if $\sigma(\pi(k) - 1) = \sigma(\pi(k))$ (i.e., $h(k)$ is the index of the group with the next higher priority at the same station as class k), and $h(k) = 0$ otherwise (i.e., class k has the highest priority at its station).

Introduce two $L \times K$ matrices: a group constituent matrix G and a higher priority group constituent matrix H. The (ℓ, k)th component of G satisfies $G_{\ell k} = 1$ if $\pi(k) = \ell$ (or equivalently $k \in g_\ell$, i.e., class k is in group ℓ), and $G_{\ell k} = 0$ otherwise. The (ℓ, k)th component of H satisfies $H_{\ell k} = 1$ if $\sigma(\ell) = \sigma(\pi(k))$ and $\pi(k) \leq \ell$ (i.e., class k is served at the same station as group ℓ and has a priority no less than that of group ℓ), and $H_{\ell k} = 0$ otherwise.

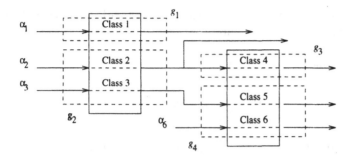

FIGURE 9.2. A two-station six-class feedforward network

To illustrate the notation, consider the network given by Figure 9.2. The network consists of $J = 2$ stations serving $K = 6$ classes of jobs with $L = 4$ priority groups. Job class 1 belongs to priority group 1; job classes 2 and 3 belong to priority group 2; job class 4 belongs to priority group 3; and job classes 5 and 6 belong to priority group 4. Priority groups 1 and 2 reside at station 1 and all the other groups reside at station 2. Then, for this network, π, g, and σ are given by

$$\pi(1) = 1, \quad \pi(2) = \pi(3) = 2,$$
$$\pi(4) = 3, \quad \pi(5) = \pi(6) = 4;$$
$$g_1 = \{1\}, \quad g_2 = \{2, 3\}, \quad g_3 = \{4\}, \quad g_4 = \{5, 6\},$$
$$\sigma(1) = \sigma(2) = 1, \quad \sigma(3) = \sigma(4) = 2.$$

The group constituent matrix G and the higher-priority group constituent matrix H, respectively, take the form

$$G = \begin{pmatrix} 1 & 0 & 0 & 0 & 0 & 0 \\ 0 & 1 & 1 & 0 & 0 & 0 \\ 0 & 0 & 0 & 1 & 0 & 0 \\ 0 & 0 & 0 & 0 & 1 & 1 \end{pmatrix}, \quad H = \begin{pmatrix} 1 & 0 & 0 & 0 & 0 & 0 \\ 1 & 1 & 1 & 0 & 0 & 0 \\ 0 & 0 & 0 & 1 & 0 & 0 \\ 0 & 0 & 0 & 1 & 1 & 1 \end{pmatrix}.$$

The queueing network is characterized by following primitive data: $2K$ sequences of nonnegative random variables $u_k = \{u_k(n), n \geq 1\}$ and $v_k = \{v_k(n), n \geq 1\}$ ($k = 1, \ldots, K$), and K sequences of K-dimensional vector $\phi^k = \{\phi^k(n), n \geq 1\}$ ($k = 1, \ldots, K$), all defined on the same probability space.

We assume that there is no job in the network at time $t = 0$. The random variable $u_k(1)$ is the time of the first exogenously arrived class k job, and $u_k(n)$, $n > 1$, is the time between the $(n-1)$-st and nth exogenous arrived class k jobs. The random variable $v_k(n)$ is the service time required for the nth class k job. The random variable ϕ^k describes the routing mechanism for class k jobs: The nth class k job after service completion turns into a class j job if $\phi^k(n) = e^j$, and leave the network if $\phi^k(n) = 0$. By our assumption on the feedforward structure, it is necessary that $\phi^k(n) \neq e^i$ for all $n \geq 1$ and $k, i = 1, \ldots, K$ with $\sigma(\pi(i)) \leq \sigma(\pi(k))$.

We introduce the summations

$$U_k(0) = 0, \quad U_k(n) = \sum_{m=1}^{n} u_k(m), n \geq 1, \qquad k = 1, \ldots, K,$$

$$V_k(0) = 0, \quad V_k(n) = \sum_{m=1}^{n} v_k(m), n \geq 1, \qquad k = 1, \ldots, K,$$

$$\Phi^k(0) = 0, \quad \Phi^k(n) = \sum_{m=1}^{n} \phi^k(m), n \geq 1, \qquad k = 1, \ldots, K.$$

Define the counting process

$$E_k(t) = \sup\{n \geq 0 : U_k(n) \leq t\}, \ k = 1, \ldots, K,$$

Let $U = (U_k)$, $V = (V_k)$, $\Phi = (\Phi^1, \ldots, \Phi^K)$, and $E = (E_k)$. We call E an exogenous arrival process, and Φ a routing sequence.

As in the single-station model above, we assume that there exist a long-run average arrival rate, an average service time, and a long-run average transition (routing) rate; namely,

$$\frac{E(t)}{t} \to \alpha \qquad \text{as } t \to \infty,$$

$$\frac{V(n)}{n} \to m \qquad \text{as } n \to \infty,$$

$$\frac{\Phi^k(n)}{n} \to P'_k \qquad \text{as } n \to \infty,$$

where P'_k is the kth row of a $K \times K$ matrix $P = (p_{kj})$. We shall call α_k, the kth coordinate of α, the (exogenous) arrival rate of the class k job, and call m_k, the kth coordinate of m, the average service time of a class k job (alternatively the mean service time of a class k job when $v_k(n)$ has the same finite mean for all $n \geq 1$). Call p_{kj}, the jth coordinate of P'_k (and the (k,j)th element of P), the average transition rate at which a class k job turns into a class j job after completing its service. When Φ^k is an i.i.d. summation, p_{kj} is the probability that a class k job turns into a class j job after its service completion. By our assumption on the routing sequence, it follows that matrix P is a strictly upper triangular matrix.

Let $\lambda = (I - P')^{-1}\alpha$, and its kth component, λ_k, is the long-run average arrival rate of class k jobs, $k = 1, \ldots, K$. Let $M = \text{diag}(m)$ and $\Lambda = \text{diag}(\lambda)$ be $K \times K$ diagonal matrices with kth diagonal elements m_k and λ_k, respectively. Let $\rho = HM\lambda$. Note that ρ is of dimension L; if g_ℓ is the lowest priority group at its station $\sigma(\ell)$, then ρ_ℓ is the traffic intensity at that station. For convenience, let $\rho_0 = 0$. For the network shown by Figure 9.2,

$$\rho_1 = \alpha_1 m_1, \qquad \rho_2 = \alpha_1 m_1 + \alpha_2 m_2 + \alpha_3 m_3,$$
$$\rho_3 = 0.5\alpha_2 m_4, \qquad \rho_4 = 0.5\alpha_2 m_4 + \alpha_3 m_5 + \alpha_6 m_6;$$

ρ_2 and ρ_4 are the traffic intensities at stations 1 and 2, respectively. We assume that $\lambda > 0$ and $m > 0$, and that the traffic intensity at all stations are no greater than one, and hence $\rho \leq e$.

9.4.2 Performance Measures and Dynamics

The performance measures of interest are the L-dimensional (*aggregated*) workload process $Z = (Z_\ell)$ with $Z_\ell = \{Z_\ell(t), t \geq 0\}$ ($\ell \in \mathcal{L}$), the K-dimensional workload process $W = (W_k)$ with $W_k = \{W_k(t), t \geq 0\}$ ($k \in \mathcal{K}$), the K-dimensional queue length process $Q = (Q_k)$ with $Q_k = \{Q_k(t), t \geq 0\}$ ($k \in \mathcal{K}$), and the L-dimensional cumulative idle time process $Y = (Y_\ell)$ with $Y_\ell = \{Y_\ell(t), t \geq 0\}$ ($\ell \in \mathcal{L}$). The process Z is nonnegative with $Z_\ell(t)$ indicating the total amount of immediate work for station $\sigma(\ell)$ embodied in jobs that are in groups g_1 to g_ℓ and that are either queued or in service at station $\sigma(\ell)$ at time t. The quantity $W_k(t)$ indicates the amount of work embodied in all class k jobs that are either queued or in service at time t. The quantity $Q_k(t)$ indicates the number of class k jobs in the network at time t. We assume that $Q(0) = 0$ and thus $Z(0) = 0$. The quantity $Y_\ell(t)$ indicates the cumulative amount of time that the server at station $\sigma(\ell)$ does not process jobs in groups g_1 to g_ℓ during $[0, t]$. (Note that if $\sigma(m) \neq \sigma(\ell)$, by definition the server at station $\sigma(\ell)$ would never serve jobs in group m.) It is clear that Y must be nondecreasing and $Y(0) = 0$.

We introduce some additional notation.

- $A_k(t)$ is the total number of class k jobs arrived at station $\sigma(\pi(k))$ during $[0, t]$ either exogenously or from other stations;

- $D_k(t)$ is the total number of service completions of class k jobs at station $\sigma(\pi(k))$ during $[0, t]$;

- $\tau_\ell(t)$ is the arrival time of the g_ℓ job that has most recently completed service at station $\sigma(\ell)$ ($\tau_\ell(t) \equiv 0$ if there have been no service completions for group ℓ).

- $S_k(t)$ is the sojourn time of class k jobs at time t at station $\sigma(\pi(k))$, denoting the time that will be spent at station $\sigma(\pi(k))$ by the first class k job that arrives at time greater than or equal to t.

- $\eta_k(t)$ is the time at which the first class k job arrives during $[t, \infty)$.

- $\mathcal{T}_k(t)$ is the time that a class k job would spent at station $\sigma(\pi(k))$ if it arrived at time t.

It is from the above definitions that we have the following dynamic relations:

$$Q(t) = A(t) - D(t), \tag{9.32}$$

$$W(t) = V(A(t)) - V(D(t)) - \nu(t), \tag{9.33}$$

$$Z(t) = HW(t), \tag{9.34}$$

$$Z(t) = HV(A(t)) - et + Y(t), \tag{9.35}$$

$$D(t) = A(G'\tau(t)), \tag{9.36}$$

$$A(t) = E(t) + \sum_{k=1}^{K} \Phi^k(D_k(t)), \tag{9.37}$$

$$S(t) = T(\eta(t)), \tag{9.38}$$

$$T_k(t) = Z_{\pi(k)}(t-) + \sum_{\substack{i:\sigma(\pi(i))=\sigma(\pi(k)) \\ \pi(i)<\pi(k)}} \left[V_i(A_i(t + T_k(t))) - V_i(A_i(t)) \right]$$
$$+ V_k(A_k(t)) - V_k(A_k(t) - 1), \tag{9.39}$$

where $\nu_k(t)$, the kth component of $\nu(t)$, is the partial service time (if any) that has been performed on a class k job during $(\tau_{\pi(k)}, t]$, which is dominated by an inequality similar to (9.6). For an understanding of the above relations it is helpful to compare them with the relations (9.1)–(9.9) for the single-station case. In particular, relation (9.35) is a workload flow balance relation (in terms of time) for service stations, and relation (9.32) is a job flow balance relation (in terms of number of jobs) for job classes. We shall assume that the work-conserving condition is in force. Hence, the pair (Z, Y) satisfies the reflection mapping relation, which implies that

$$Y(t) = \sup_{0 \le s \le t} [es - HV(A(s))]. \tag{9.40}$$

9.5   FLIL: The Network Case

This section generalizes the functional law of the iterated logarithm in Section 9.2 to the feedforward network as described in Section 9.4. The key result is that if the primitive data (the input process) have FLIL approximations, then the departure process (the output process) and the key performance measures of the queue also have FLIL approximations.

Theorem 9.11 Suppose that

$$(E(t), V(t), \Phi^1(t), \dots, \Phi^K(t)) \overset{\ell i \ell}{\approx} (\alpha t, mt, P_1' t, \dots, P_K' t).$$

Assume that the traffic intensity satisfies $\rho \le e$. Then

$$(Z(t), W(t), Q(t), S(t), D(t), \tau(t)) \overset{\ell i \ell}{\approx} (0, 0, 0, 0, \lambda t, et). \tag{9.41}$$

The proof of this theorem follows an induction argument (from stations 1 to J), and is left as an exercise. We note that it can also be proved that the FLIL approximations also hold for the following two processes:

$$A(t) \overset{\ell i \ell}{\approx} \lambda t, \tag{9.42}$$

$$\mathcal{T}(t) \overset{\ell i \ell}{\approx} 0. \tag{9.43}$$

9.6 Strong Approximation: The Network Case

This section generalizes the strong approximation in Section 9.3 to the feedforward network as described in Section 9.4. We assume that the primitive processes $(E, V, \Phi^1, \ldots, \Phi^K)$ have strong approximations, namely, we assume that they are defined on a probability space such that there exist $(K+2)$ K-dimensional r-continuous process $(\hat{E}, \hat{V}, \hat{\Phi}^1, \ldots, \hat{\Phi}^K)$ satisfying

$$E(t) \overset{r}{\approx} \alpha t + \hat{E}(t), \tag{9.44}$$

$$V(t) \overset{r}{\approx} mt + \hat{V}(t), \tag{9.45}$$

$$\Phi^k(t) \overset{r}{\approx} P_k' t + \hat{\Phi}^k(t), \qquad k = 1, \ldots, K, \tag{9.46}$$

for some $r \in (2, 4)$. Then we show that the strong approximation holds for all key performance measures and all departure processes of the network.

In (9.44)–(9.46), we note in particular that $(\hat{E}, \hat{V}, \hat{\Phi}^1, \ldots, \hat{\Phi}^K)$ may not be a Brownian motion, nor may its components $\hat{E}_k$, $\hat{V}_k$ and $\hat{\Phi}^k$, $k \in \mathcal{K}$, be mutually independent. However, suppose that u_k, v_k, and ϕ^k, $k = 1, \ldots, K$, are mutually independent i.i.d. sequences, and that u_k and v_k have finite moments of order $r \in (2, 4)$. Let $1/\alpha_k$ and m_k be the means, and $c_{0,k}$ and c_k be the coefficients of variations, of random variables, $u_k(1)$ and $v_k(1)$, respectively. Let $p_{kj} = P\{\phi^k(n) = e^j\}$, where e^j is the jth unit vector in $\Re^K$. Then by (a multidimensional generalization of) Theorem 5.14, the strong approximation assumptions (9.44)–(9.46) hold with $\hat{E}, \hat{V}, \hat{\Phi}^1, \ldots, \hat{\Phi}^K$ being mutually independent driftless Brownian motions. Their covariance matrices are respectively defined by

$$(\Gamma_E)_{i\ell} = \delta_{i\ell} \alpha_\ell c_{0,\ell}^2, \tag{9.47}$$

$$(\Gamma_V)_{i\ell} = \delta_{i\ell} m_\ell^2 c_\ell^2, \tag{9.48}$$

$$(\Gamma_\Phi^k)_{i\ell} = p_{ki}(\delta_{i\ell} - p_{k\ell}), \qquad k \in \mathcal{K}. \tag{9.49}$$

For the feedforward network (as described in Section 9.4), jobs can route from station i to station j only if $j > i$. Now we argue that given the strong approximations (9.44)–(9.46) for the primitive data, we could inductively apply Theorem 9.4 (the strong approximation theorem for a single station) to the network from stations 1 to station J. First, by Theorem 9.4 and Remark 9.5, the departure process of each job class from station 1 has a strong approximation; this, assumption (9.46) (that the routing sequence has a strong approximation), and Lemma 9.8 (v) imply that the arrival processes from station 1 to all downstream stations (if any) also have strong approximations. Since jobs arrive at station 2 either exogenously or from station 1, the total arrival process to station 2 for each job class must have a strong approximation as well. Hence, by applying Theorem 9.4 to station 2, we know in particular that the departure process of each job class from station 2 satisfies a strong approximation. Inductively, we can show that the departure process and the arrival process for each class in the network must have some strong approximations. Therefore, we could apply Theorems 9.4 to each station to obtain the strong approximations for all the performance measures, especially the workload process of each job class, the aggregated workload processes, the queue-length processes, and the sojourn time processes. The following theorem presents the strong approximations in a compact form.

Theorem 9.12 Suppose that the strong approximations (9.44)–(9.46) hold. Let

$$
\begin{aligned}
&\Delta = \Lambda G (H M \Lambda G')^{-1}, \\
&N = H M (I - P')^{-1} P' \Delta \qquad \text{and} \\
&R = (I + N)^{-1}.
\end{aligned}
$$

Then,

$$
(Z, Y, Q, W, S) \overset{r}{\approx} (\tilde{Z}, \tilde{Y}, \tilde{Q}, \tilde{W}, \tilde{S}), \tag{9.50}
$$

where

$$
\tilde{Q}(t) = \Delta \tilde{Z}(t) \quad \text{for } t \geq 0, \tag{9.51}
$$

$$
\tilde{W}(t) = M \Delta \tilde{Z}(t) = M \tilde{Q}(t), \tag{9.52}
$$

$$
\tilde{S}_k(t) = \frac{\tilde{Z}_{\pi(k)}(t) + m_k}{1 - \rho_{h(k)}}; \tag{9.53}
$$

and $(\tilde{Z}, \tilde{Y})$ are defined by

$$\tilde{Z}(t) = \theta t + \tilde{X}(t) + R\tilde{Y}(t) \geq 0 \text{ for } t \geq 0, \tag{9.54}$$

$$\theta = R(\rho - e), \tag{9.55}$$

$$\tilde{X}(t) = RH\Big[\hat{V}(\lambda t)$$

$$+ M(I - P')^{-1}[\hat{E}(t) + \sum_{k=1}^{K} \hat{\Phi}^k(\lambda_k t)]\Big]. \tag{9.56}$$

$$\tilde{Y}(\cdot) \text{ is continuous and nondecreasing with } \tilde{Y}(0) = 0, \tag{9.57}$$

$$\int_0^\infty \tilde{Z}_\ell(t)d\tilde{Y}_\ell(t) = 0 \text{ for } \ell = 1, \ldots, L, \tag{9.58}$$

Remark 9.13 It is shown in Lemma 9.17 below that the matrix Δ is well-defined. In fact, let $\beta = GM\lambda$; then the (k,ℓ)th element of the $K \times L$ matrix Δ is given by

$$\Delta_{k\ell} = \begin{cases} \lambda_k/\beta_\ell & \text{if } \ell = \pi(k), \\ -\lambda_k/\beta_{\ell+1} & \text{if } k \in g_{\ell+1} \text{ and } \sigma(\ell) = \sigma(\ell+1), \\ 0 & \text{otherwise.} \end{cases}$$

(The ℓth component of β, β_ℓ is the traffic intensity of priority group ℓ at station $\sigma(\ell)$.) It is also shown in Lemma 9.17 that the matrix N is well-defined and is strictly lower triangular. Hence, matrix R is also well-defined and is lower triangular.

Remark 9.14 Since matrix R is triangular, by inductively applying the one-dimensional reflection mapping, it is clear that for the given θ in (9.55) and $\tilde{X}$ in (9.56), relations (9.54), (9.57), and (9.58) uniquely determine the process $\tilde{Z}$ and $\tilde{Y}$. In particular, when the interarrival sequences u_k ($k = 1, \ldots, K$), the service sequences v_k ($k = 1, \ldots, K$) and the routing sequences ϕ^k ($k = 1, \ldots, K$) are mutually independent i.i.d. sequences, the process $\tilde{X}$ is a Brownian motion and the process $\tilde{Z}$ is a reflected Brownian motion with reflection matrix R. The covariance matrix of the Brownian motion $\tilde{X}$ in this case is

$$\Gamma = RH\left[\Gamma_V\Lambda + M(I - P')^{-1}\left[\Gamma_E + \sum_{k=1}^{K} \lambda_k\Gamma_\Phi^k\right](I - P)^{-1}M\right]H'R',$$

where Γ_E, Γ_V, and Γ_Φ are as given by (9.47)–(9.49). When the sequences u_k, v_k and ϕ^k, $k = 1, \ldots, K$, are i.i.d. but not mutually independent, $\tilde{X}$ given by (9.56) is still a Brownian motion but with its covariance matrix computed differently.

Remark 9.15 Also due to the fact that matrix R is triangular, it can be shown inductively that the process $\tilde{Z}$ is strong continuous.

Remark 9.16 The second equality in (9.52) is Little's law for the strong approximation limits.

Proof (of Theorem 9.12). In view of the previous discussion, we only need to show that the strong approximation limits in (9.50) are given by (9.51)–(9.58). Specifically, the starting points of our proof are the following results: the strong approximations (9.44)–(9.46), the strong approximations (9.50), and

$$A(t) \overset{r}{\approx} \lambda t + \hat{A}(t), \tag{9.59}$$

$$\nu(t) \overset{r}{\approx} 0, \tag{9.60}$$

$$\eta(t) \overset{r}{\approx} et, \tag{9.61}$$

hold. That the strong approximations in (9.50) and (9.59) hold for some limits $\tilde{Z}$, $\tilde{Y}$, $\tilde{Q}$, $\tilde{W}$, $\tilde{S}$ and $\hat{A}$ follows from an induction proof as outlined above (before the theorem). That (9.60) and (9.61) hold is an extension of Lemma 9.9. In addition, the strong approximation assumptions in (9.44)–(9.46) are clearly sufficient for the FLIL and hence Theorem 9.11 to hold; in particular, the FLIL approximations in (9.41)–(9.43) hold.

What remains now is to identify limits $\tilde{Z}$, $\tilde{Y}$, $\tilde{Q}$, $\tilde{W}$, $\tilde{S}$, and $\hat{A}$ and show that they satisfy (9.51)–(9.58). Below, we shall repeatedly use, without explicit reference, the FLIL approximations in Theorem 9.11, the strong approximations in (9.44)–(9.46), and Lemma 9.8.

First, note that $\lambda = \Lambda G' e$; from (9.33), (9.36), and (9.59)–(9.60), we have

$$
\begin{aligned}
W(t) &= V(A(t)) - V(D(t)) - \nu(t) \\
&\overset{r}{\approx} [MA(t) + \hat{V}(A(t))] - [MD(t) + \hat{V}(D(t))] \\
&\overset{r}{\approx} [M\lambda t + M\hat{A}(t) + \hat{V}(\lambda t)] - [MA(G'\tau(t)) + \hat{V}(\lambda t)] \\
&\overset{r}{\approx} [M\lambda t + M\hat{A}(t)] - [M\Lambda G'\tau(t) + M\hat{A}(G'\tau(t))] \\
&\overset{r}{\approx} M\Lambda G'(et - \tau(t));
\end{aligned}
$$

this, combined with (9.50), yields

$$\tilde{W}(t) \overset{r}{\approx} M\Lambda G'(et - \tau(t)). \tag{9.62}$$

From the above and (9.34), we have

$$\tilde{Z}(t) = H\tilde{W}(t) \overset{r}{\approx} HM\Lambda G'(et - \tau(t)); \tag{9.63}$$

substituting the above into (9.62) leads to

$$\tilde{W}(t) \overset{r}{\approx} M\Lambda G'(HM\Lambda G')^{-1}\tilde{Z}(t) = M\Delta\tilde{Z}(t);$$

this establishes the first equality in (9.52). We also note that the above, together with (9.62), implies

$$\Lambda G'(et - \tau(t)) \overset{r}{\approx} M^{-1}\tilde{W}(t) \overset{r}{\approx} \Delta\tilde{Z}(t). \tag{9.64}$$

In view of (9.32), (9.36), (9.63), and (9.64), the relation (9.51) and the second equation in (9.52) follow from

$$\begin{aligned}
Q(t) &= A(t) - D(t) \\
&\overset{r}{\approx} [\lambda t + \hat{A}(t)] - A(G'\tau(t)) \\
&\overset{r}{\approx} [\lambda t + \hat{A}(t)] - [\Lambda G'\tau(t) + \hat{A}(G'\tau(t))] \\
&\overset{r}{\approx} \Lambda G'(et - \tau(t)) \\
&\overset{r}{\approx} \Delta\tilde{Z}(t).
\end{aligned}$$

Next, in view of (9.36) and (9.37), we have

$$\begin{aligned}
A(t) &= E(t) + \sum_{k=1}^{K} \Phi^k(D_k(t)) \\
&\overset{r}{\approx} \alpha t + \hat{E}(t) + \sum_{k=1}^{K} [P'_k D_k(t) + \hat{\Phi}^k(D_k(t))] \\
&\overset{r}{\approx} \alpha t + \hat{E}(t) + P' A(G'\tau(t)) + \sum_{k=1}^{K} \hat{\Phi}^k(\lambda_k t) \\
&\overset{r}{\approx} \alpha t + \hat{E}(t) + \sum_{k=1}^{K} \hat{\Phi}^k(\lambda_k t) + P'\Lambda G'\tau(t) + P'\hat{A}(t).
\end{aligned}$$

Note that $A(t) \overset{r}{\approx} \lambda t + \hat{A}(t)$ and $\lambda = \alpha + P'\lambda = \alpha + P'\Lambda G'e$; the above leads to

$$\begin{aligned}
\hat{A}(t) &\overset{r}{\approx} (I - P')^{-1}\left[\hat{E}(t) + \sum_{k=1}^{K} \hat{\Phi}^k(\lambda_k t) - P'\Lambda G'(et - \tau(t))\right] \\
&\overset{r}{\approx} (I - P')^{-1}\left[\hat{E}(t) + \sum_{k=1}^{K} \hat{\Phi}^k(\lambda_k t) - P'\Delta\tilde{Z}(t)\right],
\end{aligned}$$

where the last approximation follows from (9.64). Using the above approximation we can rewrite (9.35) to obtain,

$$
\begin{aligned}
Z(t) &= HV(A(t)) - et + Y(t) \\
&\stackrel{r}{\approx} H[MA(t) + \hat{V}(A(t))] - et + \tilde{Y}(t) \\
&\stackrel{r}{\approx} HM\lambda t + HM\hat{A}(t) + H\hat{V}(\lambda t) - et + \tilde{Y}(t) \\
&\stackrel{r}{\approx} (\rho - e)t + H\left[\hat{V}(\lambda t) + M(I - P')^{-1}[\hat{E}(t) + \sum_{k=1}^{K} \hat{\Phi}^k(\lambda_k t)]\right] \\
&\quad - HM(I - P')^{-1}P'\Delta\tilde{Z}(t) + \tilde{Y}(t);
\end{aligned}
$$

this, together with (9.50), implies that

$$
\tilde{Z}(t) \stackrel{r}{\approx} \theta t + \tilde{X}(t) + R\tilde{Y}(t),
$$

with θ and $\tilde{X}$ as defined by (9.55) and (9.56), respectively. This establishes the relation (9.54). The relations (9.57) and (9.58) follow from the corresponding properties for the original processes Y and Z and the Lipschitz continuity of the reflection mapping; specifically, the first relation corresponds to the nondecreasing property of Y and the second relation corresponds to the work-conserving condition as stated by (9.40).

Finally, we establish (9.53). In view of (9.39) and (9.43), we have

$$
\begin{aligned}
T_k(t) &= Z_{\pi(k)}(t-) + \sum_{\substack{i:\sigma(\pi(i))=\sigma(\pi(k)) \\ \pi(i)<\pi(k)}} [V_i(A_i(t + T_k(t))) - V_i(A_i(t))] \\
&\quad + V_k(A_k(t)) - V_k(A_k(t) - 1) \\
&\stackrel{r}{\approx} Z_{\pi(k)}(t) + \sum_{\substack{i:\sigma(\pi(i))=\sigma(\pi(k)) \\ \pi(i)<\pi(k)}} \Big[m_i A_i(t + T_k(t)) + \hat{V}_i(A_i(t + T_k(t))) \\
&\quad - m_i A_i(t) - \hat{V}_i(A_i(t))\Big] + m_k \\
&\stackrel{r}{\approx} Z_{\pi(k)}(t) + \sum_{\substack{i:\sigma(\pi(i))=\sigma(\pi(k)) \\ \pi(i)<\pi(k)}} \Big[m_i\lambda_i(t + T_k(t)) + m_i\hat{A}_i(t + T_k(t)) \\
&\quad + \hat{V}_i(\lambda_i(t + T_k(t))) - m_i(\lambda_i t + \hat{A}_i(t)) - \hat{V}_i(\lambda_i t)\Big] + m_k \\
&\stackrel{r}{\approx} Z_{\pi(k)}(t) + \sum_{\substack{i:\sigma(\pi(i))=\sigma(\pi(k)) \\ \pi(i)<\pi(k)}} \Big[m_i\lambda_i T_k(t)) + m_i\hat{A}_i(t) \\
&\quad + \hat{V}_i(\lambda_i t) - m_i\hat{A}_i(t) - \hat{V}_i(\lambda_i t)\Big] + m_k \\
&\stackrel{r}{\approx} \tilde{Z}_{\pi(k)}(t) + \rho_{h(k)} T_k(t) + m_k;
\end{aligned}
$$

this establishes the relation

$$T(t) \stackrel{r}{\approx} \frac{\tilde{Z}_{\pi(k)}(t) + m_k}{1 - \rho_{h(k)}}.$$

Therefore, combining this with (9.38) and (9.61), we can conclude (9.53).
□

Lemma 9.17 Both matrices $\Delta = \Lambda G'(HM\Lambda G')^{-1}$ and $N = HM(I - P')^{-1}P'\Delta$ are well-defined, and matrix N is strictly lower triangular.

The proof of the lemma is left as Exercise 11.

9.7 Performance Analysis and Approximations

Based on the strong approximation theorem in Section 9.6, we now outline a procedure to approximate various performance measures of queueing networks. Specifically we consider the case where the interarrival time, the service time and the routing sequences are mutually independent i.i.d. sequences. In this case, we can approximate the aggregated workload process Z by an RBM $\tilde{Z}$ with drift θ, covariance matrix Γ, and reflection matrix R, which are described by

$$R = (I + HM(I - P')^{-1}P'\Delta)^{-1},$$
$$\theta = -R(e - \rho),$$
$$\Gamma = RH\left[\Gamma_V\Lambda + M(I - P')^{-1}\left[\Gamma_E + \sum_{k=1}^{K}\lambda_k\Gamma_\Phi^k\right](I - P)^{-1}M\right]H'R'.$$

Refer to Section 9.4.1 and Section 9.6 for the definition of the vectors and matrices used in the above equalities. Note that they are all from the service disciplines, the routing probability, and the mean and the variance of the interarrival and the service times. Also note that when $\rho < e$ (i.e., the traffic intensity at each station is less than one), $\tilde{Z}$ has a unique stationary distribution. In some special cases, this stationary distribution has an analytical solution, and in the general case, it can be computed numerically.

Given the estimate for the stationary distribution of the aggregated workload process, we could obtain estimates for some other performance measures of queueing networks. Let $E(\tilde{Z}_\ell)$ $(\ell = 1, \ldots, L)$ be the mean of the stationary RBM $\tilde{Z}$, which is the approximation for the stationary mean of the aggregated workload. We shall describe two alternative methods to obtain the estimates of the stationary mean queue length and mean sojourn time.

The first method is to approximate the mean queue-length by (9.51) in Section 9.6, and we have

$$E(Q) = \Delta E(\tilde{Z}).$$

Then, we use Little's law to obtain mean sojourn time as

$$E(\mathcal{S}_k) = \frac{1}{\lambda_k} E(Q_k).$$

The second method is to approximate the mean sojourn time via (9.53) by

$$E(\mathcal{S}_k) = \frac{E(\tilde{Z}_{\pi(k)}) + m_k}{1 - \rho_{h(k)}},$$

and then obtain the mean queue length by Little's law,

$$E(Q_k) = \lambda_k E(\mathcal{S}_k). \tag{9.65}$$

These two methods are summarized as Algorithms 1 and 2 below:

Algorithm 1 Computing steady-state average queue length and sojourn time

$$
\begin{aligned}
E(Q) &= \Delta E(\tilde{Z}), \\
E(\mathcal{S}_k) &= \frac{1}{\lambda_k} E(Q_k).
\end{aligned}
$$

Algorithm 2 Computing steady-state average queue length and sojourn time

$$
\begin{aligned}
E(\mathcal{S}_k) &= \frac{E(\tilde{Z}_{\pi(k)}) + m_k}{1 - \rho_{h(k)}}, \\
E(Q_k) &= \lambda_k E(\mathcal{S}_k).
\end{aligned}
$$

These two algorithms usually provide different approximations. From our numerical experiments in Section 9.8, Algorithm 2 seems to provide much more accurate estimation than Algorithm 1. Therefore, Algorithm 2 is recommended to obtain approximations for the mean stationary queue length and the mean stationary sojourn time. The numeric evidence also suggests that both algorithms are doing well and are asymptotically identical for a class k if $\beta_{\pi(k)}$ (the traffic intensity of priority group $\pi(k)$) is close to one. Intuitively, when $\beta_{\pi(k)}$ is close to one and $\rho_{\pi(k)} < 1$, $(1 - \rho_{h(k)})$ is close to or equal to one and the workload of all other priority groups at that station should almost be zero. Thus, both algorithms give

$$E(\mathcal{S}_k) \approx E(\tilde{Z}_{\pi(k)}).$$

Harrison and Williams [3] show that $\tilde{Z}$ has a product form stationary distribution if and only if

$$\Gamma_{ik} = \frac{1}{2} N_{ii}\Gamma_{kk} \quad \text{for all } 1 \le k < i \le L, \tag{9.66}$$

in which case the solution is

$$p(x) = \prod_{\ell=1}^{L} \kappa_\ell \exp(-\kappa_\ell x_\ell), \quad x \ge 0,$$

where $\kappa_1, \ldots, \kappa_L$ are the positive constants defined as

$$\kappa_\ell = \frac{2(1-\rho_\ell)R_{ll}}{\Gamma_{\ell\ell}} \quad \text{for } \ell = 1, \ldots, L.$$

However, the product-form condition is rarely satisfied for the networks discussed in this chapter (see Exercise 12).

9.8 Numerical Examples

Here we analyze two examples, both are feedforward networks as described in Section 9.4. The strong approximation is applied to obtain both RBM models. The performance estimates from the RBM approximations are compared with simulation results. The algorithm BNA/FM is used to compute the steady-state performance measures from RBM models. (Refer to Chapter 10 for the reference of the BNA/FM algorithm.)

9.8.1 A Single Station with Two Job Classes

Consider the single station network pictured in Figure 9.3. There are two job classes. Class 1 jobs have higher preemptive priority over class 2 jobs.

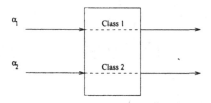

FIGURE 9.3. A two-class single-station network

We consider four versions of systems:

1. All interarrival and service times are taken to be Erlang of order 4 (SCV = 0.25).

2. All interarrival and service times are taken to be exponential (SCV=1).

3. All interarrival and service times are taken to be gamma distributed with SCV = 2.

4. All interarrival times are taken to be exponential, and all service times are taken to be Erlang of order 4.

The arrival rates of both classes are 1. For each system, we examine five cases of the mean service times:

1. $m_1 = 0.7$, $m_2 = 0.1$;

2. $m_1 = 0.5$, $m_2 = 0.3$;

3. $m_1 = 0.3$, $m_2 = 0.5$;

4. $m_1 = 0.1$, $m_2 = 0.7$;

5. $m_1 = 0.2$, $m_2 = 0.2$.

Although there is no product form solution for the joint stationary distribution of $(\tilde{Z}_1, \tilde{Z}_2)$, the marginal distributions of $\tilde{Z}_1$ and $\tilde{Z}_2$ are exponentially distributed with means

$$\mathsf{E}(\tilde{Z}_1) = \frac{m_1^2(c_{01}^2 + c_1^2)}{2(1 - \rho_1)},$$

$$\mathsf{E}(\tilde{Z}_2) = \frac{\sum_{i=1}^2 m_i^2(c_{0i}^2 + c_i^2)}{2(1 - \rho_2)},$$

respectively ($\rho_2 = \lambda_1 m_2 + \lambda_2 m_2$ and $\rho_1 = \lambda_1 m_1$). The following two analytical methods are used to obtain approximations of the mean queue lengths and mean sojourn times:

1. By algorithm 1 in Section 9.7:

$$\mathsf{E}(\mathcal{S}_1) = \mathsf{E}Q_1 = \frac{m_1(c_{01}^2 + c_1^2)}{2(1 - \rho_1)},$$

$$\mathsf{E}(\mathcal{S}_2) = \mathsf{E}Q_2 = \frac{1}{m_2}\left(\frac{\sum_{i=1}^2 m_i^2(c_{0i}^2 + c_i^2)}{2(1 - \rho_2)} - \frac{m_1^2(c_{01}^2 + c_1^2)}{2(1 - \rho_1)}\right).$$

2. By algorithm 2 in Section 9.7:

$$\mathsf{E}\mathcal{S}_1 = \mathsf{E}Q_1 = \frac{m_1^2(c_{01}^2 + c_1^2)}{2(1 - \rho_1)} + m_1,$$

$$\mathsf{E}\mathcal{S}_2 = \mathsf{E}Q_2 = \frac{1}{1 - \rho_1}\left(\frac{\sum_{i=1}^2 m_i^2(c_{0i}^2 + c_i^2)}{2(1 - \rho_2)} + m_2\right).$$

Table 9.1 summarizes the mean queue length estimates of each job class obtained from strong approximations and simulation. The columns "Algorithm 1" and "Algorithm 2" in Table 9.1 correspond to the approximations obtained by the two algorithms. The number in parentheses following each simulation estimate shows the half-width of the 95% confidence interval as a percentage of the estimate. The numbers in parentheses in the two algorithm columns indicate the relative error (against simulation estimates). This convention also applies to all subsequent tables.

As indicated in Remark 9.7 (cf. (9.30)), Algorithm 2 gives the exact mean queue lengths for Poisson arrivals (systems 1 and 2). (In these cases, we could have reported the percentage errors relative to the exact mean, but we choose to report the percentage errors relative to simulation estimates so as to be consistent with the other cases.) It appears that in almost all other cases, Algorithm 2 of the strong approximation also gives the better estimates, which are quite close to simulation results. When the SCVs of the interarrival and service times are equal to one, algorithms 1 and 2 coincide. This is true even when the arrival rates are not equal to one. Note that Algorithm 2 also performs well when the station is lightly loaded (with $\rho = 0.4$).

9.8.2   Two Stations in Tandem

Pictured in Figure 9.4 is a two-station tandem queueing network. Each sta-

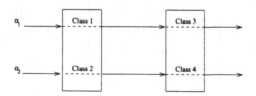

FIGURE 9.4. A two-station tandem queue

tion has two different job classes. We assume that all exogenous arrival processes and service processes are mutually independent renewal processes.

We will estimate the performance of this network under two different service disciplines:

1. The service discipline at station 1 is preemptive priority and the service discipline at station 2 is FCFS. Class 1 jobs have a higher priority over class 2 jobs at station 1.

2. The service discipline at station 1 is FCFS and the service discipline at station 2 is preemptive priority. Class 3 jobs have a higher priority over class 4 jobs at station 2.

System	m_1	m_2	Class	Algorithm 1	Algorithm 2	Simulation
1	0.7	0.1	Q_1	0.58 (42.6%)	1.11 (9.9%)	1.01 (0.5%)
			Q_2	2.17 (10.0%)	2.42 (0.4%)	2.41 (1.9%)
	0.5	0.3	Q_1	0.25 (56.1%)	0.63 (10.5%)	0.57 (0.3%)
			Q_2	1.00 (27.0%)	1.45 (5.8%)	1.37 (1.0%)
	0.3	0.5	Q_1	0.11 (65.5%)	0.33 (6.5%)	0.31 (0.2%)
			Q_2	0.79 (35.1%)	1.32 (9.1%)	1.21 (0.7%)
	0.1	0.7	Q_1	0.03 (72.0%)	0.10 (0.0%)	0.10 (0.1%)
			Q_2	0.88 (34.8%)	1.47 (8.9%)	1.35 (0.6%)
	0.2	0.2	Q_1	0.06 (70.0%)	0.21 (6.3%)	0.20 (0.1%)
			Q_2	0.10 (64.3%)	0.29 (3.6%)	0.28 (0.3%)
2	0.7	0.1	Q_1	2.33 (1.7%)	2.33 (1.7%)	2.37 (2.2%)
			Q_2	8.67 (2.9%)	8.67 (2.9%)	8.93 (4.9%)
	0.5	0.3	Q_1	1.00 (0.0%)	1.00 (0.0%)	1.00 (1.1%)
			Q_2	4.00 (2.6%)	4.00 (2.6%)	3.90 (1.8%)
	0.3	0.5	Q_1	0.43 (0.0%)	0.43 (0.0%)	0.43 (0.7%)
			Q_2	3.14 (0.6%)	3.14 (0.6%)	3.16 (1.9%)
	0.1	0.7	Q_1	0.11 (0.0%)	0.11 (0.0%)	0.11 (0.5%)
			Q_2	3.56 (0.8%)	3.56 (0.8%)	3.53 (2.5%)
	0.2	0.2	Q_1	0.25 (0.0%)	0.25 (0.0%)	0.25 (0.5%)
			Q_2	0.42 (6.8%)	0.42 (6.8%)	0.39 (0.6%)
3	0.7	0.1	Q_1	4.67 (13.3%)	3.97 (3.6%)	4.12 (3.7%)
			Q_2	17.31 (0.1%)	17.00 (3.6%)	17.23 (8.8%)
	0.5	0.3	Q_1	2.00 (25.0%)	1.50 (6.3%)	1.60 (2.0%)
			Q_2	8.00 (4.3%)	7.40 (3.5%)	7.67 (5.0%)
	0.3	0.5	Q_1	0.87 (42.6%)	0.56 (8.2%)	0.61 (1.2%)
			Q_2	6.29 (10.7%)	5.57 (1.9%)	5.68 (3.8%)
	0.1	0.7	Q_1	0.22 (70.8%)	0.12 (7.7%)	0.13 (0.6%)
			Q_2	7.12 (7.7%)	6.34 (4.1%)	6.61 (3.8%)
	0.2	0.2	Q_1	0.50 (51.0%)	0.30 (9.0%)	0.33 (0.9%)
			Q_2	0.83 (57.2%)	0.58 (10.0%)	0.53 (1.0%)
4	0.7	0.1	Q_1	1.46 (15.1%)	1.72 (0.0%)	1.72 (1.5%)
			Q_2	5.42 (2.9%)	5.54 (0.7%)	5.58 (3.9%)
	0.5	0.3	Q_1	0.63 (22.2%)	0.81 (0.0%)	0.81 (0.7%)
			Q_2	2.50 (8.8%)	2.73 (0.4%)	2.74 (2.2%)
	0.3	0.5	Q_1	0.27 (28.9%)	0.38 (0.0%)	0.38 (0.4%)
			Q_2	1.96 (12.1%)	2.23 (0.0%)	2.23 (2.0%)
	0.1	0.7	Q_1	0.07 (36.4%)	0.11 (0.0%)	0.11 (0.3%)
			Q_2	2.22 (11.9%)	2.51 (0.4%)	2.52 (2.0%)
	0.2	0.2	Q_1	0.16 (30.4%)	0.24 (4.3%)	0.23 (0.3%)
			Q_2	0.26 (25.6%)	0.35 (0.0%)	0.35 (0.5%)

TABLE 9.1. Average queue length in the single-station queue shown in Figure 9.3

System	Distribution	α_1	α_2	m_1	m_2	m_3	m_4
1	E_4	1.0	3.0	0.5	0.1	0.3	0.2
2	M	1.0	3.0	0.5	0.1	0.3	0.2
3	gamma	1.0	3.0	0.5	0.1	0.3	0.2

TABLE 9.2. System specifications of a two-station tandem queue

For each different service discipline, we consider three versions of the system with different service and interarrival time distributions. The parameters of the three systems are listed in Table 9.2. All service and interarrival time distributions are taken to be Erlang of order 4 ($SCV = 0.25$) in the first system, exponential ($SCV = 1$) in the second system, and gamma with $SCV = 2$ in the third system.

Tables 9.3 and 9.4 present the simulation estimates and BNA/FM estimates of the mean queue length for each job class and each system configuration. We use BNA/FM to get the mean aggregated workload numerically and then use Algorithm 2 of Section 9.7 to obtain the mean queue lengths. (In this case, Algorithm 1 again provides inferior estimates, so they are not presented.) The BNA/FM estimates of the mean queue lengths for this two-station network are quite accurate when compared against simulation estimates, except the case of job class 4 in the type 1 network, where BNA/FM significantly underestimates the queue length of class 4. This might be due to the large variations in the interarrival times of this class (which correspond to the departure times of class 2, the lower priority class at station 1).

9.9 Notes

Peterson [5] was the first to study a multiclass feedforward queueing network to derive a heavy traffic limit theorem. It is shown that the limit can be described by a J-dimensional reflected Brownian motion (RBM), where J equals the number of service stations in the network. In particular, a state space collapse phenomenon is observed for higher priority job classes. That is, the limiting workload or the queue length of high priority jobs is zero. Harrison and Williams [3] study the reflected Brownian motion that arises from the heavy traffic limit theorem, and obtain a necessary and sufficient condition for the existence of a product-form stationary distribution of the Brownian model.

This chapter is adapted from Chen and Shen [2]. The network presented here has a slightly more general structure than the one in Peterson [5]. As remarked in Chapters 6 and 7, the strong approximation refines the heavy traffic limit theorem in that it provides the rate of convergence and does not require the network to operate under heavy traffic. The strong

System No.	Approximation Method	$E(Q_1)$	$E(Q_2)$	$E(Q_3)$	$E(Q_4)$
1	Simulation	0.57 (0.3%)	2.71 (1.1%)	0.99 (1.3%)	4.00 (1.1%)
	BNA/FM	0.63 (10.5%)	2.70 (0.4%)	1.04 (5.1%)	2.83 (29.3%)
2	Simulation	1.00 (0.9%)	9.02 (2.6%)	3.32 (2.5%)	12.50 (1.9%)
	FNA/FM	1.01 (1.0%)	8.99 (3.3%)	3.2 (3.6%)	9.31 (25.5%)
3	Simulation	1.59 (1.6%)	17.54 (4.0%)	6.59 (4.6%)	24.41 (3.9%)
	BNA/FM	1.54 (3.1%)	17.15 (2.2%)	6.25 (5.2%)	18.45 (24.4%)

TABLE 9.3. Average queue length of network 1

System No.	Approximation Method	$E(Q_1)$	$E(Q_2)$	$E(Q_3)$	$E(Q_4)$
1	Simulation	0.78 (0.6%)	1.51 (0.8%)	0.31 (0.2%)	3.79 (1.2%)
	BNA/FM	0.85 (9.0%)	1.34 (11.3%)	0.33 (6.5%)	3.58 (5.5%)
2	Simulation	1.90 (2.1%)	4.50 (2.2%)	0.40 (0.7%)	12.17 (2.9%)
	BNA/FM	1.89 (0.5%)	4.47 (0.6%)	0.41 (3.5%)	11.80 (3.0%)
3	Simulation	3.42 (3.1%)	8.53 (3.5%)	0.55 (1.0%)	23.86 (4.4%)
	BNA/FM	3.30 (3.5%)	8.69 (1.9%)	0.51 (7.3%)	22.65 (5.1%)

TABLE 9.4. Average queue length of network 2

approximation yields appropriate approximations for the workload, queue-length, and sojourn time processes of all job classes (not just the lowest job class). The numerical examples in Chapter 10 indicate that the performance estimates based on the strong approximation have a much better accuracy than those suggested by the diffusion approximation, for both higher and lower priority classes.

That the RBM has a stationary distribution when $\rho < e$ follows from Lemma 9.17 and Lemma 3.2 of Chen [1].

9.10   Exercises

1. Prove Lemma 9.10.

2. Let $f(t) = \sqrt{t \log \log t}$ for $t \geq 0$. Show that for any given $\epsilon > 0$, there exists $t_0 > 0$ such that $f(t+s) \leq f(t) + \epsilon u$ for all $t \geq t_0$ and $s \geq 0$.

3. Establish an FLIL approximation for the single station queue, for the case $\rho > 1$.

4. Prove Lemma 9.9.

5. Suppose that the strong approximation (9.18) holds for some r-strong continuous process $\hat{V}_k$. Let

$$S_k(t) = \sup\{n \geq 0, V(n) \leq t\}.$$

Establish a strong approximation for S_k. (Refer to Exercise 27 in Chapter 5.)

6. Consider the single-station queue. Let $T_k(t)$ be the total amount of time that the server at the station has served jobs of class k during $[0, t]$, $k \in \mathcal{K}$. Obtain a FLIL approximation and a strong approximation for $T = (T_k)$.

7. Consider the single-station queue as shown in Figure 9.1. Assume that the arrival process and the service process are mutually independent i.i.d. squences, and assume that the traffic intensity $\rho \leq 1$. Write down explicitly the strong approximation for the workload process and the sojourn time process.

8. State a version of Theorem 9.4 for a sequence of networks indexed by n, and then use it to establish a diffusion approximation theorem.

9. Prove Theorem 9.11.

10. Let $\tilde{Z}$ be the reflected Brownian motion as defined in Theorem 9.12. Suppose that $R^{-1}\theta < 0$. Prove that as $T \to \infty$,

$$\sup_{0 \le t \le T} \|\tilde{Z}(t)\| \overset{a.s.}{=} O(\log T).$$

11. Prove Lemma 9.17.

12. Consider the two-station network shown by Figure 9.4.

   (a) Find a necessary and sufficient condition for the stationary distribution to have a product form.

   (b) Generalize your results for a feedforward queueing network under purely priority service disciplines.

References

[1] Chen, H. (1996). A sufficient condition for the positive recurrence of a semimartingale reflecting Brownian motion in an orthant, *Annals of Applied Probability*, **6**, 758–765.

[2] Chen, H. and X. Shen. (2000). Strong approximations for multiclass feedforward queueing networks. *Annals of Applied Probability*.

[3] Harrison, J.M. and R.J. Williams. (1992). Brownian models of feedforward queueing networks: quasireversibility and product form solutions, *Annals of Applied Probability*, **2**, 263–293.

[4] Kleinrock, L. (1976). *Queueing Systems Volume II; Computer Applications*, Wiley, New York.

[5] Peterson, W.P. (1991). A heavy traffic limit theorem for networks of queues with multiple customer types, *Mathematics of Operations Research*, **16**, 90–118.

10
Brownian Approximations

The purpose of this chapter is to develop a general approach to approximating a multiclass queueing network by a semimartingale reflected Brownian motion, (SRBM), which is a generalization of the RBM studied earlier. Our focus is *not* on proving limit theorems so as to justify *why* the network in question can be approximated by an SRBM (as we did in several previous chapters). Rather our intention is to illustrate *how* to approximate the network by an SRBM. We make no claim that the proposed approximation can always be justified by some limit theorems. To the contrary, through both analysis and numerical results, we identify cases where the SRBM may not exist, or may work poorly. (A complete characterization of *when* the proposed approximation works is a challenging and active research topic; refer to Section 10.7 for a survey on the recent advances in this research area.)

The network under consideration is more general than those considered in the previous chapters. For instance, it extends the generalized Jackson network by allowing multiple job classes, and it generalizes the feedforward network by allowing feedback routings. In addition, service stations may be subject to several types of random interruptions.

In presenting the approximation scheme, we shall rely mainly on some heuristic arguments, which draw upon the analysis and limit theorems for the simpler networks studied in the previous chapters, such as the generalized Jackson network and the feedforward network. In this regard, the proposed approximation is justified by the limit theorems (particularly the strong approximation theorems) for those simpler networks. Furthermore, we shall point out, for the more general networks, what the gaps are that have to be filled so as to justify the approximation (via limit theorems).

The rest of the chapter is organized as follows. We start with a formal description in Section 10.1 of the multiclass network under study. In particular, we specify the primitive processes and the performance measures (the derived processes), and the dynamical equations that relate the two. In Section 10.2 we present the conditions on the primitive processes, in terms of diffusion or strong approximations, and illustrate how these asymptotics hold in several example systems. Our main approximation scheme is developed in Section 10.3, where we derive the SRBM approximation for the aggregated workload process based on the conditions imposed on the primitives, and extend the approximation to other derived processes. Several issues surrounding the SRBM approximation are discussed in Section 10.4, and further illustrated through analyzing the Kumar–Seidman network. The connection to the notion of state-space collapse is also examined in that section. In Section 10.5 we provide some sufficient conditions for the existence of the stationary distribution of the SRBM; and based on the basic adjoint relation (BAR) (which characterizes the stationary distribution), we describe a numerical algorithm for computing the stationary distribution. Extensive numerical studies are presented in Section 10.6, where the SRBM approximation is compared against simulation for a variety of networks. Connections to the literature and open problems are overviewed in the notes section (Section 10.7).

10.1   The Queueing Network Model

10.1.1   Notation and Conventions

We study a queueing network that consists of J single-server stations, indexed by $j \in \mathcal{J} = \{1, \ldots, J\}$, and K job classes, indexed by $k \in \mathcal{K} := \{1, \ldots, K\}$. The K job classes are partitioned into L (nonempty) groups, $g_1, \ldots, g_L$. Group g_ℓ, or sometimes simply called group ℓ, jobs are served exclusively at station $\sigma(\ell)$. Thus, σ is a many-to-one mapping from $\mathcal{L} := \{g_1, \ldots, g_L\}$. onto $\mathcal{J}$. Necessarily, $J \leq L \leq K$.

Within each group, jobs are served in their order of arrival (i.e., FIFO). (While all job classes with a group follow the same service discipline, they are statistically different in terms of arrival times, service requirements, and routing mechanism, as shall be specified in Section 10.1.2.) Among different groups, group ℓ_1 jobs have a preemptive priority over group ℓ_2 jobs if $\ell_1 < \ell_2$, provided that they are served at the same station, i.e., $\sigma(\ell_1) = \sigma(\ell_2)$. Define a mapping π from $\mathcal{K}$ onto $\mathcal{L}$: $\pi(k) = \ell$ if and only if $k \in g_\ell$. Then the above convention implies that class k jobs are served exclusively at station $\sigma(\pi(k))$; when $\sigma(\pi(i)) = \sigma(\pi(k))$, class i jobs have a preemptive priority over class k jobs if $\pi(i) < \pi(k)$; they are served FIFO if $\pi(i) = \pi(k)$. For any $k \in \mathcal{K}$, let $h(k)$ denote the index of the group with the next higher priority at station $\sigma(\pi(k))$ (i.e., $h(k) < \pi(k)$ and $\sigma(h(k)) = \sigma(\pi(k))$; and if $h(k) < \ell < \pi(k)$, then it must be $\sigma(\ell) \neq \sigma(\pi(k))$)), and let $h(k) = 0$ if class k has the highest priority at its station.

We introduce a $J \times L$ *station constituent* matrix $C = (C_{j\ell})_{j \in \mathcal{J}, \ell \in \mathcal{L}}$: The (j, ℓ)th component of C, $C_{j\ell}$ is equal to 1 if $\sigma(\ell) = j$ (i.e., group ℓ is served at station j), and $C_{j\ell} = 0$ otherwise. Note that each row j of C specifies what job groups are served at station j. Next, we introduce two $L \times K$ matrices: *group constituent* matrix $G = (G_{\ell k})_{\ell \in \mathcal{L}, k \in \mathcal{K}}$ and *higher priority group constituent* matrix $H = (H_{\ell k})_{\ell \in \mathcal{L}, k \in \mathcal{K}}$: The (ℓ, k)th component of G, $G_{\ell k}$, is equal to 1 if $k \in g_\ell$, and $G_{\ell k} = 0$ otherwise. That is, each row ℓ of G specifies what job classes belong to the group g_ℓ. The (ℓ, k)th component of H, $H_{\ell k}$, is equal to 1 if $\sigma(\ell) = \sigma(\pi(k))$ and $\pi(k) \leq \ell$; otherwise, $H_{\ell k} = 0$. Note that the condition for $H_{\ell k} = 1$ necessarily includes $k \in g_\ell$. Hence, each row ℓ of H includes not only the classes that belong to the group g_ℓ, but also all those classes that are served at the same station $\sigma(\ell)$ but with a higher priority than those in g_ℓ. Note that while C specifies the station constituents in terms of groups and G specifies the group constituents in terms of classes, the matrix product CG spells out the station constituents in terms of classes.

Some examples are in order.

Example 10.1 Consider the generalized Jackson network, where there is a single stream of external arrivals and hence a single job class at each station. In this case, we can simply set $g_k = \{k\}$ and $\sigma(k) = k$ for $k \in \mathcal{K}$, and hence $K = J = L$ and $C = G = H = I$.

Example 10.2 Consider a multiclass network with pure priorities, i.e., each priority group contains a single class. An example of such a network is the Kumar–Seidman network, shown by Figure 8.1. Here, classes 2 and 4 have higher priorities at their respective stations. The partition of the classes can be specified as $g_1 = \{4\}$, $g_2 = \{1\}$, $g_3 = \{2\}$, and $g_4 = \{3\}$. Then, the matrices P, C, G and H can be identified as follows:

$$C = \begin{pmatrix} 1 & 1 & 0 & 0 \\ 0 & 0 & 1 & 1 \end{pmatrix}, \qquad G = \begin{pmatrix} 0 & 0 & 0 & 1 \\ 1 & 0 & 0 & 0 \\ 0 & 1 & 0 & 0 \\ 0 & 0 & 1 & 0 \end{pmatrix}$$

$$H = \begin{pmatrix} 0 & 0 & 0 & 1 \\ 1 & 0 & 0 & 1 \\ 0 & 1 & 0 & 0 \\ 0 & 1 & 1 & 0 \end{pmatrix}.$$

Example 10.3 Consider a two-station five-class network as depicted in Figure 8.11. All jobs are served FIFO. This is a variation of what is known as the Bramson network. Here, because of FIFO, we have $L = J = \{1, 2\}$,

$g_1 = \{1, 5\}$ and $g_2 = \{2, 3, 4\}$, and $\sigma(1) = 1$ and $\sigma(2) = 2$. (The priority does not play any role in this case, and we could alternatively define, e.g., $g_2 = \{1, 5\}$ and $g_1 = \{2, 3, 4\}$, and $\sigma(1) = 2$ and $\sigma(2) = 1$.) Furthermore, $C = I$,

$$G = H = \begin{pmatrix} 1 & 0 & 0 & 0 & 1 \\ 0 & 1 & 1 & 1 & 0 \end{pmatrix}. \tag{10.1}$$

10.1.2 The Primitive Processes

The network under study is driven by the following four primitive processes:

(i) *External Arrivals.* Jobs of class k arrive at the network exogenously according to a counting process $E_k = \{E_k(t), t \geq 0\}$, where $E_k(t)$ indicates the number of arrivals up to time t, with $E_k(0) = 0$.

(ii) *Service Requirements.* The nth job in class k requires $v_k(n)$ units of processing time from the server at station $\sigma(\pi(k))$. Write $v = (v_k)$ with $v_k = \{v_k(n), n \geq 1\}$, and set

$$V_k(0) = 0, \quad V_k(n) = \sum_{m=1}^{n} v_k(m); \quad V_k = \{V_k(n), n \geq 1\}.$$

(iii) *Service Capacities.* For each station j, the server is constrained by a (cumulative) service capacity process, $c_j = \{c_j(t), t \geq 0\}$, where $c_j(t)$ is the maximum amount of work that the server j can accomplish up to t. Hence, if server j is perfectly reliable, then $c_j(t) = t$. (Assume that all servers work at unit rate.) If the server is subject to random disruptions (breakdowns), then $c_j(t)$ is the cumulative up-time over the interval $[0, t]$. (More on this in Example 10.6.)

(iv) *Routing Mechanism.* After service completion a job may change classes (in particular if it is routed to another station). For each class $k \in \mathcal{K}$, define $\phi^k = \{\phi^k(n), n \geq 1\}$ such that the nth job of class k after service completion becomes a class j job if $\phi^k(n) = e^j$ (the jth unit vector), or leaves the network if $\phi^k(n) = 0$. Call $\phi = (\phi^k)$ the routing sequence. Let

$$\Phi^k(0) = 0, \qquad \Phi^k(n) = \sum_{m=1}^{n} \phi^k(m), \quad n \geq 1.$$

Set $\Phi^k(t) = \Phi^k(\lfloor t \rfloor)$, for all $t \geq 0$; and $\Phi^k = \{\Phi^k(t), t \geq 0\}$.

10.1.3 The Derived Processes

The derived processes are the performance measures of interest in the network. We shall focus on the following four processes:

(I) *the workload process:* $W = (W_k)_{k \in \mathcal{K}}$, where $W_k = \{W_k(t), t \geq 0\}$, and $W_k(t)$ denotes the amount of work embodied in all class k jobs that are either queued or in service at time t;

(II) *the queue-length process:* $Q = (Q_k)_{k \in \mathcal{K}}$, where $Q_k = \{Q_k(t), t \geq 0\}$, and $Q_k(t)$ represents the number of class k jobs either in service or in queue at time t;

(III) *the sojourn-time process:* $S = (S_k)_{k \in \mathcal{K}}$, where $S_k = \{S_k(t), t \geq 0\}$, with $S_k(t)$ denoting the time that will be spent at station $\sigma(k)$ by the first job of class k that arrives at the station at time t or afterwards;

(IV) *the aggregated workload process:* $Z = (Z_\ell)_{\ell \in \mathcal{L}}$, where $Z_\ell = \{Z_\ell(t), t \geq 0\}$ $(\ell \in \mathcal{L})$, and $Z_\ell(t)$ represents the total amount of work embodied in those jobs that are either queued or in service at station $\sigma(\ell)$ at time t and with priorities no lower than ℓ.

To specify the dynamics of the above processes, we also need some other (intermediate) performance measures:

- $A_k(t)$: total number of class k jobs arrived at station $\sigma(\pi(k))$ during $[0, t]$, either externally or from other stations;

- $D_k(t)$: total number of departures (service completions) of class k jobs from station $\sigma(\pi(k))$ during $[0, t]$;

- $Y_\ell(t)$: the cumulative amount of service capacity during $[0, t]$ that station $\pi(\ell)$ does not use to serve jobs with priority ℓ or higher. In particular, if group g_ℓ has the lowest priority at station $\sigma(\ell)$, then $Y_\ell(t)$ indicates the cumulative amount of service capacity during $[0, t]$ at station $\sigma(\ell)$ that is lost due to the lack of jobs for service (corresponding to the idle time if one interprets the service capacity as the time available for service).

Furthermore, define

- $\eta_k(t)$: the arrival epoch of the first job in class k during $[t, \infty)$;

- $\mathcal{T}_k(t)$: the time a class k job will spend at station $\sigma(k)$, given that it arrives at t;

- $\tau_\ell(t)$: the arrival time of the job in group g_ℓ that has most recently completed service at station $\sigma(\ell)$ ($\tau_\ell(t) \equiv 0$ if none of the jobs in group g_ℓ has been completed by t);

- $\nu_k(t)$: the partial service time (if any) that has been performed on a class k job during $(\tau_{\pi(k)}, t]$.

Note the following bounds for $\eta_k(t)$ and $\nu_k(t)$, $k \in \mathcal{K}$ (which will be used later):

$$0 \leq \eta_k(t) - t \leq u_k(A_k(t) + 1), \tag{10.2}$$

where $u_k(m)$ is the interarrival time between the $(m-1)$st and the mth class k jobs (jump points of A_k), and

$$0 \leq \nu_k(t) \leq \max_{1 \leq n \leq A_k(t)} v_k(n), \tag{10.3}$$

i.e., $\nu_k(t)$ is bounded by the maximum service requirement of all class k jobs that have arrived up to t.

We can now derive the following relations:

$$A(t) = E(t) + \sum_{k=1}^{K} \Phi^k(D_k(t)), \tag{10.4}$$

$$D(t) = A(G'\tau(t)), \tag{10.5}$$

$$Q(t) = A(t) - D(t), \tag{10.6}$$

$$W(t) = V(A(t)) - V(D(t)) - \nu(t), \tag{10.7}$$

$$Z(t) = HV(A(t)) - C'c(t) + Y(t), \tag{10.8}$$

$$Z(t) = HW(t), \tag{10.9}$$

$$S(t) = \mathcal{T}(\eta(t)). \tag{10.10}$$

The equation in (10.4) specifies the total arrivals (of any class) as the sum of the external arrivals and internal transitions; (10.5) is best understood in component form: $D_k(t) = A_k(\tau_{\pi(k)}(t))$, taking into account that all jobs in group $g_{\pi(k)}$ are served FIFO; (10.6) says the difference between arrivals and departures is what is left in the network (i.e., queued or in service); and (10.7) is a variation of (10.6) in terms of workload, and partially completed jobs are accounted for in the last term $\nu(t)$.

Furthermore, writing (10.8) in component form, we have

$$Z_\ell(t) = \sum_{\substack{k:\sigma(\pi(k))=\sigma(\ell) \\ \pi(k) \leq \pi(\ell)}} V_k(A_k(t)) - c_{\sigma(\ell)}(t) + Y_\ell(t), \qquad \ell \in \mathcal{L}.$$

Note that we must have, for all $\ell \in \mathcal{L}$,

$$dY_\ell \geq 0, \quad Y_\ell(0) = 0; \qquad Z_\ell dY_\ell = 0. \tag{10.11}$$

In words, $Y_\ell(t)$ is nondecreasing in t, starting from 0, and it can increase at t only if $Z_\ell(t) = 0$. These are consistent with the specification of the (preemptive) priority discipline and the work-conserving condition. The relation in (10.9) follows from the definitions of the two workload processes, W and Z, and the matrix H.

Finally, (10.10) is self-explanatory, given the definition of $\mathcal{T}$ and η. Note, however, that $\mathcal{T}$ is governed by the following recursive relation:

$$\mathcal{T}_k(t) = Z_{\pi(k)}(t-) + \sum_{\substack{i:\sigma(\pi(i))=\sigma(\pi(k)) \\ \pi(i)<\pi(k)}} [V_i(A_i(t+\mathcal{T}_k(t))) - V_i(A_i(t))]$$
$$+ \left[\mathcal{T}_k(t) - c_{\sigma(\pi(k))}(t+\mathcal{T}_k(t)) + c_{\sigma(\pi(k))}(t)\right]$$
$$+ \left[V_k(A_k(t)) - V_k(A_k(t)-1)\right], \tag{10.12}$$

where the first term on the right-hand side is the aggregated workload over all those jobs that are no lower in priority than the class k job that indexes $\mathcal{T}_k$, the second term is the sum of the work embodied in those jobs that arrive during the sojourn time of the class k job in question and have higher (preemptive) priorities, the third term is the down-time of station $\sigma(\pi(k))$ during the sojourn time of the class k job in question, and the fourth term is the job's own service time.

10.2 Two-Moment Characterization of the Primitive Processes

Recall from Section 10.1.2 that the four primitive processes that drive the dynamics of the network are:

$$E = (E_k)_{k\in\mathcal{K}}, \quad V = (V_k)_{k\in\mathcal{K}}, \quad c = (c_j)_{j\in\mathcal{J}}, \quad \Phi = (\Phi^k)_{k\in\mathcal{K}},$$

denoting, respectively, the external arrivals, the service requirements, the service capacities, and the routing mechanism.

The starting point of our approximation scheme is the assumption that (E, V, c, Φ) jointly satisfy the following:

$$E(t) \overset{d}{\approx} \alpha t + \hat{E}(t), \tag{10.13}$$

$$V(t) \overset{d}{\approx} mt + \hat{V}(t), \tag{10.14}$$

$$c(t) \overset{d}{\approx} \kappa t + \hat{c}(t), \tag{10.15}$$

$$\Phi^k(t) \overset{d}{\approx} P'_k t + \hat{\Phi}^k(t), \quad k \in \mathcal{K}. \tag{10.16}$$

Here,

$$\alpha = (\alpha_k)_{k\in\mathcal{K}}, \quad m = (m_k)_{k\in\mathcal{K}}, \quad \kappa = (\kappa_j)_{j\in\mathcal{J}}, \quad P'_k = (p_{ki})_{i\in\mathcal{K}}, \tag{10.17}$$

are nonnegative vectors (drift rates), and

$$\hat{E} = (\hat{E}_k)_{k \in \mathcal{K}}, \quad \hat{V} = (\hat{V}_k)_{k \in \mathcal{K}}, \quad \hat{c} = (\hat{c}_j)_{j \in \mathcal{J}}, \quad \hat{\Phi}^k = (\hat{\Phi}_i^k)_{i \in \mathcal{K}}, \quad (10.18)$$

are driftless Brownian motions, with covariance matrices denoted by Γ_E, Γ_V, Γ_c, and Γ_{Φ^k}, respectively. The approximation "$\overset{d}{\approx}$" means, e.g., in the case of (10.13), either the strong approximation,

$$\sup_{0 \le t \le T} \|E(t) - \alpha t - \hat{E}(t)\| \overset{\text{a.s.}}{=} o(T^{1/r}), \qquad \text{as } T \to \infty, \quad (10.19)$$

for some $r \in (2, 4)$, or the functional central limit theorem,

$$n^{-1/2} [E(n \cdot) - n\alpha \cdot] \overset{d}{\to} \hat{E}(\cdot), \qquad \text{as } n \to \infty. \quad (10.20)$$

The same holds for (10.14), (10.15), and (10.16) as well.

Note that the vectors in (10.17) have the following physical interpretation (all in the sense of long-run averages):

- α_k is the external arrival rate of class k jobs (hence, if k denotes a class that is generated only by internal transitions, then necessarily $\alpha_k = 0$);

- m_k is the mean service requirement of class k jobs;

- κ_j is the service capacity of station j, i.e., the maximum output rate from the station;

- p_{ki} is the proportion of class k jobs that transfer (upon service completion at station $\sigma(\pi(k))$) into class i (and transit to station $\sigma(\pi(i))$).

For future reference, we collect here some more notation that relates to the primitive processes: Define

- $P = (p_{ki})$, the routing matrix. Note that the kth row of P is P_k, whose transpose appears in (10.16). P is assumed to be substochastic, with $P^n \to 0$ as $n \to \infty$.

- $\lambda = (I - P')^{-1}\alpha$, whose kth component, λ_k, is the (nominal) arrival rate of class k jobs at station $\sigma(k)$, including both external arrivals (if any) and internal transfers.

- $M = \text{diag}(m)$ and $\Lambda = \text{diag}(\lambda)$, two $K \times K$ diagonal matrices with the kth diagonal elements being m_k and λ_k, respectively.

- $\beta = HM\lambda$, whose ℓth component is the summation of the group traffic intensities (which are given by $GM\lambda$) for those groups at station $\sigma(\ell)$ with the priority no less than that of group ℓ.

- $\rho = CGM\lambda$, whose jth component is the traffic intensity of station j. Note that if g_ℓ is the lowest priority group at its station, then $\beta_\ell = \rho_{\sigma(\ell)}$. Below, we assume $\rho < \kappa$, i.e., the traffic intensity at every station is strictly less than its service capacity.

Below we illustrate through some examples how to specify the parameters that characterize the approximations in (10.13)–(10.16), specifically, the drift terms and the covariance matrices (of the driftless Brownian motions).

Example 10.4 *(Renewal Arrivals, i.i.d. Service Requirement, and Markovian Routing)* These are the most common assumptions in queueing networks. Specifically, for each $k \in \mathcal{K}$, E_k is a renewal process with arrival rate α_k and squared coefficient of variation (SCV) of interarrival times $c_{0,k}^2$, and $(E_1, \ldots, E_K)$ are mutually independent. Then, the approximation in (10.13) follows from either the functional central limit theorem (Theorem 5.11) or the strong approximation theorem (Theorem 5.14), and the covariation matrix Γ_E is

$$(\Gamma_E)_{ik} = \alpha_k c_{0,k}^2 \delta_{ik}, \qquad i, k \in \mathcal{K}.$$

Similarly, the service requirements $(v_k)_{k\in\mathcal{K}}$ are K mutually independent i.i.d. sequences, with mean $(m_k)_{k\in\mathcal{K}}$ and SCV $(c_k^2)_{k\in\mathcal{K}}$. (The SCV c_k^2 should be distinguishable from the capacity process c_j from the context.) Hence, the approximation in (10.14) also holds, with the covariance matrix is

$$(\Gamma_V)_{ik} = m_k^2 c_k^2 \delta_{ik}, \qquad i, k \in \mathcal{K}.$$

The same also applies to the routing sequences $(\phi^k)_{k\in\mathcal{K}}$, which are also mutually independent i.i.d. sequences. In particular, the approximation in (10.16) holds, and the covariance matrix

$$(\Gamma_{\Phi_k})_{im} = p_{ki}(\delta_{im} - p_{km}), \qquad i, m \in \mathcal{K},$$

for all $k \in \mathcal{K}$.

Example 10.5 *(Batch Arrivals)* Suppose for each $k \in \mathcal{K}$, class k jobs arrive in batches: The interarrival times of the batches are i.i.d. with mean a_k^{-1} and SCV c_{0k}^2, and the batch sizes are i.i.d. with mean b_k and SCV c_{bk}^2. Assume that the batch sizes are independent of interarrival times and that the arrivals among different classes are mutually independent. Then the approximation in (10.13) holds with $\alpha_k = a_k b_k$ and

$$(\Gamma_E)_{ik} = a_k b_k^2 (c_{0k}^2 + c_{bk}^2)\delta_{ik}, \qquad i, k \in \mathcal{K}.$$

(See Exercise 25 in Chapter 5.)

Example 10.6 *(Perfect Service Stations)* If station j is perfectly reliable, then the jth component of the approximation in (10.15) holds with $\kappa_j = 1$ and $\hat{c}_j \equiv 0$.

Example 10.7 *(Service Stations with Breakdowns)* A station (server) j is subject to "autonomous breakdowns" if breakdowns can occur regardless of whether it is idle or working. In this case, let $\{(u_j(n), d_j(n)), n \geq 1\}$ be an i.i.d. sequence, where $u_j(n)$ and $d_j(n)$ denote the duration of the nth up-time and the nth down-time (including repair time), respectively. For simplicity, assume that $u_j(n)$ and $d_j(n)$ are independent. (The station can start in either the first up period or the first down period.) Let d_j and $c_{d_j}^2$ (u_j and $c_{u_j}^2$) denote the mean and the squared coefficient of variation of $d_j(n)$ ($u_j(n)$). Then the approximation in (10.15) holds with

$$\kappa_j = \frac{u_j}{u_j + d_j},$$

and

$$(\Gamma_c)_{jl} = \frac{u_j^2 d_j^2 (c_{u_j}^2 + c_{d_j}^2)}{u_j + d_j} \delta_{jl}.$$

(See Exercise 26 in Chapter 5.)

A station (server) is subject to "operational breakdowns" if breakdowns can occur only while the station is actively processing jobs. If a station j has an infinite supply of work and is hence always engaged in service, then the distinction between the operational breakdown and the autonomous breakdown disappears for the process c. This is the case in heavy traffic. In general, a refined approximation is proposed as follows:

$$(\Gamma_c)_{jl} = \frac{\rho_j}{\kappa_j} \cdot \frac{u_j^2 d_j^2 (c_{u_j}^2 + c_{d_j}^2)}{u_j + d_j} \delta_{jl}.$$

Either of the above modes of the breakdowns implicitly assumes that there is a dedicated repairperson for each service station. When several service stations share one repairperson, the approximation (10.15) still prevails, except that it becomes more difficult to identify the approximation parameters κ and Γ_c. See Harrison and Pich [20] for details.

10.3   The SRBM Approximation

The essence of our approximation scheme is to start with the asymptotic conditions on the primitive processes in (10.13)–(10.16), and derive the corresponding approximations for the derived processes, in particular, the four processes (W, Q, S, Z) specified in Section 10.1.3.

A key assumption we need is that the arrival process A (also defined in Section 10.1.3) satisfies the following:

$$A(t) \overset{d}{\approx} \lambda t + \hat{A}(t), \tag{10.21}$$

where $\hat{A}$ is a continuous process such that either $\hat{A}$ is r-strong continuous or the weak limit of $\hat{A}(nt)/\sqrt{n}$ (as $n \to \infty$) exists and has a continuous sample path. The specific form of $\hat{A}$ will be determined from the analysis below.

The approximation in (10.21) implies either

$$\sup_{0 \le t \le T} \|A(t) - \lambda t\| = O(\sqrt{T \log \log T}), \quad \text{as } T \to \infty,$$

or

$$\frac{1}{n} A(nt) \to \lambda t, \quad \text{u.o.c.}, \quad \text{as } n \to \infty.$$

When either of the above holds, we shall write

$$A(t) \sim \lambda t. \tag{10.22}$$

Furthermore, we can show that the above is equivalent to either one of the following two approximations

$$D(t) \sim \lambda t, \tag{10.23}$$
$$\tau(t) \sim et. \tag{10.24}$$

In view of (10.22) and (10.23)—that the departure rate equals the arrival rate—it is clear that the approximations in (10.22)–(10.24) should hold if the queueing network is stable (or weakly stable). (See Section 8.3 and Section 8.4 for a discussion of the Kumar–Seidman network as a special case.) Indeed, here we are interested in approximating only a stable network.

In deriving the approximations below, we shall repeatedly make use of the following result: Suppose two processes X and T satisfy

$$X(t) \overset{d}{\approx} at + \hat{X}(t) \quad \text{and} \quad T(t) \sim bt,$$

such that either $\hat{X}$ is r-strong continuous or the weak limit of $\hat{X}(nt)/\sqrt{n}$ (as $n \to \infty$) is continuous. Then,

$$X(T(t)) \overset{d}{\approx} aT(t) + \hat{X}(bt).$$

(Refer to Theorem 5.2 and Lemma 9.8.)

We now derive the approximations for the processes (W, Q, S, Z). First, applying (10.21) to (10.5) yield

$$D(t) = A(G'\tau(t)) \overset{d}{\approx} \Lambda G'\tau(t) + \hat{A}(t). \tag{10.25}$$

Next, applying (10.13) and (10.16) to (10.4) yields

$$A(t) \overset{d}{\approx} \alpha t + \hat{E}(t) + \sum_{k=1}^{K} \hat{\Phi}^k(\lambda_k t) + P'D(t). \tag{10.26}$$

Note that $\lambda = \alpha + P'\lambda$ and $\lambda = \Lambda G'e$. Hence, combining (10.21), (10.25), and (10.26) leads to

$$\hat{A}(t) \stackrel{d}{\approx} (I - P')^{-1}\left[\hat{E}(t) + \sum_{k=1}^{K} \hat{\Phi}^k(\lambda_k t)\right]$$
$$- (I - P')^{-1}P'\Lambda G'[et - \tau(t)]. \tag{10.27}$$

From the bound in (10.3), the residual service time satisfies $\nu(t) \stackrel{d}{\approx} 0$ (referring to Lemma 9.9). Then, applying (10.14), (10.21) and (10.25) to (10.7), we obtain

$$W(t) \stackrel{d}{\approx} M[A(t) - D(t)] \stackrel{d}{\approx} M\Lambda G'[et - \tau(t)]. \tag{10.28}$$

In view of (10.9), the above implies

$$Z(t) \stackrel{d}{\approx} HM\Lambda G'[et - \tau(t)]. \tag{10.29}$$

Making use of first (10.15), (10.14), and (10.21), and then (10.27) and (10.29), we obtain

$$Z(t) \stackrel{d}{\approx} H[MA(t) + \hat{V}(\lambda t)] - C'[\kappa t + \hat{c}(t)] + Y(t)$$
$$\stackrel{d}{\approx} HM\lambda t + HM(I - P')^{-1}\left[\hat{E}(t) + \sum_{k=1}^{K} \hat{\Phi}^k(\lambda_k t)\right]$$
$$+ H\hat{V}(\lambda t) - NZ(t) - C'[\kappa t - \hat{c}(t)] + Y(t), \tag{10.30}$$

where

$$N = HM(I - P')^{-1}P'\Delta \quad \text{and} \quad \Delta = \Lambda G'[HM\Lambda G']^{-1}. \tag{10.31}$$

We note that the above inverse matrix always exists, and the (k, ℓ)th component of the matrix Δ is as follows

$$\Delta_{k\ell} = \begin{cases} \lambda_k/(GM\lambda)_\ell & \text{if } k \in g_\ell, \\ -\lambda_k/(GM\lambda)_{\ell'} & \text{if } k \in g_{\ell'} \text{ and } \ell = h(\ell'), \\ 0 & \text{otherwise}, \end{cases}$$

where $(GM\lambda)_\ell$ is the traffic intensity of group ℓ, $\ell \in \mathcal{L}$. Assuming the existence of the inverse of $I + N$, write

$$R = (I + N)^{-1} = (HM\Lambda G')[HM(I - P')^{-1}\Lambda G']^{-1}, \tag{10.32}$$
$$\theta = R(\beta - C'\kappa), \tag{10.33}$$

and

$$\hat{X}(t) = R\Bigg[HM(I - P')^{-1}\left[\hat{E}(t) + \sum_{k=1}^{K} \hat{\Phi}^k(\lambda_k t)\right]$$
$$+ H\hat{V}(\lambda t) - C'\hat{c}(t)\Bigg]. \tag{10.34}$$

Here $\hat{X}$ is a driftless Brownian motion. To get a feel for the covariance matrix of $\hat{X}$, denoted by $\Gamma_{\hat{X}}$, consider the case where the primitive processes (E, V, c, Φ) are mutually independent. Then, we can derive

$$
\Gamma_{\hat{X}} = R\Bigg[HM(I - P')^{-1}\Bigg(\Gamma_E + \sum_{k=1}^K \lambda_k \Gamma_{\Phi_k} \Bigg)(I - P)^{-1} MH'
$$

$$
+ H\Lambda\Gamma_V H' + C'\Gamma_c C \Bigg] R', \tag{10.35}
$$

where, recall, Γ_E, Γ_V, Γ_c and Γ_{Φ_k} are, respectively, the covariance matrices of the driftless Brownian motions, $\hat{E}$, $\hat{V}$, $\hat{c}$, and $\hat{\Phi}_k$ defined in (10.13)–(10.16).

We can now rewrite the approximation in (10.30) as

$$
Z(t) \stackrel{d}{\approx} \theta t + \hat{X}(t) + RY(t).
$$

If the above, along with (10.11) and the fact that $Z \geq 0$, defines a reflection mapping that is continuous (as is the case for the generalized Jackson network where R is an M-matrix and the case for the feedforward network where R is a triangular matrix), then we will have

$$
Z(t) \stackrel{d}{\approx} \tilde{Z}(t), \tag{10.36}
$$

where $\tilde{Z}$ satisfies

$$
\tilde{Z}(t) = \theta t + \hat{X}(t) + R\tilde{Y}(t) \geq 0 \qquad \text{for } t \geq 0; \tag{10.37}
$$

$$
d\tilde{Y} \geq 0 \qquad \text{and} \qquad \tilde{Y}(0) = 0; \tag{10.38}
$$

$$
\tilde{Z}_\ell d\tilde{Y}_\ell = 0, \qquad \ell \in \mathcal{L}. \tag{10.39}
$$

The above mapping (called the "reflection mapping"), which maps the processes $\hat{X}$ and $\tilde{Y}$ to $\tilde{Z}$, defines the latter process ($\tilde{Z}$) as a "semimartingale reflected Brownian motion" (SRBM), which we shall also refer to as SRBM (θ, Γ, R). (Note that a semimartingale is a martingale plus a process of bounded variation. Here, $\hat{X}$ is a martingale, and $\theta t + R\tilde{Y}(t)$ is a linear combination of monotone functions θt and $Y(t)$ and hence is of bounded variation.)

Before proceeding any further, we must make two important technical remarks here. First, the above mapping is not always well-defined, but more on this later. Second, $\hat{X}$, being a driftless Brownian motion is a martingale with respect to the (natural) filtration generated by itself. For the above approximation to work, however, $\hat{X}$ is required to be a martingale with respect to the filtration generated by $(\hat{X}, \tilde{Y})$. Nevertheless, this usually holds under nonanticipative service disciplines such as FIFO and priority; see Williams [29] for details.

For the time being, let us assume that the above mapping uniquely defines $\tilde{Z}$ (in distribution). Then, given the approximation (10.36), from (10.28) and (10.29), we have

$$W(t) \overset{d}{\approx} \tilde{W}(t) := M\Lambda G'[HM\Lambda G']^{-1}\tilde{Z}(t) = M\Delta\tilde{Z}(t) \qquad (10.40)$$

and

$$Q(t) \overset{d}{\approx} \tilde{Q}(t) := M^{-1}\tilde{W}(t) = \Delta\tilde{Z}(t). \qquad (10.41)$$

Finally, we derive the approximation for the sojourn-time process. In view of (10.22), we can approximate the service time of a the last arrived class k job at time t by a random variable v_k^0 as follows (referring to Remark 3 after Theorem 9.4):

$$V_k(A_k(t)) - V_k(A_k(t) - 1) \overset{d}{\approx} v_k^0, \qquad k \in \mathcal{K}.$$

We assume that v_k^0 has the same distribution as the service time of a class k job and is independent of $\tilde{W}(t)$; in particular, the mean of v_k^0 is m_k. (We could replace v_k^0 by any random variable (with finite variance), but choosing v_k^0, the service time of a class k job, appears to give a better approximation.) Then, in view of (10.12), we have

$$\kappa_{\sigma(\pi(k))}\mathcal{T}_k(t) \overset{d}{\approx} \tilde{Z}_{\pi(k)}(t) + \sum_{\substack{i:\sigma(i)=\sigma(k) \\ \pi(i)<\pi(k)}} m_i\lambda_i\mathcal{T}_k(t) + v_k^0$$

$$= \tilde{Z}_{\pi(k)}(t) + \beta_{h(k)}\mathcal{T}_k(t) + v_k^0,$$

where we define $\beta_0 = 0$. Hence, we have

$$\mathcal{T}_k(t) \overset{d}{\approx} \frac{\tilde{Z}_{\pi(k)}(t) + v_k^0}{\kappa_{\sigma(\pi(k))} - \beta_{h(k)}}.$$

Note that the approximation in (10.22), along with the bound in (10.2), implies $\eta_k(t) \sim t$. Then, in view of (10.10), the above approximation yields

$$\mathcal{S}_k(t) \overset{d}{\approx} \tilde{\mathcal{S}}_k(t) := \frac{\tilde{Z}_{\pi(k)}(t) + v_k^0}{\kappa_{\sigma(\pi(k))} - \beta_{h(k)}}. \qquad (10.42)$$

To summarize, our approximation scheme is built upon the following three conditions:

(A) the primitive processes (E, V, c, Φ) satisfy the asymptotics in (10.13), (10.14), (10.15), and (10.16);

(B) the arrival process A satisfies the approximation in (10.21);

(C) the SRBM $\tilde{Z}$ is well defined (which includes conditions such as that the matrix $(I + N)$ is invertible and $\hat{X}$ is a martingale with respect to the filtration generated by $(\hat{X}, \tilde{Y})$).

Under these conditions, the aggregated workload process Z can be approximated by $\tilde{Z}$, and the other derived processes, in particular W (workload), Q (queue length), and S (sojourn time), follow the approximations in (10.40), (10.41), and (10.42), respectively.

Condition (A) is basically a modeling choice, and given a specific application, the required asymptotics are often routinely justified, as shown in the examples in Section 10.1.2. Condition (B) implies that $A(t)/t$ converges almost surely as $t \to \infty$. This requires the network to be weakly stable (and hence stable). (See Exercises 1–3.) Under some mild conditions, it appears that (B) is equivalent to what is known as the state-space collapse condition, which is a key for the heavy traffic limit theorem; refer to Section 10.4.3. An elaboration of condition (C) requires a formal definition of the SRBM, which will be provided in the next section.

10.4 Discussions and Variations

10.4.1 Issues Surrounding the SRBM

First, and most importantly, the SRBM must be well-defined. To recapitulate, assuming that $\Gamma_{\hat{X}}$ is nondegenerate, $\tilde{Z}$ is called a semimartingale reflected Brownian motion, denoted by $\mathrm{SRBM}(\theta, \Gamma_{\hat{X}}, R)$, if the relations in (10.37)–(10.39) hold and $\hat{X}$ is a martingale with respect to the filtration generated by $(\hat{X}, \tilde{Y})$; we say that $\tilde{Z}$ is well-defined if these relations uniquely determine its distribution (i.e., uniqueness in law). It is known (Reiman and Williams [27] and Taylor and Williams [28]) that the SRBM $\tilde{Z}$ is well-defined if and only if the reflection matrix R is completely-S.

Except for some rare cases (usually associated with very specially correlated primitive processes), the covariance matrix $\Gamma_{\hat{X}}$ is almost always nondegenerate. The key is therefore whether the reflection matrix R is completely-S. Some sufficient conditions are (i) R is an M-matrix; or (ii) R is a P-matrix. (Refer to Exercises 8 and 9 in Chapter 7.) As we shall see in Section 10.4.2 and Section 10.6.2, the reflection matrix R may not always be completely-S. In addition, the example in Section 10.6.2 shows that $(I + N)$ may not even be invertible, and hence R may not even be defined.

The second issue is whether or not the SRBM $\tilde{Z}$ possesses a stationary distribution, and when it does, how the stationary distribution can be computed (either analytically or numerically). This topic will be dealt with in full detail in Section 10.5. The existence of a stationary distribution of $\tilde{Z}$ is obviously motivated by the need to approximate the stationary performance measures of queueing networks. Hence, it is only natural to expect that the issue has a close relationship with the stability of queueing networks. For the latter subject, a substantial body of literature has emerged in recent years; refer to Section 8.7. It is known that the queueing network is stable if a corresponding fluid network is stable (Theorem 8.3). Note, however, that the stability of a queueing network does *not* in general imply that the SRBM $\tilde{Z}$ is even well-defined (see the examples in Section 10.4.2 and Section 10.6.2). And, the converse also remains an open problem, i.e., whether the existence of a well-defined $\tilde{Z}$ (given that the traffic intensity is less than the service capacity) implies the stability of a corresponding queueing network.

The third issue is whether or not the approximation in (10.36) is supported by a limit theorem, in the following sense (which is the strong approximation): There exists a common probability space in which both Z and $\tilde{Z}$ are defined, such that

$$\sup_{0 \leq t \leq T} \|Z(t) - \tilde{Z}(t)\| \overset{\text{a.s.}}{=} o(T^{1/r}), \qquad \text{as } T \to \infty,$$

holds for some $r > 2$. Indeed, such an approximation is established in Section 7.7 for a generalized Jackson network, and in Section 9.6 for a multiclass feedforward network. In both cases, the reflection mapping is well-defined and Lipschitz continuous. This assumption, however, appears too strong for a general multiclass network, as it would require that the relations in (10.37)–(10.39) uniquely determine the processes $\tilde{Z}$ and $\tilde{Y}$, in a pathwise sense, for any given process $\tilde{X}$ (with $\tilde{X}(0) \geq 0$). It is known that this cannot hold (see Exercises 11–13 in Chapter 7 and references in Section 7.11).

An alternative interpretation of the approximation is to view the network under study as one in a sequence of networks indexed by $n = 1, 2, \ldots$. Let ρ^n be the traffic intensity of the nth network, and assume that $\sqrt{n}[\rho^n - \kappa] \to \theta$ (which implies $\rho^n \to \kappa$ as $n \to \infty$). This is known as the heavy traffic condition. If we can show that as $n \to \infty$,

$$\frac{1}{\sqrt{n}} Z^n(n\cdot; (\rho^n - \kappa)) \overset{d}{\to} \tilde{Z}(\cdot; \theta), \tag{10.43}$$

then we may interpret the above as

$$Z^n(t; (\rho^n - \kappa)) \overset{d}{\approx} \sqrt{n}\tilde{Z}(t/n; (\rho^n - \kappa)) \overset{d}{=} \tilde{Z}(t; (\rho^n - \kappa)),$$

where the second distributional equality follows from the scaling property of the Brownian motion and the reflection mapping. Suppose the network under study is the one indexed by m. Then, we have

$$\rho = \rho^m, \quad Z(t) = Z^m(t; (\rho^m - \kappa)), \quad \tilde{Z}(t) = \tilde{Z}(t; (\rho^m - \kappa));$$

and the approximation in (10.36) holds. In this case, we can claim that (10.36) is supported by the heavy traffic limit theorem or the diffusion limit theorem. Section 7.9 provides such an example for a generalized Jackson network. In this special case, the SRBM $\tilde{Z}(\cdot; (\rho - e))$ is well-defined for any $\rho \leq e$ (where $\kappa = e$). In more general cases, however, as we shall demonstrate in Section 10.4.2 below via the Kumar–Seidman network, the situation is much more complex: $\tilde{Z}(\cdot; \rho - e)$ may not exist for some $\rho \leq e$ (although it does exist when ρ is close enough to e), even if the corresponding heavy traffic limit (10.43) does exist.

10.4.2 Kumar–Seidman Network

For the Kumar–Seidman network shown in Figure 8.1, we have

$$
\Delta = \begin{pmatrix}
-m_1^{-1} & m_1^{-1} & 0 & 0 \\
0 & 0 & m_2^{-1} & 0 \\
0 & 0 & -m_3^{-1} & m_3^{-1} \\
m_4^{-1} & 0 & 0 & 0
\end{pmatrix},
$$

$$
R = \begin{pmatrix}
1 & 0 & m_3^{-1}m_4 & -m_3^{-1}m_4 \\
0 & 1 & m_3^{-1}m_4 & -m_3^{-1}m_4 \\
m_1^{-1}m_2 & -m_1^{-1}m_2 & 1 & 0 \\
m_1^{-1}m_2 & -m_1^{-1}m_2 & 0 & 1
\end{pmatrix},
$$

$$
I + N = \begin{pmatrix}
1 & 0 & -m_3^{-1}m_4 & m_3^{-1}m_4 \\
0 & 1 & -m_3^{-1}m_4 & m_3^{-1}m_4 \\
-m_1^{-1}m_2 & m_1^{-1}m_2 & 1 & 0 \\
-m_1^{-1}m_2 & m_1^{-1}m_2 & 0 & 1
\end{pmatrix}.
$$

Note that a principal minor of R,

$$
\begin{pmatrix}
1 & -m_3^{-1}m_4 \\
-m_1^{-1}m_2 & 1
\end{pmatrix},
$$

is an S-matrix if and only if

$$
m_2 m_4 < m_1 m_3; \tag{10.44}
$$

and it is straightforward to verify that this is also a necessary and sufficient condition for R to be completely-S. Note that this condition does not depend on α_1 and α_3. Hence, by making α_1 and α_3 small enough, we can create a case in which the network is stable but the SRBM is not well-defined. Under the traffic condition $\rho < e$, the condition (10.44) is strictly stronger than

$$\alpha_1 m_2 + \alpha_3 m_4 < 1, \tag{10.45}$$

which is a sufficient condition for the stability of the Kumar–Seidman network (Theorem 8.3). On the other hand, under the heavy traffic condition $\rho = e$, (10.44) becomes equivalent to (10.45), which is a necessary and sufficient condition for the existence of a heavy traffic limit theorem (Theorem 8.11).

If the network parameters α and m are such that $\rho < e$ and $m_2 m_4 > m_1 m_3$ but (10.45) still holds, then the matrix R is not completely-S, and hence the proposed Brownian approximation does not exist. This network can clearly be considered as one particular network in a sequence of networks that approach the heavy traffic limit (if we assume that the limiting network satisfies (10.45)). This suggests that in this case the heavy traffic limit does not provide the mathematical support for the approximation. But if the particular network chosen is "closer enough" to the limiting network, i.e., the parameters α and m are chosen in a small neighborhood of

$$\{(\alpha, m) : \rho = e \text{ and } \alpha_1 m_2 + \alpha_3 m_4 < 1\}$$
$$= \{(\alpha, m) : \rho = e \text{ and } m_2 m_4 < m_1 m_3\},$$

then (10.44) must hold and the the SRBM approximation is well-defined.

10.4.3 Alternative Approximation with State-Space Collapse

Following the usual heavy traffic limit theorem, the diffusion approximation for the workload or the queue length of a higher priority group is simply zero. For instance, it is established in Theorem 8.11 for the Kumar–Seidman network that the diffusion limits for the queue-length processes of classes 2 and 4 are zero.

This phenomenon is known as a state-space collapse principle, and is critical to the heavy traffic approximation of multiclass networks. More specifically, state-space collapse means: (a) in a multiclass network that follows a static priority service discipline, only the lowest priority group has a nonempty queue under heavy traffic scaling; and (b) the queue length of each class within this lowest priority group is at a fixed proportion of the overall workload of the group under the heavy traffic scaling.

Under state-space collapse, the components of $\tilde{Z}$ corresponding to higher priority groups at each station will vanish. Hence, the approximating SRBM is of dimension J, which could be much smaller than L, resulting in a substantial computational advantage. Although zero might appear to be a rather crude approximation for the workload and the queue-length processes of a higher priority class, this approximation may work well in some cases, particularly when the traffic intensity contributed by higher priority classes is negligible.

Now, we derive the SRBM model with state-space collapse. Define

- $U = (U_j)$ with $U_j = \{U_j(t), t \geq 0\}$, $j \in \mathcal{J}$, where $U_j(t)$ indicates the total amount of workload at station j at time t; hence, if group ℓ has the lowest priority at station $\sigma(\ell)$, then $U_{\sigma(\ell)}(t) = Z_\ell(t)$;

- $I = (I_j)$ with $I_j = \{I_j(t), t \geq\}$, $j \in \mathcal{J}$, where $I_j(t)$ indicates the cumulative amount of unused capacity during $[0, t]$ at station j; hence, if group ℓ has the lowest priority at station $\sigma(\ell)$, then $I_{\sigma(\ell)}(t) = Y_\ell(t)$.

- $\tilde{\Delta} = (\tilde{\Delta}_{\ell,j})_{\ell \in \mathcal{L}, j \in \mathcal{J}}$, with

$$\tilde{\Delta}_{\ell j} = \begin{cases} 1 & \text{if group } \ell \text{ is the lowest priority group served at station } j, \\ 0 & \text{otherwise.} \end{cases}$$

Then, similar to (10.8), we have

$$U(t) = CGV(A(t)) - c(t) + I(t). \tag{10.46}$$

Under state-space collapse, the workload of all but the lowest priority group vanishes; hence, we have

$$Z(t) \overset{d}{\approx} \tilde{\Delta} U(t). \tag{10.47}$$

Combined with (10.29), this implies

$$[et - \tau(t)] \overset{d}{\approx} [HM\Lambda G']^{-1} \tilde{\Delta} U(t). \tag{10.48}$$

From (10.27), (10.48), and (10.46), we obtain

$$U(t) \overset{d}{\approx} CG[MA(t) + \hat{V}(\lambda t)] - [\kappa t + \hat{c}(t)] + I(t)$$

$$\overset{d}{\approx} CGM\lambda t + CGM(I - P')^{-1} \left[\hat{E}(t) + \sum_{k=1}^{K} \hat{\Phi}^k(\lambda_k t) \right]$$

$$+ CG\hat{V}(\lambda t) - CGM(I - P')^{-1}P'\Delta \tilde{\Delta} U(t) - [\kappa t - \hat{c}(t)] + I(t).$$

Thus, the above can be written as

$$U(t) \overset{d}{\approx} \theta t + \tilde{X}(t) + RI(t),$$

where

$$R = (I + CGM(I - P')^{-1}P'\Delta\tilde{\Delta})^{-1},$$
$$\theta = R(CGM\lambda - \kappa),$$

and

$$\tilde{X}(t) = R\left[CGM(I - P')^{-1}\left[\hat{E}(t) + \sum_{k=1}^{K}\hat{\Phi}^k(\lambda_k t)\right] + CG\hat{V}(\lambda t) - \hat{c}(t)\right].$$

Note that the jth component of $CGM\lambda$ is the traffic intensity of station j, $j \in \mathcal{J}$.

In summary, we have

$$U(t) \overset{d}{\approx} \tilde{U}(t),$$

$$Z(t) \overset{d}{\approx} \tilde{Z}(t) := \tilde{\Delta}\tilde{U}(t),$$

where $\tilde{U}$ satisfies

$$\tilde{U}(t) = \theta t + \tilde{X}(t) + R\tilde{I}(t) \geq 0,$$
$$d\tilde{I} \geq 0 \quad \text{and} \quad \tilde{I}(0) = 0,$$
$$\tilde{U}_j d\tilde{I}_j = 0, \quad j \in \mathcal{J}.$$

The approximations for the other derived processes, Q, W, and $\mathcal{S}$, remain the same as in (10.40), (10.41), and (10.42), respectively.

To conclude this part, let us revisit the Kumar–Seidman network. We can draw the same conclusion as in Section 10.4.2 by considering the approximation under state-space collapse. In particular, when $m_1 m_3 \neq m_2 m_4$, the reduced two-dimensional reflection matrix for the SRBM takes the form

$$\tilde{R}_L = \frac{1}{1 - m_1^{-1}m_2 m_3^{-1}m_4}\begin{pmatrix} 1 & m_3^{-1}m_4 \\ m_1^{-1}m_2 & 1 \end{pmatrix},$$

which is completely-S if and only if (10.44) holds, the same condition as in Section 10.4.2. We note that the reflection matrix $\tilde{R}_L$ is not defined when $m_1 m_3 = m_2 m_4$; in this case, the diffusion approximation does not exist (Theorem 8.11).

10.5   Stationary Distribution of the SRBM

Here we examine more closely the stationary distribution of the SRBM (θ, Γ, R). To simplify notation, we drop the tildes and write X, Y and Z in lieu of $\tilde{X}$, $\tilde{Y}$, and $\tilde{Z}$ in (10.37)–(10.39), respectively. We first present some sufficient conditions under which the stationary distribution exists, and then present a basic adjoint relation that characterizes the stationary distribution. Finally, we identify a special case where the stationary distribution has a product form, and derive the product form solution.

Consider a linear Skorohod problem (a special case of the Skorohod problem introduced in Section 7.2). Specifically, given $x(0) \geq 0$ and $\theta \in \Re^L$, the pair (y, z) in C^{2L} is said to be a solution to a linear Skorohod problem $\mathrm{LSP}(\theta, R)$ if the pair satisfies

$$z(t) = x(0) + \theta t + Ry(t) \geq 0 \qquad \text{for all } t \geq 0, \qquad (10.49)$$
$$dy \geq 0 \qquad \text{and} \qquad y(0) = 0, \qquad (10.50)$$
$$z_\ell dy_\ell = 0, \qquad \ell = 1, \ldots, L. \qquad (10.51)$$

A linear Skorohod problem $\mathrm{LSP}(\theta, R)$ is said to be stable with initial state $x(0)$ if every z component of all of its solutions is such that for every $\epsilon > 0$, there exists a $T < \infty$ such that $z(t) \leq \epsilon$ for all $t \geq T$. If $\mathrm{LSP}(\theta, R)$ is stable for every initial state $x(0) \geq 0$, then we simply say that $\mathrm{LSP}(\theta, R)$ is stable. The following theorem is proved by Dupuis and Williams [17].

Theorem 10.8 Assume that the matrix R is completely-S and the matrix Γ is nondegenerate. Then, the $\mathrm{SRBM}(\theta, \Gamma, R)$ is positive recurrent and has a unique stationary distribution if the corresponding linear Skorohod problem, $\mathrm{LSP}(\theta, R)$, is stable.

To characterize the stationary distribution, define the following operators:

$$\mathcal{L} = \frac{1}{2} \sum_{i,j=1}^{L} \Gamma_{ij} \frac{\partial^2}{\partial x_i \partial x_j} + \sum_{j=1}^{L} \theta_j \frac{\partial}{\partial x_j},$$

$$\mathcal{D}_j = \sum_{i=1}^{L} r_{ij} \frac{\partial}{\partial x_i} \equiv r_j' \nabla,$$

where r_j is the jth column of R. The following theorem is proved by Dai and Kurtz [13].

Theorem 10.9 Let π be a stationary distribution of $Z = \mathrm{SRBM}(\theta, \Gamma, R)$. Then, for each $j = 1, \ldots, L$, there exists a finite Borel measure ν_j on F_j (which is mutually absolutely continuous with the Lebesgue measure on the face F_j) such that a basic adjoint relation (BAR)

$$\int_{\Re_+^L} \mathcal{L}f(z) d\pi(z) + \sum_{j=1}^{L} \int_{F_j} \mathcal{D}_j f(z) d\nu_j(z) = 0 \qquad (10.52)$$

holds for every continuous and bounded function f defined on $\Re_+^J$ whose derivatives up to order 2 are also continuous and bounded.

Remark 10.10 Note that

$$\nu_j(F_j) = \frac{1}{t} \mathsf{E}_\pi[Y_j(t)], \qquad \text{for all } t > 0,$$

gives the average (over time) expected amount of push-back at the face F_j to keep the semimartingale reflected Brownian motion within its domain, the nonnegative orthant.

Next, we show how to use the basic adjoint relation to derive the stationary distribution in a special case where the stationary distribution has a product form. Let Λ_R and Λ_Γ be two $L \times L$ diagonal matrices; their jth diagonal components are respectively r_{jj} and Γ_{jj} (respectively the (j, j)th components of matrices R and Γ). The following is a natural extension of Theorem 7.12.

Theorem 10.11 Suppose the stationary distribution of $Z =$SRBM(θ, Γ, R) exists. Then, the stationary distribution has an exponential product form, namely, its density function takes the form

$$p(z) = \prod_{j=1}^{L} p_j(z_j) \qquad \text{for } z = (z_1, \ldots, z_L)' \in \Re_+^L$$

with

$$p_j(z_j) = \eta_j e^{-\eta_j z_j}, \qquad z_j \in \Re_+, \tag{10.53}$$

$j = 1, \ldots, L$, if and only if

$$\eta := -2\Lambda_\Gamma^{-1} \Lambda_R R^{-1} \theta > 0 \tag{10.54}$$

and

$$2\Gamma = R\Lambda_R^{-1}\Lambda_\Gamma + \Lambda_\Gamma\Lambda_R^{-1}R'. \tag{10.55}$$

10.6   Special Cases and Numerical Results

10.6.1   A Single-Class Queue with Breakdown

Consider a single-class, single-server queue subject to autonomous breakdown. We characterize the job arrival process by its first two moments: the arrival rate α and the SCV of interarrival time c_0^2. Similarly, let m be the mean service time and c_s^2 the SCV of the service time. Let d and c_d^2 (u and c_u^2) be the mean and the SCV of the down- (up-)time.

Following the results in Section 10.3, we can construct an SRBM $\tilde{Z}$ as an approximation to the workload process with the following data

$$\theta = \alpha m - \frac{u}{u + d},$$

$$R = 1,$$

$$\Gamma = m^2 \alpha(c_s^2 + c_0^2) + \frac{d^2 u^2 (c_u^2 + c_d^2)}{(u + d)^3}.$$

Case	α	c_0^2	m	c_s^2	u	c_u^2	d	c_d^2
1	4.0	1	0.15	1	12	1	4	1
2	4.0	1	0.15	1	12	2	4	2
3	4.0	2	0.15	2	12	1	4	1
4	4.0	2	0.15	2	12	2	4	2
5	4.0	1	0.15	1	18	1	2	1
6	4.0	1	0.15	1	18	2	2	2
7	4.0	2	0.15	2	18	1	2	1
8	4.0	2	0.15	2	18	2	2	2
9	1.0	1	0.75	1	18	1	2	1
10	1.0	1	0.75	1	18	2	2	2
11	1.0	2	0.75	2	18	1	2	1
12	1.0	2	0.75	2	18	2	2	2

TABLE 10.1. Parameters of a single-station queue with breakdown

For one-dimensional SRBM, the steady-state distribution is known to be exponentially distributed with mean $E(\tilde{Z}) = \Gamma/(2\theta)$. From (10.42), we obtain the steady-state average system time

$$E(\tilde{S}) = \left(\frac{\Gamma}{2\theta} + m\right) \Big/ \kappa,$$

where $\kappa = u/(u + d)$.

In Table 10.1 we list several different sets of parameters of the queue. In simulation, the random variables (representing service times, up-times, and down-times) are fitted with Erlang distributions, exponential distributions, or gamma distributions corresponding to the SCV being less than 1, equal to 1, or larger than 1. This scheme is used for simulations in all subsequent examples as well. Table 10.2 presents both simulation and SRBM estimates for $E(\tilde{S})$. In this table, and in all subsequent tables, the number in parentheses after each simulation result is the half-width of the 95% confidence interval, expressed as a percentage of the simulation estimate. The number in parentheses after each SRBM estimate represents the percentage error from the simulation result.

10.6.2 A Variation of the Bramson Network

For the network shown in Figure 8.11, we have

$$\Delta = \lambda \begin{pmatrix} \rho_1^{-1} & 0 \\ 0 & \rho_2^{-1} \\ 0 & \rho_2^{-1} \\ 0 & \rho_2^{-1} \\ \rho_1^{-1} & 0 \end{pmatrix}$$

Case	Simulation Results	SRBM Estimates
1	5.924 (6.8%)	6.000 (1.3%)
2	10.017 (7.7%)	11.00 (9.8%)
3	6.972 (7.4%)	6.800 (2.5%)
4	12.003 (7.3%)	11.800 (1.7%)
5	1.070 (3.0%)	1.100 (2.8%)
6	1.464 (5.3%)	1.700 (16.1%)
7	1.471 (2.6%)	1.433 (2.6%)
8	1.861 (4.2%)	2.033 (9.2%)
9	6.230 (6.2%)	6.200 (0.5%)
10	7.501 (6.9%)	7.400 (1.3%)
11	10.282 (5.0%)	10.367 (0.8%)
12	12.245 (7.3%)	11.567 (5.5%)

TABLE 10.2. Average sojourn time in the single-class single-server queue with breakdown

and

$$R = \frac{1}{\delta} \begin{pmatrix} 1 + \lambda(m_3 + 2m_4)\rho_2^{-1} & -3\lambda m_5 \rho_2^{-1} \\ -\rho_2 \rho_1^{-1} & 1 + \lambda m_5 \rho_1^{-1} \end{pmatrix},$$

where

$$\delta = [1 + \lambda m_5 \rho_1^{-1}][1 + \lambda(m_3 + 2m_4)\rho_2^{-1}] - 3\lambda m_5 \rho_1^{-1}.$$

Note that $\delta > 0$ if and only if

$$m_1(m_2 + m_3 + m_4) > (m_2 - m_4)m_5; \tag{10.56}$$

and it is easy to see that this condition is a necessary and sufficient condition for the matrix R to be completely-S (which in this case is equivalent to R being an M-matrix). (Note that if (10.56) holds with equality, then the matrix $(I + N)$ in (10.32) does not exist and matrix R is not well-defined.) Notice that this condition does not depend on the arrival rate λ ($\lambda = \alpha_1$). This implies that similar to the Kumar–Seidman network in Section 10.4.2, we can create a case in which the network is stable (by making λ small enough) but the SRBM approximation is not well-defined. For instance, $m = (0.02, 0.8, 0.05, 0.05, 0.88)$ provides such a case where (10.56) is violated. In this case, if we let $\lambda = 1$, the traffic intensities at both stations are 0.9. The simulation results presented in Figures 8.12 and 8.13 suggest that the total number of jobs in the system grows with time and that the network cannot be stable. In this sense, the existence of the SRBM appears to be closely related to the stability of the queueing network.

Next, we fix the arrival rate at $\lambda = 1$, and vary the mean service times (m, see Tables 10.3) and 10.4, which all satisfy (10.56), and hence, the corresponding SRBM exists. In Cases 1, 2, 8, and 9, the m values are very close to the region where the SRBM does not exist; and interestingly, the

No.	m	Method	Station 1	Station 2
1	(0.40, 0.80, 0.05, 0.05, 0.45)	Simulation	5.759 (5.7%)	9.898 (9.2%)
		SRBM	11.937 (107.3%)	12.030 (21.5%)
2	(0.40, 0.80, 0.05, 0.05, 0.40)	Simulation	3.544 (5.2%)	7.866 (7.0%)
		SRBM	9.760 (175.4%)	14.388 (82.9%)
3	(0.40, 0.80, 0.05, 0.05, 0.20)	Simulation	0.748 (2.3%)	6.370 (10.3%)
		SRBM	1.012 (35.3%)	6.879 (8.0%)
4	(0.40, 0.30, 0.30, 0.30, 0.45)	Simulation	2.375 (3.9%)	2.797 (6.9%)
		SRBM	2.432 (2.4%)	2.743 (1.9%)
5	(0.02, 0.30, 0.30, 0.30, 0.88)	Simulation	8.659 (6.9%)	5.767 (7.4%)
		SRBM	8.959 (3.5%)	4.838 (16.1%)
6	(0.10, 0.30, 0.30, 0.30, 0.80)	Simulation	6.349 (5.9%)	4.645 (5.6%)
		SRBM	7.156 (12.7%)	4.164 (10.4%)
7	(0.20, 0.30, 0.30, 0.30, 0.70)	Simulation	4.922 (4.8%)	3.719 (4.1%)
		SRBM	5.552 (12.8%)	3.549 (4.6%)
8	(0.02, 0.425, 0.05, 0.425, 0.80)	Simulation	9.499 (4.9%)	6.311 (4.4%)
		SRBM	9.911 (4.3%)	5.432 (13.9%)
9	(0.02, 0.43, 0.05, 0.42, 0.88)	Simulation	9.484 (4.3%)	6.074 (4.0%)
		SRBM	10.034 (5.8%)	5.529 (9.0%)

TABLE 10.3. Average waiting times in the Bramson network

No.	Method	$E(S_1)$	$E(S_2)$	$E(S_3)$	$E(S_4)$	$E(S_5)$
1	Simulation	5.68 (5.4%)	10.89 (7.4%)	10.25 (7.8%)	9.45 (8.4%)	6.69 (5.2%)
	SRBM	12.34 (117.2%)	12.83 (17.8%)	12.08 (17.8%)	12.08 (27.8%)	12.39 (85.2%)
2	Simulation	3.55 (4.9%)	8.83 (6.8%)	8.17 (7.3%)	7.49 (7.9%)	4.34 (4.6%)
	SRBM	10.16 (186.2%)	15.19 (72.0%)	14.44 (76.7%)	14.44 (92.8%)	10.16 (134.1%)
3	Simulation	1.02 (1.5%)	7.31 (7.8%)	6.62 (8.5%)	6.08 (9.1%)	1.08 (2.0%)
	SRBM	1.41 (38.2%)	7.68 (5.1%)	6.93 (4.7%)	6.93 (14.0%)	1.21 (12.0%)
4	Simulation	2.78 (3.4%)	3.09 (6.8%)	3.10 (6.9%)	3.10 (6.8%)	2.83 (3.3%)
	SRBM	2.83 (1.8%)	3.04 (1.6%)	3.04 (1.9%)	3.04 (1.9%)	2.88 (1.8%)
5	Simulation	8.65 (6.9%)	5.91 (6.4%)	6.15 (6.3%)	6.14 (6.3%)	9.57 (6.2%)
	SRBM	8.98 (3.8%)	5.14 (13.0%)	5.14 (16.4%)	5.14 (16.3%)	9.84 (2.8%)
6	Simulation	6.44 (5.8%)	4.85 (5.2%)	5.00 (5.2%)	4.99 (5.1%)	7.16 (5.2%)
	SRBM	7.26 (12.7%)	4.46 (8.0%)	4.46 (10.8%)	4.46 (10.6%)	7.96 (11.2%)
7	Simulation	5.13 (4.6%)	3.96 (4.0%)	4.05 (3.9%)	4.05 (3.9%)	5.61 (4.2%)
	SRBM	5.75 (12.1%)	3.85 (2.8%)	3.85 (4.9%)	3.85 (4.9%)	6.25 (11.4%)
8	Simulation	9.44 (4.9%)	6.74 (4.4%)	6.36 (4.6%)	6.74 (4.3%)	10.46 (4.5%)
	SRBM	9.93 (5.2%)	5.86 (13.1%)	5.48 (13.8%)	5.86 (13.1%)	10.71 (2.4%)
9	Simulation	9.42 (4.3%)	6.50 (3.4%)	6.14 (3.6%)	6.49 (3.4%)	10.44 (4.0%)
	SRBM	10.05 (6.7%)	5.96 (8.3%)	5.58 (9.1%)	5.95 (8.3%)	10.91 (4.5%)

TABLE 10.4. Average sojourn time of each job class in the Bramson network

Case	m_1	m_2	m_3	m_4	m_5	m_6	ρ_1	ρ_2
1	0.5	0.5	1.0	0.50	0.50	1.0	0.90	0.65
2	1.0	0.5	0.4	1.00	1.00	0.4	0.80	0.90
3	0.5	0.5	0.4	0.25	0.25	2.0	0.60	0.70

TABLE 10.5. Mean service times of three cases of the complex queueing network

associated SRBM approximation gives very poor estimates in Cases 1 and 2 but reasonably good estimates in Case 8 and 9. The best approximation appears in Case 4, where the mean service times are identical at station 2 and nearly identical at station 1.

10.6.3   A More Complex Multiclass Network

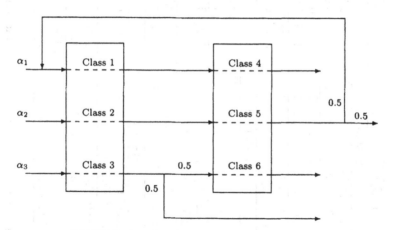

FIGURE 10.1. A two-station multiclass network

Consider the two-station network shown in Figure 10.1. Job classes 1, 2, and 3 are processed at station 1, where the service discipline is FIFO; classes 4, 5, and 6 are processed at station 2, where class 4 has a higher priority over classes 5 and 6 (which are at the same priority rank, and hence served FIFO). Thus, there exist three priority job groups; jobs in classes 1, 2, and 3 belong to group 1; class 4 jobs belong to group 2; class 5 and 6 jobs belong to group 3. Classes 1, 2, and 3 are associated with external arrivals, which are Poisson processes with rates 0.2, 0.4, and 0.5 respectively. Service times of all six classes are i.i.d. with the SCV denoted by c_s^2. Consider three versions of the network, labeled as systems A, B, and C, with $c_s^2 = 0.25, 1$, and 2, respectively. In each system there are 3 different subcases, as shown in Table 10.5, where m_i is the mean service time of class i, $i = 1, \ldots, 6$; ρ_1 and ρ_2 denote the traffic intensities at the two stations. The routing matrix

Case	Method	$E(S_1)$	$E(S_2)$	$E(S_3)$	$E(S_4)$	$E(S_5)$	$E(S_6)$
				System A			
1	Simu.	4.88 (4.2%)	4.86 (4.2%)	5.36 (3.0%)	0.55 (0.3%)	1.33 (0.8%)	1.70 (0.6%)
	SRBM	4.70 (3.7%)	4.70 (3.3%)	5.30 (1.1%)	0.57 (3.6%)	1.29 (3.0%)	1.91 (12.4%)
2	Simu.	2.54 (1.8%)	2.13 (2.1%)	2.02 (2.2%)	1.14 (0.4%)	9.23 (5.3%)	8.30 (5.3%)
	SRBM	2.52 (0.8%)	2.02 (5.2%)	1.92 (5.0%)	1.15 (0.9%)	8.77 (5.0%)	7.77 (6.4%)
3	Simu.	1.00 (0.7%)	0.96 (0.8%)	0.86 (0.7%)	0.26 (0.3%)	2.46 (2.0%)	4.45 (1.5%)
	SRBM	0.98 (2.0%)	0.98 (21.%)	0.88 (2.3%)	0.25 (3.8%)	2.57 (4.5%)	4.51 (1.3%)
				System B			
1	Simu.	7.07 (5.3%)	7.15 (5.6%)	7.64 (4.8%)	0.65 (0.5%)	2.48 (1.7%)	2.81 (1.5%)
	SRBM	7.26 (2.7%)	7.26 (1.5%)	7.76 (1.6%)	0.63 (3.1%)	2.31 (6.9%)	2.93 (4.3%)
2	Simu.	3.76 (1.8%)	3.28 (1.9%)	3.19 (2.0%)	1.46 (0.7%)	14.15 (5.8%)	13.34 (6.4%)
	SRBM	3.85 (2.4%)	3.35 (2.1%)	3.25 (1.9%)	1.45 (0.7%)	14.61 (3.3%)	13.61 (2.0%)
3	Simu.	1.28 (1.1%)	1.23 (1.2%)	1.13 (1.6%)	0.28 (0.4%)	4.01 (3.6%)	5.95 (2.7%)
	SRBM	1.26 (1.6%)	1.26 (2.4%)	1.16 (2.7%)	0.27 (3.6%)	4.09 (2.0%)	6.03 (1.3%)
				System C			
1	Simu.	9.90 (5.4%)	10.17 (5.3%)	10.61 (5.5%)	0.77 (1.5%)	3.95 (2.4%)	4.26 (2.4%)
	SRBM	10.18 (2.8%)	10.18 (0.1%)	10.67 (0.6%)	0.71 (7.8%)	3.58 (9.4%)	4.20 (1.4%)
2	Simu.	5.42 (3.4%)	4.90 (3.8%)	4.80 (3.5%)	1.87 (1.4%)	22.99 (7.9%)	22.09 (8.2%)
	SRBM	5.62 (3.7%)	5.12 (4.5%)	5.02 (4.6%)	1.85 (1.1%)	22.82 (0.7%)	21.82 (1.2%)
3	Simu.	1.60 (1.5%)	1.57 (1.4%)	1.46 (1.3%)	0.30 (0.7%)	5.95 (4.2%)	7.90 (3.4%)
	SRBM	1.62 (1.3%)	1.62 (3.2%)	1.52 (4.1%)	0.30 (0.0%)	6.12 (2.3%)	8.07 (2.2%)

TABLE 10.6. Average sojourn time of each job class in the complex queueing network

is

$$P = \begin{pmatrix} 0 & 0 & 0 & 1 & 0 & 0 \\ 0 & 0 & 0 & 0 & 1 & 0 \\ 0 & 0 & 0 & 0 & 0 & 0.5 \\ 0 & 0 & 0 & 0 & 0 & 0 \\ 0.5 & 0 & 0 & 0 & 0 & 0 \\ 0 & 0 & 0 & 0 & 0 & 0 \end{pmatrix}.$$

The performance measure of interest is the average sojourn times experienced by each job class at the two stations, denoted by $E(S_i)$, $i = 1, \ldots, 6$. From (10.42), we obtain

$$E(S_i) = m_i + E(\tilde{Z}_1), \qquad i = 1, 2, 3,$$
$$E(S_i) = m_i + E(\tilde{Z}_2), \qquad i = 4,$$
$$E(S_i) = \frac{1}{1 - 0.4m_4}(m_i + E(\tilde{Z}_3)), \qquad i = 5, 6.$$

The numerical comparisons against simulation are summarized in Table 10.6. The quality of the approximation in this example is very good.

Next we present some numerical results on the stationary *distribution* for Case 2 of System A. In Figures 10.2 and 10.3 the cumulative stationary distributions are plotted for the waiting times of job groups 1 and 3, respectively. The SRBM approximations of the waiting times are by the approximations of the workload process. Specifically, $\tilde{W}_1 = \tilde{W}_2 = \tilde{W}_3$ for job group 1 and $\tilde{W}_5 = \tilde{W}_6$ for job group 3; their stationary distributions are generated by the BNA/FM algorithm. From these two figures, we can see that SRBM provides a fairly good approximation for the stationary distribution as well.

To approximate the stationary distribution of the sojourn time by (10.42) would involve numerically computing the convolution of two random variables, which is not reported here.

10.7 Notes

This chapter is adapted from Chen et al. [5], with the exception of Section 10.5, which is adapted from an extension of Harrison and Williams [21].

In the performance evaluation of queueing networks, the Brownian approximation refers to approximating certain processes of interest, such as workload, queue lengths, and sojourn times, by the SRBM or its special case, the RBM. Based on the heavy traffic limit theorem by Reiman [25] (refer to Section 7.9), Harrison and Williams [21] propose an RBM approximation for the generalized Jackson network as given in Section 7.5, and in [23] another RBM approximation—a version with the state-space

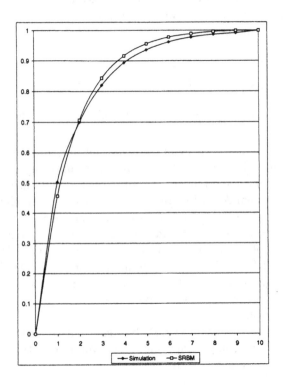

FIGURE 10.2. Cumulative stationary distributions of waiting time in the complex queueing network: group 1 jobs

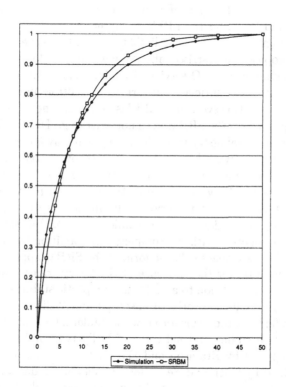

FIGURE 10.3. Cumulative stationary distributions of waiting time in the complex queueing network: group 3 jobs

collapse—for the feedforward network, based on the heavy traffic limit theorem of Peterson [24]. (The term "state-space collapse" was first used in
Reiman [25, 26], in the context of a single multiclass station with a static
priority discipline.) In Chapters 7 and 9 we have shown that the RBM approximations are better supported by the strong approximations for these
networks. In another special case, the reentrant line under a first-buffer-
first-served (FBFS) discipline and a last-buffer-first-served (LBFS) discipline, Dai, et al. [16] propose an RBM approximation, which is supported
by the heavy traffic limit theorem of Chen and Zhang [8] for the FBFS
discipline, and by the heavy traffic limit theorem of Bramson and Dai [2]
and of Chen and Ye [7] for the LBFS discipline.

In anticipation of a heavy traffic limit theorem, this type of approximation is also proposed in Harrison and Nguyen [18, 19] for a general multiclass network under a FIFO service discipline, and in Harrison and Pich
[20] for a four-station multiclass network with multiple unreliable servers
and other features. However, for multiclass queueing networks, there are
known examples where the SRBM does not exist (e.g., Dai and Wang [15]),
implying that the usual heavy traffic limit theorem may not exist. Studying
when a multiclass network has a heavy limit theorem has been an active
research area. Chen and Zhang [9, 11] establish some sufficient conditions
for the existence of a heavy traffic limit theorem for a multiclass network
under a FIFO discipline and a priority discipline, respectively. Williams
[29] establishes that (under some other mild conditions) a state-space collapse condition, together with the completely-S condition for the reflection
mapping (in the state-space collapse form of the SRBM approximation), is
a sufficient condition for the existence of the heavy traffic limit theorem.
Relating the latter condition to a uniform asymptotic stability condition of
the corresponding fluid network, Bramson [1] and Williams [29] establish
the heavy traffic limit theorems for networks under a head-of-line processor
sharing service discipline and networks of Kelly type under FIFO. Their
approach is applied by Bramson and Dai [2] to some network examples,
and is simplified by Chen and Ye [7] for the network under a priority service discipline to a more explicit condition that requires the higher priority
classes in the corresponding fluid network to be stable.

One of the foundations for the heavy traffic limit theorem and the SRBM
approximation discussed in this chapter is the necessary and sufficient condition for the existence of the SRBM, by Reiman and Williams [27] and by
Taylor and Williams [28], that the reflection matrix be completely-S. Given
the existence of an SRBM, we want to know whether it has a stationary
distribution, which can be used to approximate the stationary distribution
of the performance of the queueing network. Dupuis and Williams [17] relate the existence of the SRBM stationary distribution to the stability of
a linear Skorohod problem. Applying a linear Lyapunov function to this
Skorohod problem, Chen [3] finds certain more explicit conditions, which,
in particular, imply the sufficient conditions for the RBM associated with

approximating the generalized Jackson network in Chapter 7 and the feed-forward network in Chapter 9.

A necessary and sufficient condition for the existence of an exponential product-form solution to the stationary distribution of the SRBM (Theorem 10.11) follows from Harrison and Williams [22]. But except for some rare cases, the stationary distribution does not have an exponential product form, and analytical solutions are difficult to find. Hence, numerical computation is required. For this purpose, an algorithm was developed and implemented in Dai [12] and Dai and Harrison [14], which uses a global polynomial basis. Chen and Shen [4] provide an alternative implementation, known as BNA/FM , based on the finite element basis, which improves the numerical stability. This is the algorithm used to obtain the numerical results in Section 10.6, as well as in the previous chapters.

10.8 Exercises

1. Consider the Kumar–Seidman network as described in Chapter 8. Suppose that the virtual station condition in (8.2) is violated strictly (i.e., the condition in (8.12) holds). Show that the limit of $A_k(t)/t$ as $t \to \infty$ does not exist for $k = 2, 4$, even if the traffic condition holds. Conclude that the assumption in (10.21) cannot hold in this case.

2. Consider the Kumar–Seidman network under the heavy traffic condition in (8.3). If the virtual station condition in (8.2) holds, show that the assumption in (10.21) holds and identify λ and $\hat{A}$.

3. Consider the variation of Bramson network as shown by Figure 8.11. With the same set of parameters as discussed there, repeat Exercise 1 above.

4. For the generalized Jackson network, show that the assumption in (10.21) holds and identify λ and $\hat{A}$.

5. Consider the three-station variation of the Kumar–Seidman network shown by Figure 8.6.

   (a) Derive a 5-dimensional SRBM for this network. Under what condition is this SRBM well-defined?

   (b) Derive a 3-dimensional SRBM for this network (using the state-space collapse). Under what condition is this SRBM well-defined?

   (c) Compare the existence conditions in the above with the stability condition of this network (refer to Exercise 5).

6. Consider the three-station variation of the Bramson network shown in Figure 8.14.

(a) Derive an SRBM for this network. Under what condition is this SRBM well-defined?

(b) With the assistance of simulation, study the relationship between the above condition and the stability condition for this queueing network.

7. Consider a queueing network operating under a work-conserving discipline. (Clearly, FIFO or priority disciplines discussed in this chapter belong to work-conserving disciplines.) Suppose that its corresponding fluid network is weakly stable. (You are required to identify this fluid network.)

(a) Prove (10.22).

(b) For the network considered in this chapter (with the specific mix of FIFO and priority service discipline), prove (10.24).

References

[1] Bramson, M. (1998). State space collapse with application to heavy traffic limits for multiclass queueing networks. *Queueing Systems*. **30**, 89–148.

[2] Bramson, M. and J.G. Dai. (1999). Heavy traffic limits for some queueing networks. Preprint.

[3] Chen, H. (1996). A sufficient condition for the positive recurrence of a martingale reflecting Brownian motion in an orthant. *Annals of Applied Probability*, **6**, 3, 758–765.

[4] Chen, H. and X. Shen. (2000). Computing the stationary distribution of SRBM in an orthant. In preparation.

[5] Chen, H., X. Shen., and D.D. Yao. (2000). Brownian approximations of multiclass queueing networks. Preprint.

[6] Chen, H. and W. Whitt. (1993). Diffusion approximations for open queueing networks with service interruptions. *Queueing Systems, Theory and Applications*, **13**, 335–359.

[7] Chen, H. and H. Ye. (1999). Existence condition for the diffusion approximations of multiclass priority queueing networks. Preprint.

[8] Chen, H. and H. Zhang. (1996a). Diffusion approximations for reentrant lines with a first-buffer-first-served priority discipline. *Queueing Systems, Theory and Applications*, **23**. 177–195.

[9] Chen, H. and H. Zhang. (1996b). Diffusion approximations for some multiclass queuing networks with FIFO service discipline. *Mathematics of Operations Research* (to appear).

[10] Chen, H. and H. Zhang. (1998). Diffusion approximations for Kumar–Seidman network under a priority service discipline. *Operations Research Letters*, **23**, 171–181.

[11] Chen, H. and H. Zhang. (1999). A sufficient condition and a necessary condition for the diffusion approximations of multiclass queuing networks under priority service disciplines. *Queuing Systems, Theory and Applications* (to appear).

[12] Dai, J.G. (1990). *Steady-State Analysis of Reflected Brownian Motions: Characterization, Numerical Methods and Queuing Applications*, Ph.D. dissertation, Stanford University.

[13] Dai, J.G. and T.G. Kurtz. (1997). Characterization of the stationary distribution for a semimartingale reflecting Brownian motion in a convex polyhedron. Preprint.

[14] Dai, J.G. and J.M. Harrison. (1992). Reflected Brownian motion in an orthant: numerical methods for steady-state analysis. *Annals of Applied Probability*, **2**, 65–86.

[15] Dai, J.G. and Wang, Y. (1993). Nonexistence of Brownian models for certain multiclass queuing networks. *Queuing Systems, Theory and Applications*, **13**, 41–46.

[16] Dai, J.G., D.H. Yeh, and C. Zhou. (1997). The QNET method for re-entrant queuing networks with priority disciplines. *Operations Research*, **45**, 610–623.

[17] Dupuis, P. and R.J. Williams. (1994). Lyapunov functions for semimartingale reflecting Brownian motions. *Annals of Probability*, **22**, 680–702.

[18] Harrison, J.M. and V. Nguyen. (1990). The QNET method for two-moment analysis of open queuing networks. *Queuing Systems, Theory and Applications*, **6**, 1–32.

[19] Harrison, J.M. and V. Nguyen. (1993). Brownian models of multiclass queuing networks: current status and open problems. *Queuing Systems*, **13**, 5–40.

[20] Harrison, J.M. and M.T. Pich. (1996). Two-moment analysis of open queuing networks with general workstation capabilities. *Operations Research*, **44**, 936–950.

[21] Harrison, J.M. and R.J. Williams. (1987). Brownian models of open queueing networks with homogeneous customer populations. *Stochastics and Stochastic Reports*, **22**, 77–115.

[22] Harrison, J.M. and R.J. Williams. (1987). Multidimensional reflected Brownian motions having exponential stationary distributions. *Annals of Probability*, **15**, 115–137.

[23] Harrison, J.M. and R.J. Williams. (1992). Brownian models of feedforward queueing networks: quasireversibility and product form solutions, *Annals of Applied Probability*, **2**, 263–293.

[24] Peterson, W.P. (1991). A heavy traffic limit theorem for networks of queues with multiple customer types, *Mathematics of Operations Research*, **16**, 90–118.

[25] Reiman, M.I. (1984). Open queueing networks in heavy traffic. *Mathematics of Operations research*, **9**, 441–458.

[26] Reiman, M.I. (1988). A multiclass feedback queue in heavy traffic. *Advances in Applied Probability*, **20**, 179–207.

[27] Reiman, M.I. and R.J. Williams. (1988). A boundary property of semimartingale reflecting Brownian motions. *Probability Theory and Related Fields*, **77**, 87–97. and **80**(1989), 633.

[28] Taylor, L.M. and R.J. Williams. (1993). Existence and uniqueness of semimartingale reflecting Brownian motions in an orthant. *Probability Theory Related Fields*, **96**, 283–317.

[29] Williams, R.J. (1998). Diffusion approximations for open multiclass queueing networks: sufficient conditions involving state space collapse. *Queueing Systems, Theory and Applications*, **30**, 27–88.

11

Conservation Laws

Conservation laws belong to the most fundamental principles that govern the dynamics, or law of motion, of a wide range of stochastic systems. Under conservation laws, the performance space of the system becomes a *polymatroid*, that is, a polytope with a matroid-like structure, with all the vertices corresponding to the performance under priority rules, and all the vertices are easily identified. Consequently, the optimal control problem can be translated into an optimization problem. When the objective is a linear function of the performance measure, the optimization problem becomes a special linear program, for which the optimal solution is a vertex that is directly determined by the relative order of the cost coefficients in the linear objective. This implies that the optimal control is a priority rule that assigns priorities according to exactly the order of the cost coefficients.

In a more general setting, conservation laws extend to so-called generalized conservation laws (GCL), under which the performance space becomes an *extended polymatroid* (EP). Although the structure of the performance polytope that is an EP is not as strong as in the polymatroid case, the above facts still apply, the only modification being that the priority indices become more involved and have to be recursively generated.

The rest of the chapter is organized as follows. In Section 11.1 we present the basics of polymatroid and related optimization results. These are then connected to conservations laws in Section 11.2 and related stochastic scheduling problems, with many queueing examples and applications. In Section 11.3 we extend the conservations laws to GCL, using Klimov's model and the branching bandit problems as motivating examples. Equivalent definitions and properties of the EP are presented in Section 11.4, where we also bring out the connection between EP and GCL. Finally, in Section 11.5 we discuss the optimization of a linear objective over an EP, and its relationship with the dynamic scheduling problem in which the system performance satisfies GCL.

Throughout the chapter, "increasing" and "decreasing" are in the non-strict sense, meaning "nondecreasing" and "nonincreasing", respectively.

11.1   Polymatroid

11.1.1   Definitions and Properties

As we mentioned earlier, a polymatroid, simply put, is a polytope that has matroid-like properties. Below, we present three equivalent definitions of a polymatroid. Definition 11.1 is the most standard one; Definition 11.2 will later motivate the definition for EP; Definition 11.3 is new, and will provide a contrast against the structure of the EP later in Section 11.4 (refer to Definition 11.22).

Throughout, let $E = \{1, \ldots, n\}$ be a finite set; let A^c denote the complement of set A: $A^c = E \setminus A$.

Definition 11.1 (Welsh [47], Chapter 18) The polytope

$$\mathcal{P}(f) = \left\{ x \geq 0 : \sum_{i \in A} x_i \leq f(A), \ A \subseteq E \right\} \tag{11.1}$$

is termed a polymatroid if the function $f : 2^E \mapsto \Re_+$ satisfies the following properties:

(i)   (*normalized*) $f(\emptyset) = 0$;
(ii)   (*increasing*)   if $A \subseteq B \subseteq E$, then $f(A) \leq f(B)$;
(iii)   (*submodular*)   if $A, B \subseteq E$, then $f(A) + f(B) \geq f(A \cup B) + f(A \cap B)$

In matroid parlance, a function f that satisfies the above properties is termed a "rank function." Also note that a companion to submodularity is *supermodularity*, meaning that the inequality in (iii) holds in the opposite direction ($\leq$).

We now present the second definition for polymatroid. Given a set function $f : 2^E \mapsto \Re_+$, with $f(\emptyset) = 0$, and a permutation π of $\{1, 2, \ldots, n\}$, the elements of the set E, we define a vector x^π with the following components (to simplify notation, x_{π_i} below is understood to be $x_{\pi_i}^{\pi}$):

$$x_{\pi_1} = f(\{\pi_1\})$$
$$x_{\pi_2} = f(\{\pi_1, \pi_2\}) - x_{\pi_1} = f(\{\pi_1, \pi_2\}) - f(\{\pi_1\})$$
$$\vdots$$
$$x_{\pi_n} = f(\{\pi_1, \pi_2, \ldots, \pi_n\}) - f(\{\pi_1, \pi_2, \ldots, \pi_{n-1}\})$$

x^{π} is termed a "vertex" of the polytope $\mathcal{P}(f)$ in (11.1). Note, however, that this terminology could be misleading, since a priori there is no guarantee that x^{π} necessarily belongs to the polytope, since we simply do not know as yet whether x^{π} defined as above satisfies the set of inequalities that define $\mathcal{P}(f)$ in (11.1). In fact, this is the key point in the second definition of polymatroid below.

Definition 11.2 $\mathcal{P}(f)$ of (11.1) is a polymatroid if $x^{\pi} \in \mathcal{P}(f)$ for all permutations π.

Here is a third definition.

Definition 11.3 $\mathcal{P}(f)$ of (11.1) is a polymatroid if for any $A \subset B \subseteq E$, there exists a point $x \in \mathcal{P}(f)$, such that

$$\sum_{i \in A} x_i = f(A) \qquad \text{and} \qquad \sum_{i \in B} x_i = f(B).$$

Below we show that the three definitions are equivalent.

Theorem 11.4 The above three definitions for polymatroid are equivalent.

Proof. (Definition 11.1 $\Longrightarrow$ Definition 11.2)
That $x_{\pi_i} \geq 0$ for all i follows directly from the increasing property of f. For any $A \subseteq E$ and $\pi_i \in A$, since f is submodular, we have

$$f(A \cap \{\pi_1, \ldots, \pi_i\}) + f(\{\pi_1, \ldots, \pi_{i-1}\})$$
$$\geq f(A \cap \{\pi_1, \ldots, \pi_{i-1}\}) + f((A \cap \{\pi_1, \ldots, \pi_i\}) \cup \{\pi_1, \ldots, \pi_{i-1}\})$$
$$= f(A \cap \{\pi_1, \ldots, \pi_{i-1}\}) + f(\{\pi_1, \ldots, \pi_i\}),$$

which implies

$$f(\{\pi_1, \ldots, \pi_i\}) - f(\{\pi_1, \ldots, \pi_{i-1}\})$$
$$\leq f(A \cap \{\pi_1, \ldots, \pi_i\}) - f(A \cap \{\pi_1, \ldots, \pi_{i-1}\}).$$

Summing over $\pi_i \in A$, we have

$$\sum_{\pi_i \in A} x_{\pi_i} = \sum_{\pi_i \in A} (f(\{\pi_1, \ldots, \pi_i\}) - f(\{\pi_1, \ldots, \pi_{i-1}\}))$$

$$\leq \sum_{\pi_i \in A} f(A \cap \{\pi_1, \ldots, \pi_i\}) - f(A \cap \{\pi_1, \ldots, \pi_{i-1}\})$$

$$= f(A).$$

Hence, $x^\pi \in \mathcal{P}(f)$, and Definition 11.2 follows.

(Definition 11.2 $\Longrightarrow$ Definition 11.3)

For any given $A \subset B \subseteq E$, from Definition 11.2 it suffices to pick a vertex x^π such that its first $|A|$ components constitute the set A, and its first $|B|$ components constitute the set B.

(Definition 11.3 $\Longrightarrow$ Definition 11.1)

Taking $A = \emptyset$ in Definition 11.3 yields $f(\emptyset) = 0$. Monotonicity is trivial, since $x_i \geq 0$. For submodularity, take any $A, B \subseteq E$, $A \neq B$; then there exists $x \in \mathcal{P}(f)$ such that

$$\sum_{A \cup B} x_i = f(A \cup B) \quad \text{and} \quad \sum_{A \cap B} x_i = f(A \cap B),$$

since $A \cap B \subset A \cup B$. Therefore,

$$f(A \cup B) + f(A \cap B) = \sum_{A \cup B} x_i + \sum_{A \cap B} x_i$$

$$= \sum_{i \in A} x_i + \sum_{i \in B} x_i$$

$$\leq f(A) + f(B),$$

where the inequality follows from $x \in \mathcal{P}(f)$.   □

11.1.2 Optimization

Here we consider the optimization problem of maximizing a linear function over the polymatroid $\mathcal{P}(f)$:

$$\textbf{(P)} \quad \max \quad \sum_{i \in E} c_i x_i$$

$$\text{s.t.} \quad \sum_{i \in A} x_i \leq f(A), \quad \text{for all } A \subseteq E,$$

$$x_i \geq 0, \quad \text{for all } i \in E.$$

Assume

$$c_1 \geq c_2 \geq \cdots \geq c_n \geq 0, \tag{11.2}$$

without loss of generality, since any negative c_i clearly results in the corresponding $x_i = 0$. Let $\pi = (1, 2, \ldots, n)$. Then we claim that the vertex x^π in Definition 11.2 is the optimal solution to $(\mathbf{P})$.

To verify the claim, we start with writing down the dual problem as follows:

$$(\mathbf{D}) \quad \min \quad \sum_{A \subseteq E} y_A \, f(A)$$

$$\text{s.t.} \quad \sum_{A \ni i} y_A \geq c_i, \quad \text{for all } i \in E,$$

$$y_A \geq 0, \quad \text{for all } A \subseteq E.$$

Define y^π, a candidate dual solution, componentwise as follows:

$$y^\pi_{\{1\}} = c_1 - c_2,$$

$$y^\pi_{\{1,2\}} = c_2 - c_3,$$

$$\vdots$$

$$y^\pi_{\{1,\ldots,n-1\}} = c_{n-1} - c_n,$$

$$y^\pi_{\{1,\ldots,n\}} = c_n;$$

and set $y^\pi_A = 0$, for all other $A \subseteq E$.

Now the claimed optimality follows from

(1) primal feasibility: x^π is a vertex of the polymatroid $\mathcal{P}(f)$, and hence is feasible by definition (refer to Definition 11.2);

(2) dual feasibility: that y^π is feasible is easily checked (in particular, nonnegativity follows from (11.2));

(3) complementary slackness: also easily checked, in particular, the n binding constraints in $(\mathbf{P})$ that define the vertex x^π correspond to the n nonzero (not necessarily zero, to be precise) components of y^π listed above.

It is also easy to verify that the primal and the dual objectives are equal: letting $c_{n+1} := 0$, we have

$$\sum_{i \in E} c_i \, x^\pi_i = \sum_{i \in E} c_i \, [f(\{1, \ldots, i\}) - f(\{1, \ldots, i-1\})]$$

$$= \sum_{i=1}^{n} (c_i - c_{i+1}) f(\{1, \ldots, i\})$$

$$= \sum_{A \subseteq E} y^\pi_A f(A).$$

To summarize, x^π is optimal for $(\mathbf{P})$ and y^π is optimal for $(\mathbf{D})$. And it is important to note that

(a) Primal feasibility is always satisfied, by definition of the polymatroid.

(b) It is the dual feasibility that determines the permutation π, which, by way of complementary slackness, points to a vertex of $\mathcal{P}(f)$ that is optimal.

More specifically, the sum of the dual variables yields the cost coefficients:

$$y^\pi_{\{1,...,i\}} + \cdots + y^\pi_{\{1,...,n\}} = c_i, \qquad i = 1, \ldots, n; \tag{11.3}$$

the order of which [cf. (11.2)] decides the permutation π.

11.2   Conservation Laws

11.2.1   Polymatroid Structure

To relate to the last section, here $E = \{1, 2, \ldots, n\}$ denotes the set of all job classes, and x denotes the vector of performance measures of interest. For instance, x_i is the (long-run) average delay or throughput of job class i.

The conservation laws defined below were first formalized in Shanthikumar and Yao [39], where the connection to polymatroids was made. In [39], as well as subsequent papers in the literature, these laws are termed "strong conservation laws." Here, we shall simply refer to these as conservation laws.

Verbally, conservation laws can be summarized into the following two statements:

(i) the total performance (i.e., the sum) over all job classes in E is invariant under any admissible policy;

(ii) the total performance over any given subset, $A \subset E$, of job classes is minimized (or maximized) by offering priority to job classes in this subset over all other classes.

As a simple example, consider a system of two job classes. Each job (of either class) brings a certain amount of "work" (service requirement) to the system. Suppose the server serves (i.e., depletes work) at unit rate. Then it is not difficult to see that (i) the total amount of work, summing over all jobs of both classes that are present in the system, will remain invariant regardless of the actual policy that schedules the server, as long as it is nonidling; and (ii) if class 1 jobs are given preemptive priority over class 2 jobs, then the amount of work in the system summing over class 1 jobs is minimized, namely, it cannot be further reduced by any other admissible policy.

We now state the formal definition of conservation laws. For any $A \subseteq E$, denote by $|A|$ the cardinality of A. Let $\mathcal{A}$ denote the space of all *admissible* policies: all nonanticipative and nonidling policies (see more details below), and x^u the performance vector under an admissible policy $u \in \mathcal{A}$. As before, let π denote a permutation of the integers $\{1, 2, \ldots, n\}$. In particular, $\pi = (\pi_1, \ldots, \pi_n)$ denotes a priority rule that is admissible and in which class π_1 jobs are assigned the highest priority, and class π_n jobs the lowest priority.

Definition 11.5 (Conservation Laws) The performance vector x is said to satisfy conservation laws if there exists a set function b (or respectively f): $2^E \mapsto \Re_+$, satisfying

$$b(A) = \sum_{i \in A} x_{\pi_i}, \quad \forall \pi : \{\pi_1, \ldots, \pi_{|A|}\} = A, \quad \forall A \subseteq E, \qquad (11.4)$$

or respectively

$$f(A) = \sum_{i \in A} x_{\pi_i}, \quad \forall \pi : \{\pi_1, \ldots, \pi_{|A|}\} = A, \quad \forall A \subseteq E, \qquad (11.5)$$

(when $A = \emptyset$, by definition, $b(\emptyset) = f(\emptyset) = 0$); such that for all $u \in \mathcal{A}$ the following is satisfied:

$$\sum_{i \in A} x_i^u \geq b(A), \quad \forall A \subset E; \qquad \sum_{i \in E} x_i^u = b(E), \qquad (11.6)$$

or respectively

$$\sum_{i \in A} x_i^u \leq f(A), \quad \forall A \subset E; \qquad \sum_{i \in E} x_i^u = f(E). \qquad (11.7)$$

Note that whether the function b or the function f applies in a particular context is determined by whether the performance in question is minimized or maximized by the priority rules. (For instance, b applies to delay, and f applies to throughput.) It is important to note that this minimal (or maximal) performance is required to be independent of the priority assignment among the classes within the subset A on the one hand and the priority assignment among the classes within the subset $E \setminus A$ on the other hand, as long as any class in A has priority over any class in $E \setminus A$. This requirement is reflected in the qualifications imposed on π in defining $b(A)$ and $f(A)$ in (11.4) and (11.5). In particular, the definition requires that $b(A)$ and $f(A)$ be respectively the minimal and the maximal total performances summing over all job classes in the subset A that are given priority over all the other classes.

For the time being, ignore the b part of Definition 11.5. It is clear that when x satisfies the conservation laws, the performance space, as defined by the polytope in (11.7), is a polymatroid. This is because following (11.5) and (11.7), all the vertices x^π indeed, *by definition*, belong to the polytope.

In fact, the polytope in (11.7) is the polymatroid $\mathcal{P}(f)$ of (11.1) restricted to the hyperplane $\sum_{i \in E} x_i = f(E)$ (instead of the half-plane $\sum_{i \in E} x_i \leq f(E)$), and is hence termed the *base* of the polymatroid $\mathcal{P}(f)$, denoted by $\mathcal{B}(f)$ below. Furthermore, following Theorem 11.4, we know that when x satisfies conservation laws, the function $f(\cdot)$ as defined in (11.5) is increasing and submodular.

Next, consider the b part of Definition 11.5. Note that subtracting the inequality constraint from the equality constraint in (11.6), we can express these constraints in the same form as in (11.7), by letting $f(A) := b(E) - b(E \setminus A)$, or equivalently, $b(A) := b(E) - f(E \setminus A)$. Hence, the polytope in (11.6) is also (the base of) a polymatroid. Furthermore, the increasingness and submodularity of f translate into the increasingness and supermodularity of b.

To sum up the above discussion, we have the following result.

Theorem 11.6 If the performance vector x satisfies conservation laws, then its feasible space (i.e., the achievable performance region) constitutes the base polytope of a polymatroid, $\mathcal{B}(f)$ or $\mathcal{B}(b)$, of which the vertices correspond to the *priority rules*. Furthermore, the functions f and b, which are the performance functions corresponding to priority rules, are, respectively, increasing and submodular, and increasing and supermodular.

11.2.2 Examples

Consider a queueing system with n different job classes, which are indexed by the set E. Let u be the control or scheduling rule that governs the order of service among different classes of jobs. Let $\mathcal{A}$ denote the class of admissible controls, which are required to be nonidling and nonanticipative. That is, no server is allowed to be idle when there are jobs waiting to be served, and the control is only allowed to make use of past history and the current state of the system. Neither can an admissible control affect the arrival processes or the service requirements of the jobs. Otherwise, we impose no further restrictions on the system. For instance, the arrival processes and the service requirements of the jobs can be arbitrary. Indeed, since the control cannot affect the arrival processes and the service requirements, all the arrival and service data can be viewed as generated a priori following any given (joint) distribution and with any given dependence relations. We allow multiple servers, and multiple stages (e.g., tandem queues or networks of queues). We also allow the control to be either preemptive or nonpreemptive. (Some restrictions will be imposed on individual systems to be studied below.)

Let x_i^u be a performance measure of class i ($i \in E$) jobs under control u. This need not be a steady-state quantity or an expectation; it can very well be a sample-path realization over a finite time interval, for instance the delay (sojourn time) of the first m class i jobs, the number of class i jobs in the system at time t, or the number of class i job completions by time t. Let $x^u := (x_i^u)_{i \in E}$ be the performance vector.

For any given permutation $\pi \in \Pi$, let x^π denote the performance vector under a priority scheduling rule that assigns priority to the job classes according to the permutation π, i.e., class π_1 has the highest priority, $\dots$, class π_n has the lowest priority. Clearly, any such priority rule belongs to the admissible class.

In all the queueing systems studied below, the service requirements of the jobs are mutually independent, and are also independent of the arrival processes. (One exception to these independence requirements is Example 11.7 below, where these independence assumptions are not needed.) No independence assumption, however, is required for the arrival processes, which can be arbitrary. When a performance vector satisfies conservation laws, whether its state space is $\mathcal{B}(b)$ (11.6) or $\mathcal{B}(f)$ (11.7) depends on whether the performance of a given subset of job classes is minimized or maximized by giving priority to this subset. This is often immediately evident from the context.

Example 11.7 Consider a $G/G/1$ system that allows preemption. For $i \in E$, let $V_i(t)$ denote the amount of work (processing requirement) in the system at time t due to jobs of class i. (Note that for any given t, $V_i(t)$ is a random quantity, corresponding to some sample realization of the workload process.) Then it is easily verified that for any t, $x := [V_i(t)]_{i \in E}$ satisfies conservation laws.

Example 11.8 Continue with the last example. For all $i \in E$, let $N_i(t)$ be the number of class i jobs in the system at time t. When the service times follow exponential distributions, with mean $1/\mu_i$ for class i jobs, we have $\mathsf{E}N_i(t) = \mu_i \mathsf{E}V_i(t)$. Let W_i be the steady-state sojourn time in the system for class i jobs. From Little's law we have $\mathsf{E}W_i = \mathsf{E}N_i/\lambda_i = \mathsf{E}V_i/\rho_i$, where λ_i is the arrival rate of class i jobs, $\rho_i := \lambda_i/\mu_i$, N_i and V_i are the steady-state counterparts of $N_i(t)$ and $V_i(t)$, respectively. Hence, the following x also satisfies conservation laws:

(i) for any given t, $x := [\mathsf{E}N_i(t)/\mu_i]_{i \in E}$;

(ii) $x := [\rho_i \mathsf{E}W_i]_{i \in E}$.

Example 11.9 In a $G/M/c$ ($c > 1$) system that allows preemption, if all job classes follow the same exponential service-time distribution (with mean $1/\mu$), then it is easy to verify that for any t, $x := [\mathsf{E}N_i(t)]_{i \in E}$ satisfies conservation laws. In this case, $\mathsf{E}V_i(t) = \mathsf{E}N_i(t)/\mu$ and $\mathsf{E}W_i = \mathsf{E}N_i/\lambda_i$. Hence, x defined as follows satisfies conservation laws:

(i) for any given t, $x := [\mathsf{E}N_i(t)]_{i \in E}$, $x := [\mathsf{E}V_i(t)]_{i \in E}$;

(ii) $x := [\lambda_i \mathsf{E}W_i]_{i \in E}$.

(If the control is restricted to be nonpreemptive, the results here still hold true. See Example 11.12 below.)

Example 11.10 The results in Example 11.9 still hold when the system is a network of queues, provided that all job classes follow the same exponential service-time distribution and the same routing probabilities at each node (service-time distributions and routing probabilities can, however, be node dependent); (external) job arrival processes can be arbitrary and can be different among the classes.

Example 11.11 Another variation of Example 11.9 is the finite-buffer queue, $G/M/c/K$, where $K \geq c$ denotes the upper limit on the total number of jobs allowed in the system at any time. In this system, higher priority jobs can preempt lower priority jobs not only in service but also in occupancy. That is, whenever a higher priority job finds (on its arrival) a fully occupied system, a lower priority job within the system (if any) will be removed from the system and its occupancy given to the higher priority job. If there is no lower priority job, then the arriving job is rejected and lost. As in Example 11.9, all jobs follow the same exponential service-time distribution. Let $R_i(t)$ and $D_i(t)$ $(i \in E)$ denote, respectively, the (cumulated) number of rejected/removed class i jobs and the (cumulated) number of class i departures (service completions) up to time t. Then, for any given t, (i) $x := [ER_i(t)]_{i \in E}$ and (ii) $x := [ED_i(t)]_{i \in E}$ satisfy conservation laws.

We next turn to considering cases where the admissible controls are restricted to be nonpreemptive.

Example 11.12 Consider the $G/G/c$ system, $c \geq 1$. If all job classes follow the same service-time distribution, then it is easy to see that the scheduling of the servers will not affect the departure epochs of jobs (in a pathwise sense); although it *will* affect the identity (class) of the departing jobs at those epochs. (See Shanthikumar and Sumita [38], §2, for the $G/G/1$ case; the results there also hold true for the $G/G/c$ case.) Hence, for any given t, $x := [N_i(t)]_{i \in E}$ satisfies conservation laws.

Example 11.13 Comparing the above with Example 11.9, we know that the results there also hold for nonpreemptive controls. However, in contrast to the extension of Example 11.9 to the network case in Example 11.10, the above can only be extended to queues in tandem, where overtaking is excluded. Specifically, the result in Example 11.12 also holds for a series of $G/G/c$ queues in tandem, where at each node all job classes have the same service-time distribution, which, however, can be node dependent. External job arrival processes can be arbitrary and can be different among classes. The number of servers can also be node dependent.

Example 11.14 With nonpreemptive control, there is a special case for the $G/G/1$ system with only two job classes $(n = 2)$ that may follow *different* service-time distributions: for any given t, $x := [V_i(t)]_{i \in E}$ satisfies conservation laws.

For steady-state measures, from standard results in $GI/G/1$ queues (see, e.g., Asmussen [1], Chapter VIII, Proposition 3.4), we have

$$EV_i = \mu_i^{-1}[EN_i - \rho_i] + \rho_i \mu_i m_i/2$$

and

$$EV_i = \rho_i[EW_i - \mu_i^{-1} + \mu_i m_i/2],$$

where m_i is the second moment of the service time of class i jobs. Hence, following the above, we know that $x = [EN_i/\mu_i]_{i \in E}$ and $x = [\rho_i EW_i]_{i \in E}$ also satisfy conservation laws.

Example 11.15 Two more examples that satisfy conservation laws:

(i) for the $G/G/1$ system with preemption,

$$x := \left[\int_0^t \exp(-\alpha \tau) V_i(\tau) d\tau \right]_{i \in E} ;$$

(ii) for the $G/M/1$ system with preemption,

$$x := \left[E \int_0^t \exp(-\alpha \tau) N_i(\tau) d\tau / \mu_i \right]_{i \in E} ,$$

where in both (i) and (ii) $\alpha > 0$ is a discount rate, and t is any given time.

Finally, note that in all the above examples, with the exception of Example 11.11, whenever $[EN_i(t)]_{i \in E}$ satisfies conservation laws, $[ED_i(t)]_{i \in E}$ also satisfies conservation laws, since in a no-loss system the number of departures is the difference between the number of arrivals (which is independent of the control) and the number in the system.

Evidently, based on the above discussions, the state space of the performance vectors in each of the examples above is a polymatroid.

11.2.3 Optimal Scheduling

Theorem 11.16 Consider the optimal control (scheduling) of n jobs classes in the set E:

$$\max_{u \in \mathcal{A}} \sum_{i \in E} c_i x_i^u \quad [\text{or} \quad \min_{u \in \mathcal{A}} \sum_{i \in E} c_i x_i^u],$$

where x is a performance measure that satisfies conservation laws, and the cost coefficients c_i $(i \in E)$ satisfy, without loss of generality, the ordering in (11.2). Then, this optimal control problem can be solved by solving the following linear program (LP):

$$\max_{x \in \mathcal{B}(f)} \sum_{i \in E} c_i x_i \quad [\text{or} \quad \min_{x \in \mathcal{B}(b)} \sum_{i \in E} c_i x_i].$$

The optimal solution to this LP is simply the vertex $x^\pi \in \mathcal{B}(f)$, with $\pi = (1, \ldots, n)$ being the permutation corresponding to the decreasing order of the cost coefficients in (11.2). And the optimal control policy is the corresponding priority rule, which assigns the highest priority to class 1 jobs, and the lowest priority to class n jobs.

Example 11.17 ($c\mu$-rule)   Consider one of the performance vectors in Example 11.8, $x := [\mathsf{E}(N_i)/\mu_i]_{i \in E}$, where N_i is the number of jobs of class i in the system (or, "inventory") in steady state, and μ_i is the service rate. Suppose our objective is to minimize the total inventory cost

$$\min \sum_{i \in E} c_i \mathsf{E}(N_i),$$

where c_i is the inventory holding cost rate for class i jobs. We then rewrite this objective as

$$\min \sum_{i \in E} c_i \mu_i x_i.$$

(Note that $(N_i)_{i \in E}$ does not satisfy conservation laws; $(x_i)_{i \in E}$ does.) Then, we know from the above theorem that the optimal policy is a priority rule, with the priorities assigned according to the $c_i \mu_i$ values: The larger the value, the higher the priority. This is what is known as the "$c\mu$-rule." When all jobs have the same cost rate, the priorities follow the μ_i values, i.e., the faster the processing rate (or the shorter the processing time), the higher the priority, which is the so-called SPT (shortest processing time) rule.

The connection between conservation laws and polymatroid, as specified in Theorem 11.6, guarantees that any admissible control will yield a performance vector that belongs to the polymatroid. Furthermore, the converse is also true: any performance vector that belongs to the polymatroid can be realized by an admissible control. This is because since $\mathcal{B}(f)$ (or $\mathcal{B}(b)$) is a convex polytope, any vector in the performance space can be expressed as a convex combination of the vertices. Following Carathéodory's theorem (refer to, e.g., Chvátal [7]), any vector in the performance space can be expressed as a convex combination of no more than $n + 1$ vertices. In other words, any performance vector can be realized by a control that is a *randomization* of at most $n + 1$ priority rules, with the convex combination coefficients being the probabilities for the randomization.

In terms of implementation, however, randomization can be impractical. First, computationally, there is no easy way to derive the randomization coefficients. Second, in order to have an unbiased implementation, randomization will have to be applied at the beginning of each regenerative cycle, e.g., a busy period. In heavy traffic, busy periods could be very long, making implementation extremely difficult, and also creating large variance of the performance.

In fact, one can do better than randomization. It is known (e.g., Federgruen and Groenevelt [15]) that any interior point of the performance space can be realized by a particular dynamic scheduling policy, due originally to Kleinrock [30, 31], in which the priority index of each job present in the system grows proportionately to the time it has spent waiting in queue, and the server always serves the job that has the highest index. This scheduling policy is completely specified by the proportionate coefficients associated with the job classes, which, in turn, are easily determined by the performance vector (provided that it is at the interior of the performance space). In terms of practical implementation, there are several versions of this scheduling policy, refer to [17, 18].

11.3 Generalized Conservation Laws

11.3.1 Definition

Although conservation laws apply to the many examples in the last section, there are other interesting and important problems that do not fall into this category. A primary class of such examples includes systems with feedback, i.e., jobs may come back after service completion. For example, consider the so-called Klimov problem: a multiclass $M/G/1$ queue in which jobs, after service completion, may return and switch to another class, following a Bernoulli mechanism. Without feedback, we know that this is a special case of Example 11.7, and the work in the system, $[V_i(t)]_{i \in E}$, satisfies conservation laws. With feedback, however, the conservation laws as defined in Definition 11.5 need to be modified.

Specifically, with the possibility of feedback, the work of a particular job class, say class i, should not only include the work associated with class i jobs that are present in the system, but also take into account the *potential* work that will be generated by feedback jobs, which not only include class i jobs but also all other classes that may feed back to become class i. With this modification, the two intuitive principles of conservation laws listed at the beginning of Section 11.2.1 will apply.

To be concrete, let us paraphrase here the simple example at the beginning of Section 11.2.1 with two job classes, allowing the additional feature of feedback. As before, suppose the server serves at unit rate. Then it is not difficult to see that (i) the total amount of potential work, summing over both classes, will remain invariant regardless of the actual schedule that the server follows, as long as it is a nonidling schedule; and (ii) if class 1 jobs are given (preemptive) priority over class 2 jobs, then the amount of potential work due to class 1 jobs is minimized, namely, it cannot be further reduced by any other scheduling rule. And the same holds for class 2 jobs, if given priority over class 1 jobs.

Another way to look at this example is to let T be the first time there are no class 1 jobs left in the system. Then, T is minimized by giving class 1 jobs (preemptive) priority over class 2 jobs. In particular, T is no smaller than the potential work of class 1 generated by class 1 jobs (only); T is equal to the latter if and only if class 1 jobs are given priority over class 2 jobs.

Therefore, with this modification, the conservation laws in Definition 11.5 can be generalized. The net effect, as will be demonstrated in the examples below, is that the variables x_i in Definition 11.5 will have to be multiplied with different coefficients a_i^A that depend on both the job classes (i) and the subsets (A). In particular, when x_i is, for instance, the average number of jobs of class i, a_i^A denotes the rate of potential work of those classes in set A that is generated by class i jobs.

We now state the formal definition of generalized conservation laws (GCL), using the same notation wherever possible as in Definition 11.5.

Definition 11.18 (Generalized Conservation Laws) The performance vector x is said to satisfy generalized conservation laws (GCL) if there exists a set function b (or respectively f): $2^E \mapsto \Re_+$, and a matrix $(a_i^S)_{i \in E, S \subseteq E}$ (which is in general different for b and f, but we will not make this distinction below for notational simplicity), satisfying

$$a_i^S > 0, \ i \in S \quad \text{and} \quad a_i^S = 0, \ i \notin S; \quad \forall S \subseteq E,$$

such that

$$b(A) = \sum_{i \in A} a_{\pi_i}^A x_{\pi_i}, \quad \forall \pi : \{\pi_1, \ldots, \pi_{|A|}\} = A, \quad \forall A \subseteq E, \qquad (11.8)$$

or respectively,

$$f(A) = \sum_{i \in A} a_{\pi_i}^A x_{\pi_i}, \quad \forall \pi : \{\pi_1, \ldots, \pi_{|A|}\} = A, \quad \forall A \subseteq E, \qquad (11.9)$$

such that for all $u \in \mathcal{A}$ the following is satisfied:

$$\sum_{i \in A} a_i^A x_i^u \geq b(A), \quad \forall A \subset E, \quad \sum_{i \in E} a_i^E x_i^u = b(E); \qquad (11.10)$$

or respectively,

$$\sum_{i \in A} a_i^A x_i^u \leq f(A), \quad \forall A \subset E; \quad \sum_{i \in E} a_i^E x_i^u = f(E). \qquad (11.11)$$

It is obvious from the above definition that GCL reduces to the conservation laws if $a_i^A = 1$ for all $i \in A$, and all $A \subseteq E$.

11.3.2 Examples

Example 11.19 (Klimov's problem [32]) This concerns the optimal control of a system in which a single server is available to serve n classes of jobs. Class i jobs arrive according to a Poisson process with rate α_i, which is independent of other classes of jobs. The service times for class i jobs are independent and identically distributed with mean μ_i. When the service of a class i job is completed, it either returns to become a class j job, with probability p_{ij}, or leaves the system with probability $1 - \sum_j p_{ij}$. Set $\alpha = (\alpha_i)_{i \in E}$, $\mu = (\mu_i)_{i \in E}$, and $P = [p_{ij}]_{i,j \in E}$.

Consider the class of nonpreemptive policies. The performance measure is

$$x_i^u = \text{long-run average number of class } i \text{ jobs in system under policy } u.$$

The objective is to find the optimal policy that minimizes $\sum_j c_j x_j^u$. Klimov proved that a priority policy is optimal and gave a recursive procedure for obtaining the priority indices.

Tsoucas [42] showed that the performance space of Klimov's problem is the following polytope:

$$\left\{ x \geq 0 : \sum_{i \in S} a_i^S x_i \geq b(S), S \subset E; \sum_{i \in E} a_i^E x_i = b(E) \right\},$$

where the coefficients are given as $a_i^S = \lambda_i \beta_i^S$, with $\lambda = (\lambda)_{i \in E}$ and $\beta^S = (\beta)_{i \in S}$ obtained as

$$\lambda = (I - P')^{-1} \alpha \qquad \text{and} \qquad \beta^S = (I - P_{SS})^{-1} \mu_S,$$

where P_{SS} and μ_S are, respectively, the restriction of P and μ to the set S. Note that here, λ_i is the overall arrival rate of class i jobs (including both external arrivals and feedback jobs), β_i^S is the amount of potential work of the classes in S generated by a class i job. (Hence, this potential work is generated at rate α_i in the system.) Summing over $i \in S$ yields the total amount of potential work of the classes in S (generated by the same set of jobs), which is minimized when these jobs are given priority over other classes. This is the basic intuition as to why x satisfies GCL.

Example 11.20 (Branching bandit process) There are m projects at time 0. They are of K classes, labeled $k = 1, \ldots, K$. Each class k project can be in one of a finite number of states, with E_k denoting the state space. Classifying different project classes or projects of the same class but in different states as different "classes," we set $E = \cup_k E_k = \{1, \ldots, n\}$ as the set of all project classes. A single server works on the projects one at a time. Each class i project keeps the server busy for a duration of v_i time units. Upon completion, the class i project is replaced by N_{ij} projects of class

j. The server then has to decide which project to serve next, following a scheduling rule (control) u. The collection $\{(v_i, N_{ij}), j \in E\}$ follows a general joint distribution, which is independent and identically distributed for all $i \in E$.

Given $S \subseteq E$, the S-descendants of a project of class $i \in S$ constitute all of its immediate descendants that are of classes belonging to S, as well as the immediate descendants of those descendents, and so on. (If a project in S transfers into a class that is not in S, and later transfers back into a class in S, it will not be considered as an S-descendant of the original project.) Given a class i project, the union of the time intervals in which its S-descendants are being served is called an (i, S) period. Let T_i^S denote the length of an (i, S) period. It is the "potential work" of the classes in the set S generated by the class i project. And we use T_m^S to denote the time until the system has completely cleared all classes of projects in S class—under a policy that gives priority to those classes in S over other classes. Note, in particular, that T_m^E represents the length of a busy period.

In the discounted case, the expected reward associated with the control u is $\sum_{i \in E} c_i x_i^u$, where

$$x_i^u = \mathsf{E}_u \left[\int_0^\infty e^{-\alpha t} I_i^u(t) dt \right]$$

$\alpha > 0$ is the discount rate and

$$I_i^u(t) = \begin{cases} 1, & \text{if a class } i \text{ project is being served at time } t, \\ 0, & \text{otherwise.} \end{cases}$$

Bertsimas and Niño-Mora [3] showed that $x^u = (x_i^u)_{i \in E}$, as defined above, satisfy the GCL (see also Example 11.28 below), with coefficients

$$a_i^S = \frac{\mathsf{E}\left[\int_0^{T_i^{S^c}} e^{-\alpha t} dt \right]}{\mathsf{E}\left[\int_0^{v_i} e^{-\alpha t} dt \right]}, \quad i \in S \subseteq E,$$

and

$$b(S) = \mathsf{E}\left[\int_0^{T_m^E} e^{-\alpha t} dt \right] - \mathsf{E}\left[\int_0^{T_m^{S^c}} e^{-\alpha t} dt \right].$$

Intuitively, the GCL here says that the time until all the S^c-descendents of all the projects in S are served is minimized by giving project classes in S^c priority over those in S.

An undiscounted version is also available in [3]. (This includes Klimov's problem, the last example above, as a special case.) The criterion here is to minimize the total expected cost incurred under control u during the first busy period (of the server) $[0, T]$, $\sum_{i \in E} c_i x_i^u$, with

$$x_i^u = \mathsf{E}_u \left[\int_0^\infty t I_i^u(t) dt \right].$$

Following [3], the x^u satisfy GCL with coefficients

$$a_i^S = \mathsf{E}[T_i^{S^c}]/\mathsf{E}[v_i], \quad i \in S \subseteq E,$$

and

$$b(S) = \frac{1}{2}\mathsf{E}\left[(T_m^E)^2\right] - \frac{1}{2}\mathsf{E}\left[(T_m^{S^c})^2\right] + \sum_{i \in S} b_i(S),$$

where

$$b_i(S) = \frac{\mathsf{E}[v_i]\mathsf{E}[v_i^2]}{2} \left(\frac{\mathsf{E}[T_i^{S^c}]}{\mathsf{E}[v_i]} - \frac{\mathsf{E}[(T_i^{S^c})^2]}{\mathsf{E}[v_i^2]}\right), \quad i \in S.$$

The intuition is similar to the discounted case.

11.4 Extended Polymatroid

11.4.1 Equivalent Definitions

Recall that the space of any performance measure that satisfies conservation laws is a polymatroid. Analogously, one can ask what is the structure of the performance space under GCL, i.e., what is the structure of the following polytopes:

$$\mathcal{EP}(b) = \left\{ x \geq 0 : \sum_{i \in S} a_i^S x_i \geq b(S), \ S \subseteq E \right\}, \tag{11.12}$$

$$\mathcal{EP}(f) = \left\{ x \geq 0 : \sum_{i \in S} a_i^S x_i \leq f(S), \ S \subseteq E \right\}. \tag{11.13}$$

The most natural route to approach this issue appears to be mimicking Definition 11.2 of polymatroid (and this is indeed the route taken in [3]). As with the definition of x^π preceding Definition 11.2, here, given a permutation π (of $\{1, 2, \ldots, n\}$), we can generate a vertex x^π as follows:

$$x_{\pi_1} = f(\{\pi_1\})/a_{\pi_1}^{\{\pi_1\}}$$

$$x_{\pi_2} = \left(f(\{\pi_1, \pi_2\}) - a_{\pi_1}^{\{\pi_1, \pi_2\}} x_{\pi_1}\right) \Big/ a_{\pi_2}^{\{\pi_1, \pi_2\}}$$

$$\vdots$$

$$x_{\pi_n} = \left(f(\{\pi_1, \ldots, \pi_n\}) - \sum_{i=1}^{n-1} a_{\pi_i}^{\{\pi_1, \ldots, \pi_n\}} x_{\pi_i}\right) \Big/ a_{\pi_n}^{\{\pi_1, \ldots, \pi_n\}}.$$

Same as in the polymatroid case, we should emphasize here that as yet, x^π does not necessarily belong to the polytope in (11.13). The vertices for $\mathcal{EP}(b)$ are analogously generated, with $f(\cdot)$ replaced by $b(\cdot)$.

Definition 11.21 $\mathcal{EP}(f)$ (respectively $\mathcal{EP}(b)$) is an extended polymatroid (EP) if x^π as generated above (respectively with b replacing f) belongs to the polytope $\mathcal{EP}(f)$, (respectively $\mathcal{EP}(b)$), for any permutation π.

(The term "extended polymatroid" was previously used to refer to a polymatroid without the requirement that $x \geq 0$; e.g., see [26], page 306. Since [3, 42] and other works in the queueing literature, it has been used to refer to the polytopes defined above. Also, in [3], the EP corresponding to the b function is termed "extended contra-polymatroid," with the term "extended polymatroid" reserved for the f function. For simplicity, we do not make such a distinction here and below.)

With the above definition for EP, the right-hand side functions b and f are not necessarily increasing and supermodular/submodular. In other words, we do not have a counterpart of Definition 11.1 for EP (more on this later). On the other hand, the counterpart for Definition 11.3 does apply.

Definition 11.22 $\mathcal{EP}(f)$ is an extended polymatroid if the following is satisfied: For any $A \subset B \subset E$, there exists a point $x \in \mathcal{EP}(f)$ such that

$$\sum_{i \in A} a_i^A x_i = f(A) \quad \text{and} \quad \sum_{i \in B} a_i^B x_i = f(B).$$

Theorem 11.23 The two definitions of EP in Definitions 11.21 and 11.22 are equivalent.

Proof. If $\mathcal{EP}(f)$ is EP, then the stated condition in Definition 11.22 is obviously satisfied: Just pick the vertex x^π such that the first $|A|$ components in π constitute the set A, and the first $|B|$ components constitute the set B.

For the other direction, i.e., if the stated condition in Definition 11.22 holds, then $\mathcal{EP}(f)$ is EP, and we use induction on $n = |E|$. That this holds for $n = 1$ is trivial.

Suppose this holds for $n = k$, i.e. for a polytope of the kind in (11.13) with k variables. Now consider such a polytope with $k + 1$ variables, i.e., $|E| = k + 1$. Without loss of generality, consider the permutation $\pi = (1, 2, \ldots, k+1)$. We want to show that the corresponding x^π (i.e., generated from the triangulation above) is in the polytope $\mathcal{EP}(f)$.

Since $x_1^\pi = f(\{1\})/a_1^{\{1\}}$, we substitute it into the other x_i^π expressions, $i \neq 1$, to arrive at the following polytope of k variables:

$$\mathcal{EP}(\tilde{f}) = \left\{ x \geq 0 : \sum_{i \in S, i \neq 1} a_i^S x_i \leq \tilde{f}(S), \{1\} \in S \subseteq E \right\},$$

where

$$\tilde{f}(S) := f(S) - \frac{f(\{1\})}{a_1^{\{1\}}} a_1^S.$$

Clearly, since the stated condition in Definition 11.22 is assumed to hold for $\mathcal{EP}(f)$ (the one with $k+1$ variables), it also holds for $\mathcal{EP}(\tilde{f})$ (the one with k variables), since the equations in question all differ by an amount $f(\{1\})a_1^{\{1\}}/a_1^S$ on both sides. Hence, the induction hypothesis confirms that $\mathcal{EP}(\tilde{f})$ is an EP. This implies that $(x_2^\pi, \ldots, x_n^\pi) \in \mathcal{EP}(\tilde{f})$, which is equivalent to $x^\pi = (x_1^\pi, x_2^\pi, \ldots, x_n^\pi)$ satisfying all the constraints in $\mathcal{EP}(f)$ that involve $S \subseteq E$ with $1 \in S$.

We still need to check that x^π satisfies all the other constraints in $\mathcal{EP}(f)$ corresponding to $S \subseteq E$ with $1 \notin S$. To this end, consider the following polytope:

$$\left\{ x \geq 0 : \sum_{i \in S} a_i^S x_i \leq f(S), \, S \subseteq E \setminus \{1\} \right\}. \tag{11.14}$$

The above is another polytope with k variables. Obviously, the stated condition in Definition 11.22, which is assumed to hold for the polytope $\mathcal{EP}(f)$, holds for the above polytope as well (since the defining inequalities in the latter are just part of those in $\mathcal{EP}(f)$). Hence, based on the induction hypothesis, the polytope in (11.14) is also an EP. This implies that $(x_2^\pi, \ldots, x_n^\pi)$, and hence x^π, satisfies all the inequalities involved in (11.14).

Hence, we have established that given the stated condition in Definition 11.22, x^π does satisfy all the constraints in $\mathcal{EP}(f)$, for each permutation π. Therefore, $\mathcal{EP}(f)$ is an EP. □

The above theorem leads immediately to the following:

Corollary 11.24 If $\mathcal{EP}(f)$ is an extended polymatroid, then

$$\mathcal{EP}^-(f) := \left\{ x \geq 0 : \sum_{i \in S} a_i^S x_i \leq f(S), \, S \subseteq E \setminus E_0 \right\}$$

is also an extended polymatroid, for any $E_0 \subset E$.

Proof. Just verify Definition 11.22. Since $\mathcal{EP}(f)$ is an EP, we can pick any $A \subset B \subseteq E \setminus E_0 \subset E$, and there exists an $x \in \mathcal{EP}(f)$, such that $\sum_{i \in A} a_i^A x_i = f(A)$ and $\sum_{i \in B} a_i^B x_i = f(B)$. But this is exactly what is required for $\mathcal{EP}^-(f)$ to be EP.

11.4.2 Connections to Sub/Supermodularity

As mentioned earlier, the right-hand-side functions f and b of the EP are not necessarily increasing and submodular/supermodular as in the polymatroid case. It turns out that in order for them to have these properties, we need conditions on the coefficients a_i^S. These coefficients are called increasing (decreasing) if $a_i^A \leq (\geq)a_i^B$ for all $i \in E$ and all $A \subseteq B \subseteq E$. They are called submodular (supermodular) if

$$a_i^A + a_i^B \geq (\leq) a_i^{A \cup B} + a_i^{A \cap B}, \tag{11.15}$$

for all $i \in E$ and all $A, B \subseteq E$.

Theorem 11.25 Suppose $\mathcal{EP}(b)$ in (11.12) is an EP, with the coefficients $\{a_i^S\}$ being increasing and supermodular as defined above. Then, $b(\cdot)$ is an increasing and supermodular function.

Proof. For the increasingness of $b(\cdot)$, as in the proof of Theorem 11.4, consider two sets $A \subset B \subseteq E$: We can pick a π, such that the vertex x^π satisfies

$$\sum_{\pi_j \in A} a_{\pi_j}^A x_{\pi_j} = b(A) \quad \text{and} \quad \sum_{\pi_j \in B} a_{\pi_j}^B x_{\pi_j} = b(B). \qquad (11.16)$$

Since $A \subset B$ implies $a_{\pi_j}^A \leq a_{\pi_j}^B$, and since both x_{π_j} and its coefficients are nonnegative, we have $b(A) \leq b(B)$.

For the supermodularity of $b(\cdot)$, consider two subsets of E, A and B. Since $A \cap B \subseteq A \subseteq E$, we can, as above, pick a vertex x^π such that

$$\sum_{\pi_j \in A \cap B} a_{\pi_j}^{A \cap B} x_{\pi_j} = b(A \cap B),$$

$$\sum_{\pi_j \in A} a_{\pi_j}^A x_{\pi_j} = b(A), \qquad (11.17)$$

$$\sum_{\pi_j \in A \cup B} a_{\pi_j}^{A \cup B} x_{\pi_j} = b(A \cup B).$$

Making use of the increasingness and supermodularity of a_i^S, we have

$$\sum_{\pi_j \in A \cap B} a_{\pi_j}^{A \cap B} x_{\pi_j} + \sum_{\pi_j \in A \cup B} a_{\pi_j}^{A \cup B} x_{\pi_j} - \sum_{\pi_j \in A} a_{\pi_j}^A x_{\pi_j} \geq \sum_{\pi_j \in B} a_{\pi_j}^B x_{\pi_j}, \qquad (11.18)$$

since from the increasingness and supermodularity of a_i^S we have

$$\begin{aligned} \text{for } \pi_j \in A \backslash B : & \quad a_{\pi_j}^{A \cup B} \geq a_{\pi_j}^A, \\ \text{for } \pi_j \in B \backslash A : & \quad a_{\pi_j}^{A \cup B} \geq a_{\pi_j}^B, \\ \text{for } \pi_j \in A \cap B : & \quad a_{\pi_j}^{A \cup B} + a_{\pi_j}^{A \cap B} \geq a_{\pi_j}^A + a_{\pi_j}^B. \end{aligned}$$

Since $\mathcal{EP}(b)$ is an EP, we have $x^\pi \in \mathcal{EP}(b)$ following Definition 11.21. Hence,

$$\sum_{\pi_j \in B} a_{\pi_j}^B x_{\pi_j} \geq b(B).$$

This, together with (11.17) and (11.18), yields the desired supermodularity:

$$b(A \cup B) + b(A \cap B) \geq b(A) + b(B).$$

□

The above theorem almost has a converse, except that the requirements on the coefficients have to be changed: from increasing and supermodular to decreasing and submodular.

Theorem 11.26 Consider the polytope $\mathcal{EP}(b)$ in (11.12), not necessarily an EP. Suppose the coefficients $\{a_i^S\}$ are decreasing and submodular, and the function $b(\cdot)$ is supermodular. Then, $\mathcal{EP}(b)$ is an EP.

Proof. We verify that Definition 11.22 is satisfied. The submodularity of the coefficients implies

$$\sum_{i \in A \cap B} (a_i^{A \cup B} x_i + a_i^{A \cap B} x_i) \le \sum_{i \in A \cap B} (a_i^A x_i + a_i^B x_i);$$

and as in the proof of Theorem 11.25, with the decreasingness of the coefficients, the above inequality can be extended to the following:

$$\sum_{i \in A \cup B} a_i^{A \cup B} x_i + \sum_{i \in A \cap B} a_i^{A \cap B} x_i \le \sum_{i \in A} a_i^A x_i + \sum_{i \in B} a_i^B x_i, \qquad (11.19)$$

for any $x \in \mathcal{EP}(b)$. Now, consider $x^* \in \mathcal{EP}(b)$ that satisfies

$$\sum_{i \in A} a_i^A x_i^* = b(A) \quad \text{and} \quad \sum_{i \in B} a_i^B x_i^* = b(B), \qquad (11.20)$$

for some $A, B \subset E$. Then, based on the inequality in (11.19), we have

$$\sum_{i \in A \cup B} a_i^{A \cup B} x_i^* \le \sum_{i \in A} a_i^A x_i^* + \sum_{i \in B} a_i^B x_i^* - \sum_{i \in A \cap B} a_i^{A \cap B} x_i^*$$
$$\le b(A) + b(B) - b(A \cap B)$$
$$\le b(A \cup B),$$

where the last inequality follows from the supermodularity of $b(\cdot)$, and the second inequality combines (11.20) with

$$\sum_{i \in A \cap B} a_i^{A \cap B} x_i^* \ge b(A \cap B),$$

which, in turn, follows from $x^* \in \mathcal{EP}(b)$.

On the other hand, since $x^* \in \mathcal{EP}(b)$, we must also have

$$\sum_{i \in A \cup B} a_i^{A \cup B} x_i^* \ge b(A \cup B).$$

Hence,

$$\sum_{i \in A \cup B} a_i^{A \cup B} x_i^* = b(A \cup B)$$

follows, which is what is desired.   □

Note that the above theorem does not require the increasingness of b. In fact, the conclusion that $\mathcal{EP}(b)$ is an EP does not necessarily imply that b is an increasing function, since now the coefficients are decreasing. For instance, letting $A \subset B$ in (11.20), we cannot conclude that $b(A) \leq b(B)$.

The same twist will affect $\mathcal{EP}(f)$ in the following theorem, which parallels the last two theorems for $\mathcal{EP}(b)$.

Theorem 11.27 (i) Suppose $\mathcal{EP}(f)$ in (11.13) is an EP, with the coefficients $\{a_i^S\}$ being decreasing and submodular. Then, $f(\cdot)$ is a submodular function.

(ii) Conversely, suppose in (11.13) the coefficients $\{a_i^S\}$ are increasing and supermodular, and the function $f(\cdot)$ is submodular. Then, $\mathcal{EP}(f)$ is an EP.

Notice that in the first part above, which parallels Theorem 11.25, we cannot conclude that f is increasing, again because the coefficients are decreasing. In the second part, as in Theorem 11.26, we do not need the increasingness of f; but in contrast to Theorem 11.26, the conclusion that $\mathcal{EP}(f)$ is an EP does imply the increasingness of f, since the coefficients in this case are increasing.

Example 11.28 As an application of Theorem 11.26, let us revisit the branching bandit problem in Example 11.20. Recall that given a class i project, an (i, S) period constitutes the union of the time intervals in which the class i project's S-descendants are being served; and T_i^S denotes the length of an (i, S) period, which is the potential work of all the classes in the set S generated by the class i project. It is not difficult to verify that T_i^S is supermodular in S, for each i. That is, for any $A, B \subseteq E$, we have

$$T_i^{A \cup B} + T_i^{A \cap B} \geq T_i^A + T_i^B. \tag{11.21}$$

Intuitively, this is because when A and B are grouped together, then the potential work that can be generated by all of the project classes and their descendents will include those intergroup transfers between A and B, which are excluded when A and B are treated as separate groups.

Recall that the first (discounted) class of performance vector in Example 11.20 constitutes a polytope of $\mathcal{EP}(b)$ as in (11.12), with

$$a_i^S = \frac{\mathsf{E}\left[\int_0^{T_i^{S^c}} e^{-\alpha t} dt\right]}{\mathsf{E}\left[\int_0^{v_i} e^{-\alpha t} dt\right]}, \quad i \in S \subseteq E,$$

(recall that S^c denotes the complement set of S), and

$$b(S) = \mathsf{E}\left[\int_0^{T_m^E} e^{-\alpha t} dt\right] - \mathsf{E}\left[\int_0^{T_m^{S^c}} e^{-\alpha t} dt\right].$$

The numerator of a_i^S (which is the only part that involves S) is a function

$$g(x) := \int_0^x e^{-\alpha t} dt = \frac{1}{\alpha}(1 - e^{-\alpha x}),$$

which is increasing and concave in x. Write $h(S) := T_i^{S^c}$. Then, $h(S)$ is decreasing and submodular in S, following (11.21). It is then easy to verify that $g(h(S))$ is decreasing and submodular in S; and hence, so is a_i^S. In particular, for $A, B \subseteq E$,

$$g(h(A)) + g(h(B)) \geq g(h(A \cap B)) + g(h(A \cup B))$$

follows from the increasing concavity of $g(x)$, along with

$$h(A \cap B) \geq h(A), h(B) \geq h(A \cup B),$$

since h is increasing, and

$$h(A) + h(B) \geq h(A \cap B) + h(A \cup B),$$

since h is submodular.

Similarly, for $b(S)$, only the second expectation involves S, and it is equal to $-g(h(S))$ using the above notation. Hence, $b(S)$ is supermodular.

Therefore, following Theorem 11.26, we can conclude that the performance polytope $\mathcal{EP}(b)$ is an EP. This, of course, is a known result, since the performance vector satisfies GCL ([3]). However, the above provides a direct proof that the performance polytope is an EP (and hence, GCL is satisfied). It also establishes that the right-hand-side function $b(\cdot)$ of the polytope is a supermodular function, which is a new result. (By the way, in this particular case, $b(\cdot)$ also happens to be increasing, as is evident from the above discussion.)

To conclude this section we summarize below the relationship between GCL and EP.

Theorem 11.29 If the performance vector x satisfies GCL, then the performance polytope is an EP, of which the vertices correspond to the performance under priority rules, and the functions $b(A)$ and $f(A)$ correspond to the performance of job classes in the set A when A is given priority over all other classes in $E \setminus A$. Furthermore, if the coefficients $\{a_i^S\}$ are decreasing and submodular, then $b(A)$ and $f(A)$ are, respectively, supermodular and submodular functions.

11.5 Optimization over EP

Here we consider the optimization problem of maximizing a linear function over the EP, $\mathcal{EP}(f)$, defined in (11.13):

$$
\textbf{(PG)} \quad \max \quad \sum_{i \in E} c_i \, x_i
$$

$$
\text{s.t.} \quad \sum_{i \in A} a_i^A x_i \leq f(A), \quad \text{for all } A \subseteq E,
$$

$$
x_i \geq 0, \quad \text{for all } i \in E.
$$

The dual problem can be written as follows:

$$
\textbf{(DG)} \quad \min \quad \sum_{A \subseteq E} y_A \, f(A)
$$

$$
\text{s.t.} \quad \sum_{A \ni i} y_A a_i^A \geq c_i, \quad \text{for all } \in E,
$$

$$
y_A \geq 0, \quad \text{for all } A \subseteq E.
$$

Let us start with $\pi = (1, 2, \ldots, n)$, and consider x^π, the vertex defined at the beginning of the last section. Below we write out the objective function of **(PG)** at x^π, and use the expression, along with complementary slackness, to identify a candidate for the dual solution. From dual feasibility, we then identify the conditions under which π is the optimal permutation. Collectively, these steps constitute an algorithm that finds the optimal π.

For simplicity, write x for x^π below. We first write out x_n in the objective function

$$
\sum_{i=1}^{n} c_i x_i = c_n \left(f(\{1, \ldots, n\}) - \sum_{i=1}^{n-1} a_i^{\{1,\ldots,n\}} x_i \right) \Big/ a_n^{\{1,\ldots,n\}} + \sum_{i=1}^{n-1} c_i x_i
$$

$$
= y_{\{1,\ldots,n\}} f(\{1, \ldots, n\}) + \sum_{i=1}^{n-1} \left(c_i - y_{\{1,\ldots,n\}} a_i^{\{1,\ldots,n\}} \right) x_i,
$$

where we set

$$
y_{\{1,\ldots,n\}} = c_n / a_n^{\{1,\ldots,n\}}.
$$

Next, we write out x_{n-1} in the summation above, and set

$$
y_{\{1,\ldots,n-1\}} = (c_{n-1} - y_{\{1,\ldots,n\}} a_{n-1}^{\{1,\ldots,n\}}) / a_{n-1}^{\{1,\ldots,n-1\}},
$$

to reach the following expression:

$$\sum_{i=1}^{n} c_i x_i = y_{\{1,\ldots,n\}} f(\{1,\ldots,n\}) + y_{\{1,\ldots,n-1\}} f(\{1,\ldots,n-1\})$$

$$+ \sum_{i=1}^{n-2} \left(c_i - y_{\{1,\ldots,n\}} a_i^{\{1,\ldots,n\}} - y_{\{1,\ldots,n-1\}} a_i^{\{1,\ldots,n-1\}} \right) x_i^{\pi}.$$

This procedure can be repeated to yield the following:

$$\sum_{i=1}^{n} c_i x_i = y_{\{1,\ldots,n\}} f(\{1,\ldots,n\}) + y_{\{1,\ldots,n-1\}}) f(\{1,\ldots,n-1\})$$

$$+ \cdots + y_{\{1,2\}} f(\{1.2\}) + y_{\{1\}} f(\{1\}), \tag{11.22}$$

where

$$y_{\{1,\ldots,k\}} = \left(c_k - \sum_{j=k+1}^{n} y_{\{1,\ldots,j\}} a_k^{\{1,\ldots,j\}} \right) \Big/ a_k^{\{1,\ldots,k\}}, \tag{11.23}$$

for $k = 1, \ldots, n$. (When $k = n$, the vacuous summation in (11.23) vanishes.) Furthermore, set $y_A := 0$ for all other $A \subseteq E$.

With the above choice of x and y, it is easy to check that complementary slackness is satisfied. Also, primal feasibility is automatic, guaranteed by the definition of EP, since x is a vertex. Hence, we only need to check dual feasibility.

From the construction of y in (11.23), we have

$$\sum_{j=i}^{n} y_{\{1,\ldots,j\}} a_i^{\{1,\ldots,j\}} = c_i, \qquad i \in E,$$

satisfying the first set of constraints in **(DG)**. So it suffices to show that the n nonzero dual variables in (11.23) are nonnegative. To this end, we need to be specific about the construction of the permutation $\pi = (1,\ldots,n)$.

Let us start from the last element in π. Note that from (11.23), we have

$$y_{\{1,\ldots,n\}} = \frac{c_n}{a_n^{\{1,\ldots,n\}}} \geq 0.$$

Next, to ensure $y_{\{1,\ldots,n-1\}} \geq 0$, the numerator of its expression in (11.23) must be nonnegative, i.e.,

$$\frac{c_{n-1}}{a_{n-1}^{\{1,\ldots,n\}}} \geq y_{\{1,\ldots,n\}} = \frac{c_n}{a_n^{\{1,\ldots,n\}}}.$$

Therefore, the index n has to be

$$n = \arg\min_i \frac{c_i}{a_i^{\{1,\ldots,n\}}}.$$

Note that this choice of n guarantees $y_{\{1,\ldots,n-1\}} \geq 0$, independent of the ordering of the other $n-1$ elements in the permutation.

Similarly, to ensure $y_{\{1,\ldots,n-2\}} \geq 0$, from (11.23), we must have

$$c_{n-2} - y_{\{1,\ldots,n-1\}}a_{n-2}^{\{1,\ldots,n-1\}} - y_{\{1,\ldots,n\}}a_{n-2}^{\{1,\ldots,n\}} \geq 0,$$

or

$$\frac{c_{n-2} - y_{\{1,\ldots,n\}}a_{n-2}^{\{1,\ldots,n\}}}{a_{n-2}^{\{1,\ldots,n-1\}}} \geq y_{\{1,\ldots,n-1\}}.$$

Hence, the choice of $n-1$ has to be

$$n - 1 = \arg\min_{i \leq n-1} \frac{c_i - y_{\{1,\ldots,n\}}a_i^{\{1,\ldots,n\}}}{a_i^{\{1,\ldots,n-1\}}}.$$

This procedure can be repeated until all elements of the permutation are determined. In general, the index k is chosen in the order of $k = n, n-1, \ldots, 1$, and it has to satisfy

$$k = \arg\min_{i \leq k} \frac{c_i - \sum_{j=k+1}^n y_{\{1,\ldots,j\}}a_i^{\{1,\ldots,j\}}}{a_i^{\{1,\ldots,k\}}}.$$

Formally, the following algorithm solves the dual problem (DG) in terms of generating the permutation π, along with the dual solution y^π. The optimal primal solution is then the vertex x^π, corresponding to the permutation π.

Algorithm 11.30 [for (DG)]

(i) Initialization: $S(n) = E$, $k = n$;

(ii) If $k = 1$, stop, and output $\{\pi, S(k); y^\pi(S(k))\}$; otherwise, set

$$\pi_k := \arg\min_i \frac{c_i - \sum_{j=k+1}^n y_{S(j)}^\pi a_i^{S(j)}}{a_i^{S(k)}}$$

$$y_{S(k)}^\pi := \min_i \frac{c_i - \sum_{j=k+1}^n y_{S(j)}^\pi a_i^{S(j)}}{a_i^{S(k)}};$$

(iii) $k \leftarrow k - 1$, $S(k) = S(k+1) \setminus \{\pi_k\}$; goto (ii).

Theorem 11.31 Given an extended polymatroid $\mathcal{EP}(f)$, the above algorithm solves the primal and dual LPs **(PG)** and **(DG)** in $O(n^2)$ steps, with x^π and y^π being the optimal primal–dual solution pair.

Proof. Following the discussions preceding the algorithm, it is clear that we only need to check $y_{S(k)}^\pi \geq 0$, for $k = 1, \ldots, n$.

When $k = n$, following the algorithm, we have $S(n) = E$, and

$$\pi_n = \arg\min_i \{c_i/a_i^E\}, \qquad y_E^\pi = c_{\pi_n}/a_{\pi_n}^E \geq 0.$$

Inductively, suppose $y_{S(j)}^\pi \geq 0$, for $j = k+1, \ldots, n$, have all been determined. The choice of π_{k+1} and hence $y_{S(k+1)}^\pi$ in the algorithm guarantees

$$c_k - \sum_{j=k}^n y_{S(j)}^\pi a_k^{S(j)} \geq 0,$$

and hence $y_{S(k)}^\pi \geq 0$.

That the optimal solution is generated in $O(n^2)$ steps is evident from the description of the algorithm. $\qquad\square$

To summarize, the two remarks at the end of Section 11.1.2 for the polymatroid optimization also apply here: (i) Primal feasibility is automatic, by way of the definition of EP; and (ii) dual feasibility, along with complementary slackness, identifies the permutation π that defines the (primal) optimal vertex.

Furthermore, there is also an analogy to (11.3), i.e., the sum of dual variables yields the priority index. To see this, for concreteness consider Klimov's problem, with the performance measure x_i being the (long-run) average number of class i jobs in the system. (For this example, we are dealing with a minimization problem over the EP $\mathcal{EP}(b)$. But all of the above discussions, including the algorithm, still apply, mutatis mutandis, such as changing f to b and max to min.) The optimal policy is a priority rule corresponding to the permutation π generated by the above algorithm, with the jobs of class π_1 given the highest priority, and jobs of class π_n, the lowest priority. Let y^* be the optimal dual solution generated by the algorithm. Define

$$\gamma_i := \sum_{S \ni i} y_S^*, \qquad i \in E.$$

Then, we have

$$\gamma_{\pi_i} = y_{\{\pi_1,\ldots,\pi_i\}}^* + \cdots + y_{\{\pi_1,\ldots,\pi_n\}}^*, \quad i \in E. \tag{11.24}$$

Note that γ_{π_i} is decreasing in i, since the dual variables are nonnegative. Hence, the order of the γ_{π_i} is in the same direction as the priority assignment. In other words, (11.24) is completely analogous to (11.3): Just as with the indexing role played by the cost coefficients in the polymatroid case, in the EP case here $\{\gamma_i\}$ is also a set of indices upon which the priorities are assigned: At each decision epoch the server chooses to serve, among all waiting jobs, the job class with the highest γ index.

Finally, we can synthesize all the above discussions on GCL and its connection to EP, and on optimization over an EP, to come up with the following generalization of Theorem 11.16.

Theorem 11.32 Consider the optimal control problem in Theorem 11.16:

$$\max_{u \in \mathcal{A}} \sum_{i \in E} c_i x_i^u \quad [\text{or} \quad \min_{u \in \mathcal{A}} \sum_{i \in E} c_i x_i^u].$$

Suppose x is a performance measure that satisfies GCL. Then, this optimal control problem can be solved by solving the following LP:

$$\max_{x \in \mathcal{EP}(f)} \sum_{i \in E} c_i x_i \quad [\text{or} \quad \min_{x \in \mathcal{EP}(b)} \sum_{i \in E} c_i x_i].$$

The optimal solution to this LP is simply the vertex $x^\pi \in \mathcal{B}(f)$, with π being the permutation identified by Algorithm 11.30; and the optimal policy is the corresponding priority rule, which assigns the highest priority to class π_1 jobs, and the lowest priority to class π_n jobs.

Applying the above theorem to Klimov's model we can generate the optimal policy, which is a priority rule dictated by the permutation π, which, in turn, is generated by Algorithm 11.30.

11.6   Notes

A standard reference to matroids, as well as polymatroids, is Welsh [47]. The equivalence of the first two definitions of the polymatroid, Definitions 11.1 and 11.2, is a classical result; refer to, e.g., Edmonds [12], Welsh [47], and Dunstan and Welsh [11].

The original version of conservation laws, due to Kleinrock [31], takes the form of a single equality constraint, $\sum_{i \in E} x_i = b(E)$ or $= f(E)$. In the works of Coffman and Mitrani [8], and Gelenbe and Mitrani [20], the additional inequality constraints were introduced, which, along with the equality constraint, give a full characterization of the performance space. In a sequence of papers, Federgruen and Groenevelt [14, 15, 16] established the polymatroid structure of the performance space of several queueing systems, by showing that the RHS (right-hand-side) functions are increasing and submodular.

Shanthikumar and Yao [39] revealed the *equivalence* between conservations laws and the polymatroid nature of the performance polytope. In other words, the increasingness and submodularity of the RHS functions are not only sufficient but also necessary conditions for conservation laws. This equivalence is based on two key ingredients: On the one hand, the polymatroid Definition 11.2 asserts that if the "vertex" x^π—generated through a triangular system of n linear equations (made out of a total of $2^n - 1$ inequalities that define the polytope)—belongs to the polytope (i.e., if it satisfies all the other inequalities), for every permutation, π, then the polytope is a polymatroid. On the other hand, in conservation laws the RHS functions that characterize the performance polytope can be defined in such a way that they correspond to those "vertices." This way, the vertices will automatically belong to the performance space, since they are achievable by priority rules.

The direct implication of the connection between conservation laws and polymatroids is the translation of the scheduling (control) problem into an optimization problem. And in the case of a linear objective, the optimal solution follows immediately from examining the primal–dual pair: Primal feasibility is guaranteed by the polymatroid property: All vertices belong to the polytope, and dual feasibility, along with complementary slackness, yields the priority indices.

Motivated by Klimov's problem, Tsoucas [42], and Bertsimas and Niño-Mora [3] extended conservation laws and related polymatroid structure to GCL and EP. The key ingredients in the conservation laws/polymatroid theory of [39] are carried over to GCL/EP. In particular, EP is defined completely analogously to the polymatroid Definition 11.2 mentioned above, via the "vertex" x^π, whereas GCL is such that for every permutation π, x^π corresponds to a priority rule, and thereby guarantees its membership to the performance polytope.

Very little was known about the properties of EP, until two doctoral dissertations by Lu [34] and Zhang [52] (also see [51]), from which much of the materials in Section 11.4 are drawn, including the equivalent definition for EP in Definition 11.22, the connections between EP and submodularity and supermodularity in Theorems 11.25, 11.26, and 11.27, and Example 11.28.

Dynamic scheduling of a multiclass stochastic network is a complex and difficult problem that has continued to attract much research effort. A sample of more recent works shows a variety of different approaches to the problem, from Markov decision programming (e.g., Harrison [27], Weber and Stidham [45]) and monotone control of generalized semi-Markov processes (Glasserman and Yao [24, 25]) to asymptotic techniques via diffusion limits (Harrison [28], and Harrison and Wein [29]). This chapter presents yet another approach, which is based on polymatroid optimization. It exploits, in the presence of conservation laws and GCL, the polymatroid or EP structure of the performance polytope and turns the dynamic control problem into a static optimization problem.

The $c\mu$-rule in Example 11.17 is a subject with a long history that can be traced back to Smith [40] and the monograph of Cox and Smith [9]; also see, e.g., [5, 6]. More ambitious examples of applications that are based on Theorem 11.16 include scheduling in a Jackson network ([36]), scheduling and load balancing in a distributed computer system ([37]), and scheduling multiclass jobs in a flexible manufacturing system ([50]).

Klimov's problem generalizes the $c\mu$-rule model by allowing completed jobs to feed back and change classes. Variations of Klimov's model have also been widely studied using different techniques, e.g., Harrison [27], Tcha and Pliska [41]. The optimal priority policy is often referred to as the "Gittins index" rule, as the priority indices are closely related to those indices in dynamic resource allocation problems that are made famous by Gittins ([21, 22, 23]).

Klimov's model, in turn, belongs to the more general class of branching bandit problems (refer to Section 11.3), for which scheduling rules based on Gittins indices are optimal. There is a vast literature on this subject; refer to, e.g., Lai and Ying [33], Meilijson and Weiss [35], Varaiya et al. [43], Weber [44], Weiss [46], Whittle [48, 49], as well as Gittins [21, 22] and Gittins and Jones [23].

GCL corresponds to the so-called indexable class of stochastic systems, including Klimov's model and branching bandits as primary examples; refer to [3, 4]. Beyond this indexable class, however, the performance space is not even an EP. There have been recent studies that try to bound such performance spaces by more structured polytopes (e.g., polymatroid and EP), e.g., Bertsimas [2], Bertsimas et al. [4], Dacre et al. [10], and Glazebrook and Garbe [19].

11.7   Exercises

1. Verify that the workload process in a $G/G/1$ system that allows preemption satisfies conservation laws; refer to Example 11.7.

2. Verify that the queue-length process (i.e., number of jobs in the system) in the $G/M/c$ queue that allows preemption satisfies conservation laws, provided that all job classes follow a common exponential service-time distribution; refer to Example 11.9.

3. Verify that assuming nonpreemption will extend the result in the above problem to general service-time distributions, i.e., the $G/G/c$ system; refer to Example 11.12.

4. Verify the conservation laws claimed for the performance measures in Example 11.14 and Example 11.15.

5. Prove *directly* that Definition 11.2 implies Definition 11.1 of the poly-matroid.

6. Prove the relation in (11.21) for T_i^S, the length of the (i, S) period, of the branching bandit problem.

7. Prove the results in Theorem 11.27.

8. In Theorems 11.25, 11.26, and 11.27 the coefficients of the EP, a_i^S, are either increasing and supermodular or decreasing and submodular. Now, suppose the coefficients are increasing and submodular. Then, show that we must have

$$a_i^A = a_i^{A \cup B}, \qquad i \in A \backslash B,$$

for any $A, B \subseteq E$.

9. Consider the polytope $\mathcal{EP}(b)$ in (11.12), as yet not necessarily an EP. Suppose for each $i \in S \subset E$, $a_i^S = a_i \cdot \alpha^S$, where $a_i > 0$ and $\alpha^S > 0$ are two constants (depending, respectively, on i and on S only). Then, prove that $\mathcal{EP}(b)$ is an EP if and only if $b(S)/\alpha^S$ is increasing and supermodular in S. Similarly, $\mathcal{EP}(f)$ is an EP if and only if $f(S)/\alpha^S$ is increasing and submodular in S.

10. Let $\{E_1, \ldots, E_K\}$ be a partition of the set $E = \{1, \ldots, n\}$. Suppose the coefficients of the EP, a_i^S, satisfy

$$a_i^S = a_i^{S \cup E_k}, \qquad \forall i \in S \subseteq E.$$

Show that $b(\cdot)$ (respectively $f(\cdot)$) is supermodular (respectively sub-modular), *with respect to the partition* $\{E_1, \ldots, E_K\}$, meaning that for any Σ that is the union of some E_k's, and Σ^c, its complement (neither is empty), and any

$$A \subseteq B \subseteq \Sigma \qquad \text{and} \qquad A' \subseteq B' \subseteq \Sigma^c,$$

the following holds: For supermodularity

$$b(A \cup A') + b(B \cup B') \geq b(A \cup B') + b(B \cup A'),$$

or, for submodularity,

$$f(A \cup A') + f(B \cup B') \leq f(A \cup B') + f(B \cup A').$$

References

[1] ASMUSSEN, S., *Applied Probability and Queues*. Wiley, Chichester, U.K., 1987.

[2] BERTSIMAS, D., The Achievable Region Method in the Optimal Control of Queueing Systems; Formulations, Bounds and Policies. *Queueing Systems*, **21** (1995), 337–389.

[3] BERTSIMAS, D. AND NIÑO-MORA, J., Conservation Laws, Extended Polymatroid and Multi-Armed Bandit Problems: A Unified Approach to Indexable Systems. *Mathematics of Operations Research*, **21** (1996), 257–306.

[4] BERTSIMAS, D. PASCHALIDIS, I.C. AND TSITSIKLIS, J.N., Optimization of Multiclass Queueing Networks: Polyhedral and Nonlinear Characterization of Achievable Performance. *Ann. Appl. Prob.*, **4** (1994), 43–75.

[5] BARAS, J.S., DORSEY, A.J. AND MAKOWSKI, A.M., Two Competing Queues with Linear Cost: the μc Rule Is Often Optimal. *Adv. Appl. Prob.*, **17** (1985), 186–209.

[6] BUYUKKOC, C., VARAIYA, P. AND WALRAND, J., The $c\mu$ Rule Revisited. *Adv. Appl. Prob.*, **30** (1985), 237–238.

[7] CHVÁTAL, V., *Linear Programming*. W.H. Freeman, New York, 1983.

[8] COFFMAN, E. AND MITRANI, I., A Characterization of Waiting Time Performance Realizable by Single Server Queues. *Operations Research,* **28** (1980), 810–821.

[9] COX, D.R. AND SMITH, W.L., *Queues.* Methunen, London, 1961.

[10] DACRE, K.D., GLAZEBROOK, K.D., AND NINÕ-MORA, J., The Achievable Region Approach to the Optimal Control of Stochastic Systems. *J. Royal Statist. Soc.* (1999).

[11] DUNSTAN, F.D.J. AND WELSH, D.J.A., A Greedy Algorithm for Solving a Certain Class of Linear Programmes. *Math. Programming,* **5** (1973), 338–353.

[12] EDMONDS, J., Submodular Functions, Matroids and Certain Polyhedra. *Proc. Int. Conf. on Combinatorics (Calgary),* Gordon and Breach, New York, 69–87, 1970.

[13] FEDERGRUEN, A. AND GROENEVELT, H., The Greedy Procedure for Resource Allocation Problems: Necessary and Sufficient Conditions for Optimality. *Operations Res.,* **34** (1986), 909–918.

[14] FEDERGRUEN, A. AND GROENEVELT, H., The Impact of the Composition of the Customer Base in General Queueing Models. *J. Appl. Prob.,* **24** (1987), 709–724.

[15] FEDERGRUEN, A. AND GROENEVELT, H., M/G/c Queueing Systems with Multiple Customer Classes: Characterization and Control of Achievable Performance under Non-Preemptive Priority Rules. *Management Science,* **34** (1988), 1121–1138.

[16] FEDERGRUEN, A. AND GROENEVELT, H., Characterization and Optimization of Achievable Performance in Queueing Systems. *Operations Res.,* **36** (1988), 733–741.

[17] FONG, L.L. AND SQUILLANTE, M.S., Time-Function Scheduling: A General Approach to Controllable Resource Management. IBM Research Report RC-20155, IBM Research Division, T.J. Watson Research Center, Yorktown Hts., New York, NY 10598, 1995.

[18] FRANASZEK, P.A. AND NELSON, R.D., Properties of Delay Cost Scheduling in Timesharing Systems. IBM Research Report RC-13777, IBM Research Division, T.J. Watson Research Center, Yorktown Hts., New York, NY 10598, 1990.

[19] GLAZEBROOK, K.D. AND GARBE, R., Almost Optimal Policies for Stochastic Systems which Almost Satisfy Conservation Laws. Preprint, 1996.

[20] GELENBE, E. AND MITRANI, I., *Analysis and Synthesis of Computer Systems*. Academic Press, London, 1980.

[21] GITTINS, J.C., Bandit Processes and Dynamic Allocation Indices (with discussions). *J. Royal Statistical Society, Ser. B*, **41** (1979), 148–177.

[22] GITTINS, J.C., *Multiarmed Bandit Allocation Indices*. Wiley, Chichester, 1989.

[23] GITTINS, J.C. AND JONES, D.M., A Dynamic Allocation Index for the Sequential Design of Experiments. In: *Progress in Statistics: European Meeting of Statisticians, Budapest, 1972*, J. Gani, K. Sarkadi and I. Vince (eds.), North-Holland, Amsterdam, 1974, 241–266.

[24] GLASSERMAN, P. AND YAO, D.D., *Monotone Structure in Discrete-Event Systems*. Wiley, New York, 1994.

[25] GLASSERMAN, P. AND YAO, D.D., Monotone Optimal Control of Permutable GSMP's. *Mathematics of Operations Research*, **19** (1994), 449–476.

[26] GRÖTSCHEL, M., LOVÁSZ, L., AND SCHRIJVER, A., *Geometric Algorithms and Combinatorial Optimization*, second corrected edition. Springer-Verlag, Berlin, 1993.

[27] HARRISON, J.M., Dynamic Scheduling of a Multiclass Queue: Discount Optimality. *Operations Res.*, **23** (1975), 270–282.

[28] HARRISON, J.M., The BIGSTEP Approach to Flow Management in Stochastic Processing Networks. In: *Stochastic Networks: Theory and Applications*, Kelly, Zachary, and Ziedens (eds.), Royal Statistical Society Lecture Note Series, #4, 1996, 57–90.

[29] HARRISON, J.M. AND WEIN, L., Scheduling Networks of Queues: Heavy Traffic Analysis of a Simple Open Network. *Queueing Systems*, **5** (1989), 265–280.

[30] KLEINROCK, L., A Delay Dependent Queue Discipline. *Naval Research Logistics Quarterly*, **11** (1964), 329–341.

[31] KLEINROCK, L., *Queueing Systems*, Vol. 2. Wiley, New York, 1976.

[32] KLIMOV, G.P., Time Sharing Service Systems, *Theory of Probability and Its Applications*, **19** (1974), 532–551 (Part I) and **23** (1978), 314–321 (Part II).

[33] LAI, T.L. AND YING, Z., Open Bandit Processes and Optimal Scheduling of Queueing Networks. *Adv. Appl. Prob.*, **20** (1988), 447–472.

[34] LU, Y., *Dynamic Scheduling of Stochastic Networks with Side Constraints.* Ph.D. thesis, Columbia University, 1998.

[35] MEILIJSON, I. AND WEISS, G., Multiple Feedback at a Single-Server Station. *Stochastic Proc. and Appl.,* **5** (1977), 195–205.

[36] ROSS, K.W. AND YAO, D.D., Optimal Dynamic Scheduling in Jackson Networks. *IEEE Transactions on Automatic Control,* **34** (1989), 47–53.

[37] ROSS, K.W. AND YAO, D.D., Optimal Load Balancing and Scheduling in a Distributed Computer System. *Journal of the Association for Computing Machinery,* **38** (1991), 676–690.

[38] SHANTHIKUMAR, J.G. AND SUMITA, U., Convex Ordering of Sojourn Times in Single-Server Queues: Extremal Properties of FIFO and LIFO Service Disciplines. *J. Appl. Prob.,* **24** (1987), 737–748.

[39] SHANTHIKUMAR J.G. AND YAO D.D., Multiclass queueing systems: polymatroid structure and optimal scheduling control. *Operation Research,* **40** (1992), Supplement 2, S293–299.

[40] SMITH, W.L., Various Optimizers for Single-Stage Production. *Naval Research Logistics Quarterly,* **3** (1956), 59–66.

[41] TCHA, D. AND PLISKA, S.R., Optimal Control of Single-Server Queueing Networks and Multiclass M/G/1 Queues with Feedback. *Operations Research,* **25** (1977), 248–258.

[42] TSOUCAS, P., The Region of Achievable Performance in a Model of Klimov. IBM Research Report RC-16543, IBM Research Division, T.J. Watson Research Center, Yorktown Hts., New York, NY 10598, 1991.

[43] VARAIYA, P., WALRAND, J., AND BUYYOKOC, C., Extensions of the Multiarmed Bandit Problem: The Discounted Case. *IEEE Trans. Automatic Control,* **30** (1985), 426–439.

[44] WEBER, R., On the Gittins Index for Multiarmed Bandits. *Annals of Applied Probability,* (1992), 1024–1033.

[45] WEBER, R. AND STIDHAM, S., JR., Optimal Control of Service Rates in Networks of Queues. *Adv. Appl. Prob.,* **19** (1987), 202–218.

[46] WEISS, G., Branching Bandit Processes. *Probability in the Engineering and Informational Sciences,* **2** (1988), 269–278.

[47] WELSH, D., *Matroid Theory,* (1976), Academic Press, London.

[48] WHITTLE, P., Multiarmed Bandits and the Gittins Index. *J. Royal Statistical Society, Ser. B,* **42** (1980), 143–149.

[49] WHITTLE, P., *Optimization over Time: Dynamic Programming and Stochastic Control,* vols. I, II, Wiley, Chichester, 1982.

[50] YAO, D.D. AND SHANTHIKUMAR, J.G., Optimal Scheduling Control of a Flexible Machine. *IEEE Trans. on Robotics and Automation,* **6** (1990), 706–712.

[51] YAO, D.D. AND ZHANG, L., Stochastic Scheduling and Polymatroid Optimization, *Lecture Notes in Applied Mathematics,* **33**, G. Ying and Q. Zhang (eds.), Springer-Verlag, 1997, 333–364.

[52] ZHANG, L., *Reliability and Dynamic Scheduling in Stochastic Networks.* Ph.D. thesis, Columbia University, 1997.

12

Scheduling of Fluid Networks

Here we study the optimal scheduling of a multiclass fluid network. We start with model formulation in Section 12.1, followed by developing a solution procedure based on linear programming in Section 12.2, and establishing several key properties of the procedure in Section 12.3. In particular, we show that the procedure involves up to 2^K iterations (K being the total number of types), and that the so-called global optimality, i.e., optimality of the objective function over every time point, is not guaranteed. In this sense, the solution procedure is termed "myopic" (or greedy). On the other hand, we show that the procedure is guaranteed to lead to the stability of the fluid network. In addition, we derive the minimum "clearing time" as the time it takes to drive all fluid levels down to zero.

We then focus on the single-station model, which is essentially the deterministic equivalent of Klimov's problem (refer to Section 11.3.2). In Section 12.4 we present an algorithm that solves a sequence of linear programs following the myopic procedure. In this special case, the myopic solution procedure takes no more than K iterations, and the global optimality is guaranteed, as shown in Section 12.5.

12.1 Problem Formulation and Global Optimality

Consider a fluid network that consists of a set of stations, or buffers, $\mathcal{J} = \{1, \ldots, J\}$, where different classes of fluid are processed. Let $\mathcal{K} = \{1, \ldots, K\}$ denote the set of all fluid classes. Let the mapping $\sigma : \mathcal{K} \mapsto \mathcal{J}$

map each fluid class to the station where it is processed. For instance, $\sigma(i)$ denotes the station where class i fluid is processed. Note that under this notation, every fluid class is processed at one and only one station. A station, however, may in general process more than one fluid class.

For each fluid class i, after it is processed at station $\sigma(i)$, a proportion p_{ij} of it becomes class j fluid and proceeds to station $\sigma(j)$ for processing, while the remaining fraction, $1 - \sum_{j \in \mathcal{K}} p_{ij}$, flows to the outside world (the environment external to the network). The source of any fluid j can be either from the outside world or from the internal transition of other fluid classes within the network. Let $P = (p_{ij})$, $i, j \in \mathcal{K}$, denote the switching (or routing) matrix, which is a nonnegative matrix, with row sums less than or equal to unity for each row. Furthermore, we assume that P has a spectral radius less than unity; namely, we consider an open network.

Let λ_i, $i = 1, \ldots, K$, denote the exogenous inflow rate for class i fluid into the network. Let μ_i, $i = 1, \ldots, K$, denote the *potential* processing rate of class i fluid, i.e., when station $\sigma(i)$ devotes its entire capacity to processing class i fluid. The inflow rate λ_i is assumed to be nonnegative and the processing rate μ_i is assumed to be positive for all $i \in \mathcal{K}$.

Let $Q_i(t)$ denote the fluid level (volume) of class i at time t; let $Q(t) := (Q_i(t))_{i \in \mathcal{K}}$; and $q := Q(0)$, the vector of initial fluid levels. To specify the dynamics governing $Q(t)$, we need first to specify the allocation of the processing capacity at each station. Let $T_i(t)$ denote the *cumulative* amount of time that class i fluid has been processed, at station $\sigma(i)$ over the time interval $[0, t]$. Let $T(t) := (T_i(t))_{i \in \mathcal{K}}$; We have

$$Q_i(t) = Q_i(0) + \lambda_i t + \sum_{j \in \mathcal{K}} \mu_j T_j(t) p_{ji} - \mu_i T_i(t).$$

The right-hand side above is simply the difference between the input (from both external and internal sources) up to t and the fluid that is depleted up to t, plus the initial fluid level. Note that the more complex aspect of the dynamics is hidden in $T(t)$. For instance, when the fluid level is at zero, the allocation of processing capacity has to be zero.

In matrix notation, the above can be expressed as

$$Q(t) = Q(0) + \lambda t - (I - P')DT(t), \tag{12.1}$$

where $\lambda := (\lambda_i)_{i \in \mathcal{K}}$, P' is the transpose of P, and D is a diagonal matrix with diagonal elements the fluid processing rates μ_i, $i \in \mathcal{K}$.

By definition, we must have $T(t)$ nondecreasing in t, and $T(0) = 0$. Furthermore, we also require the cumulative idle time (or unused capacity) $U_j(t)$, at each station $j \in \mathcal{J}$, to be nondecreasing in t. Let $U(t) := (U_j(t))_{j \in \mathcal{J}}$. Define a $J \times K$ (station) constituent matrix $C = (c_{jk})$, with $c_{jk} = 1$ if $\sigma(k) = j$, and $c_{jk} = 0$ otherwise. The jth row of C specifies which fluid classes are processed at station j. Then, we have

$$U(t) = et - CT(t).$$

Hence, the allocation (or scheduling) policy $T = \{T(t), t \geq 0\}$ is termed *admissible* if it satisfies

$$Q(t) = Q(0) + \lambda t - (I - P')DT(t) \geq 0; \qquad (12.2)$$
$$T(t) \text{ nondecreasing}, \qquad T(0) = 0;$$
$$U(t) = et - CT(t) \text{ nondecreasing}.$$

Let $\mathcal{A}$ denote the class of all admissible allocation policies. We note that in addition to (λ, μ, P, C), $\mathcal{A}$ implicitly depends on $Q(0)$.

Write (12.1) as $Q(t) = \Psi(T)(t)$, where $\Psi \colon T \mapsto Q$ is the mapping that represents the linear system in (12.1). Let $h = (h_k)_{k \in \mathcal{K}} \geq 0$ denote the vector of inventory cost rates charged for holding each unit of fluid. The dynamic scheduling problem we want to solve can be expressed as:

$$\min_{T \in \mathcal{A}} \{h'\Psi(T)(t), \ \forall t \geq 0\}. \qquad (12.3)$$

An allocation process T^* is termed *globally optimal* if $h'\Psi(T^*)(t) \leq h'\Psi(T)(t)$ for all $t \geq 0$ and all $T \in \mathcal{A}$. Naturally, the globally optimal allocation in general depends on (or varies with) the network parameters such as (λ, μ, P, C); on the other hand, it follows from our definition that the globally optimal allocation also *depends on (or varies with) the initial fluid level* $Q(0)$.

To conclude this section, we illustrate through an example where a globally optimal allocation as defined above does *not* exist.

Example 12.1 Consider a network with two stations in series and each station processing one fluid class. Specifically, we have $\mathcal{K} = \{1, 2\}$, $\mathcal{J} = \{1, 2\}$, $\sigma(1) = 1$, $\sigma(2) = 2$, $p_{12} = 1$, and $p_{ij} = 0$ for all other elements in the flow-transfer matrix. The inflow rate is $\lambda = (0, 0)'$ and the outflow rate is $\mu = (1, 2)'$. The initial fluid level is $Q_1(0) = Q_2(0) = 1$, and the unit holding costs for classes 1 and 2 are $h_1 = 1$ and $h_2 = 2$, respectively.

Now, suppose T^* is globally optimal; we will reach a contradiction. First consider another allocation $\tilde{T}(t) := (t, t)'$ for $t \in [0, 1]$; then at $t = 1$, $Q_1(1) = Q_2(1) = 0$ and $h'\Psi(\tilde{T})(1) = 0$. On the other hand, since T^* is globally optimal, it has to be optimal over the initial period, say $[0, t_1]$. Since the unit holding cost of class 2 is larger than that of class 1, it is clear that at least initially, one should not process class 1 until the level of class 2 fluid drops to zero; for otherwise, the outflow of class 1 (from station 1) becomes class 2 (inventory at station 2), incurring a higher holding cost. Indeed, it is not difficult to derive that we must have $T_1^*(t) = 0$ and $T_2^*(t) = t$ for $0 \leq t \leq 0.5$. At $t = 0.5$, we have $Q_1(0.5) = 1$ and $Q_2(0.5) = 0$. After this, no matter what one does, at $t = 1$ one always has $Q_1(1) \geq 0.5$, and hence $h'\Psi(T^*)(1) \geq 0.5 > h'\Psi(\tilde{T})(1)$, contradicting the optimality of T^* at $t = 1$.

12.2   The Myopic Procedure via LP

Suppose there exists a finite time epoch t_1, and T^1 is the optimal allocation over $[0, t_1]$, i.e., $h'\Psi(T^1)(t) \leq h'\Psi(T)(t)$ for all $T \in \mathcal{A}$ and $t \in [0, t_1]$. Suppose T^2 is a globally optimal allocation given the initial state (which is the end inventory under T^1)

$$Q^1(t_1) := Q(0) + \lambda t_1 - (I - P')DT^1(t_1).$$

Construct an allocation $\{T^*(t), t \geq 0\}$ through "pasting" together T^1 and T^2 as follows:

$$T^*(t) = \begin{cases} T^1(t), & 0 \leq t \leq t_1; \\ T^1(t_1) + T^2(t - t_1), & t \geq t_1. \end{cases}$$

Then we have

$$h'\Psi_{Q(0)}(T^*)(t) = h'\Psi_{Q(0)}(T^1)(t) \leq h'\Psi_{Q(0)}(T)(t)$$

for all $T \in \mathcal{A}$ and $t \in [0, t_1]$, where the subscript $Q(0)$ emphasizes the dependence on the initial state, and for $t \geq t_1$, we have

$$\begin{aligned} Q(t) &= Q(0) + \lambda t - (I - P')D[T^1(t_1) + T^2(t - t_1)] \\ &= Q^1(t_1) + \lambda(t - t_1) - -(I - P')DT^2(t - t_1). \end{aligned}$$

Since the right-hand side above equals $\Psi_{Q^1(t_1)}(T^2)(t - t_1)$ for $t \geq t_1$, we also have

$$h'\Psi_{Q(0)}(T^*)(t) \leq h'\Psi_{Q^1(t_1)}(T)(t - t_1)$$

for all $T \in \mathcal{A}$ and $t \geq t_1$.

Note, however, that we cannot claim that T^* constructed as above is necessarily a globally optimal allocation with respect to the initial state $Q(0)$. This is because there may exist another allocation T^3, under which the fluid level at t_1 reaches $Q^3(t_1) \neq Q^1(t_1)$, and after t_1, it is quite possible for T^3 to outperform T^*, since T^3 acts on the fluid network that starts from a different initial state (at t_1).

Hence, the pasting mechanism described above is a "myopic" procedure in solving the control problem in (12.3). Specifically, we start with the (given) initial state $Q(0)$, and find an allocation $T^1(t)$ and a time epoch t_1 such that $T^1(t)$ is optimal for $t \in [0, t_1]$. Next, we roll back the clock so that t_1 becomes the new time zero, let $Q(t_1)$ be the initial state, and find a second allocation $T^2(t)$ that is optimal for $t \in [0, t_2]$, for some t_2 and with respect to the initial state $Q(t_1)$. We continue this procedure until at some step n the corresponding allocation $T^n(t)$ becomes optimal for the entire (remaining) time horizon, i.e., $t_n = +\infty$. Later we will show that under a certain mild regulation this n is guaranteed to be finite; i.e., the myopic procedure is guaranteed to terminate within a finite number of steps.

It turns out that at each step in the myopic procedure, the allocation can be identified through solving a linear program (LP). (Refer to, e.g., Chvátal [7] for preliminaries of LP.) Set

$$A := (a_{kj}) = (I - P')D,$$

and hence

$$(Ax - \lambda)_k = \sum_{j \in \mathcal{K}} a_{kj} x_j - \lambda_k.$$

Set $c := A'h$. Let $T(t) = xt$, with x being a vector of decision variables. Then, from (12.2), we have

$$
\begin{aligned}
h'Q(t) &= h'[Q(0) + \lambda t] - h'(I - P')DT(t) \\
&= h'[Q(0) + \lambda t] - c'xt.
\end{aligned}
$$

Hence, the objective in (12.3) can be replaced by maximizing $cT(t) = c'xt$.

Now consider the constraints associated with the control problem in (12.3). Set $q = Q(0)$ and

$$\Pi(q) := \{k : q_k = 0\}.$$

Note that initially, i.e, shortly after $t = 0$, we need be concerned only with those fluid classes in $\Pi(q)$, in terms of enforcing the nonnegativity constraint (12.2). In other words, if a fluid class, say class i, has an initial inventory level that is positive, then regardless of the allocation policy, we can always follow the policy over a sufficiently short time interval so as to guarantee that Q_i will stay nonnegative. (This is due to the continuity of $Q(t)$.)

Therefore, we can translate (12.3) into the following LP:

$$
\begin{aligned}
\max \quad & c'x & (12.4) \\
\text{s.t.} \quad & (Ax - \lambda)_k \le 0, \quad k \in \Pi(q) \\
& Cx \le e, \\
& x \ge 0.
\end{aligned}
$$

Observe that the above LP depends on the initial state q only through the index set $\Pi(q)$, i.e., only on the identity of those classes with a zero fluid level. In other words, the actual positive values of the initial fluid levels have no bearings on the LP solution. That is, the dependency of the LP solution on q is only through the index set $\Pi(q)$. Let x denote an LP solution; the corresponding allocation policy is $T(t) = xt$. To determine the range for which T is feasible, we need consider only the the nonnegativity constraint in (12.2). This leads to

$$\tau(q; x) := \min_{k \notin \Pi(q)} \frac{q_k}{(Ax - \lambda)_k^+},$$

where $\tau(q; x) = +\infty$ if $(Ax - \lambda)_k \leq 0$ for all $k \notin \Pi(q)$.

Note that the LP in (12.4) in general will have multiple optimal solutions, in which case we pick a solution that yields the largest τ value. Specifically, let $X^{\Pi(q)}$ denote the set of optimal solutions to the LP in (12.4). As a *regulation*, we pick the solution $x^{\Pi(q)} \in X^{\Pi(q)}$ that yields the largest τ value:

$$\tau(q) := \tau(q; x^{\Pi(q)}) = \sup_{x \in X^{\Pi(q)}} \tau(q; x). \qquad (12.5)$$

Note that the above "sup" can be achieved regardless of whether it is finite or infinite; this is due to the compactness of $X^{\Pi(q)}$ and the continuity of $\tau(q; \cdot)$. Also note that to identify $x^{\Pi(q)} \in X^{\Pi(q)}$ in (12.5) amounts to solving another LP (see Exercise 1).

If $\tau(q) = \infty$, the myopic procedure is stopped, and $T(t) = x^{\Pi(q)}t$ is the globally optimal allocation, with respect to the initial state q. Otherwise, at $\tau(q)$ we reset the time zero and the initial inventory level, update the set $\Pi(q)$, and repeat the above myopic procedure via solving the LP in (12.4).

The procedure can be summarized as follows.

The Myopic Solution Procedure

- STEP 0. Set $n = 0$, $t_0 = 0$, $q = Q(0)$. Identify $\Pi(q)$.

- STEP 1. Solve (12.4) to obtain $x^{\Pi(q)}$ (following the regulation if the set of solutions is not unique) and $\tau(q)$.

- STEP 2. Set

$$t_{n+1} = t_n + \tau(q);$$

set

$$T(t) = T(t_n) + x^{\Pi(q)}(t - t_n), \qquad t_n < t \leq t_{n+1}.$$

- STEP 3. If $t_{n+1} = \infty$, stop; otherwise, set

$$q \leftarrow q + [\lambda - Ax^{\Pi(q)}]\tau(q), \qquad n \leftarrow n + 1;$$

go to STEP 1.

12.3   Properties of the Myopic Procedure

12.3.1   Termination Rules

Lemma 12.2 The objective value of the LP in (12.4) in any two consecutive iterations of the myopic solution procedure either decreases or remains unchanged.

Proof. Consider any two consecutive iterations in the myopic solution procedure. Denote the corresponding LP's by (P_1) and (P_2), and their optimal solutions by x^1 and x^2. Note that (P_2) is formulated as follows: First, the nonbinding constraints in (P_1), if any, are deleted; denote the resulting LP by (P_0) and its optimal solution by x^0. Second, a new constraint (which did not appear in (P_1)) is added to (P_0) to yield (P_2). It is then clear that $c'x^1 = c'x^0 \geq c'x^2$: The equality can be verified, for instance, through examining the corresponding dual problems; the inequality is obvious, since x^2 is also feasible for (P_0). $\qquad\square$

As a result of the above lemma, we next show that the myopic procedure stops within no more than 2^K iterations.

Theorem 12.3 Following the regulation stipulated in (12.5), the myopic procedure will terminate within at most 2^K iterations. More specifically, there exists an $n \leq 2^K$ such that $t_{n+1} = \infty$ in Step 2 of the myopic procedure.

Proof. First observe that there are only a maximum of 2^K different LPs involved in the myopic procedure, one corresponding to each subset $\Pi(q)$. From Lemma 12.2, we know as the procedure iterates, the objective value of the LP either strictly decreases or stays the same. Hence, the only possible scenario for the procedure *not* to terminate is that it loops forever among a number of LPs that all have the same objective value. Therefore, it suffices to show that if the objective value of any two consecutive iterations is the same then the corresponding τ value becomes infinity.

Denote the two consecutive LPs under consideration by (P_1) and (P_2). Denote the corresponding Π sets by Π^1 and Π^2, and the optimal solutions [observing (12.5)] as x^1 and x^2, respectively. Let q^1 denote the initial fluid level corresponding to (P_1). Below we show that a contradiction will be reached if $\tau(q^1) < \infty$ and $c'x^1 = c'x^2$ hold simultaneously.

Take the convex combination of x^1 and x^2, and define

$$x(\theta) = (1-\theta)x^1 + \theta x^2, \qquad \theta \in (0,1).$$

Below, we show that

(i) for a sufficiently small θ, $x(\theta)$ is another optimal solution for (P_1);

(ii) $\tau(q^1; x(\theta)) > \tau(q^1; x^1)$; hence, a contradiction, since the regulation in (12.5) is violated.

To show (i), note that $c'x(\theta) = c'x^1 = c'x^2$, for any $\theta \in (0,1)$; hence, it remains to argue that $x(\theta)$ is feasible to (P_1). Consider the set Π^1. Feasibility to (P_1) requires that $x(\theta)$ satisfy, in particular, the first inequality in (12.4). (The other two are obviously satisfied.) Consider the inequality regarding Π^1. We need

$$(\lambda - Ax(\theta))_\ell = (1-\theta)(\lambda - Ax^1)_\ell + \theta(\lambda - Ax^2)_\ell \leq 0, \quad \forall \ell \in \Pi^1. \quad (12.6)$$

If $(\lambda - Ax^2)_\ell \leq 0$, then the above holds, since $(\lambda - Ax^1)_\ell \leq 0$, due to $\ell \in \Pi^1$. If $(\lambda - Ax^2)_\ell > 0$, then we must have $\ell \in \Pi^1 \setminus \Pi^2$. This implies that we must have $(\lambda - Ax^1)_\ell < 0$, i.e., it is a nonbinding constraint in (P_1), which was subsequently removed from (P_2). Therefore, by choosing $\theta > 0$ small enough, (12.6) will hold.

To show (ii), recall the following:

$$\tau(q^1; x^1) = \min_{k \notin \Pi^1} \frac{q_k}{(Ax^1 - \lambda)_k^+}, \tag{12.7}$$

$$\tau(q^1; x(\theta)) = \min_{k \notin \Pi^1} \frac{q_k}{(Ax(\theta) - \lambda)_k^+}. \tag{12.8}$$

Since $\tau(q^1; x^1) = \tau(q^1) < \infty$ as assumed, we also have $\tau(q^1; x(\theta)) < \infty$, for $\theta > 0$ sufficiently small. Suppose the minima of (12.7) and (12.8) are attained by classes k and j, respectively. (It is possible that $k = j$.) Then, we argue that

$$\frac{q_k}{(Ax^1 - \lambda)_k} < \frac{q_j}{(1 - \theta)(Ax^1 - \lambda)_j + \theta(Ax^2 - \lambda)_j}, \tag{12.9}$$

and hence the claimed contradiction that $\tau(q^1; x^1) < \tau(q^1; x(\theta))$. Note that we have, by the definition of k,

$$\frac{q_k}{(Ax^1 - \lambda)_k} \leq \frac{q_j}{(Ax^1 - \lambda)_j}.$$

If the above holds as a strict inequality, then (12.9) holds for a sufficiently small $\theta > 0$. If the above holds as an equality, then class j fluid, like class k, becomes binding, and hence $j \in \Pi^2$, which implies $(Ax^2 - \lambda)_j \leq 0$. Therefore, (12.9) still holds.   □

12.3.2 Stability and Clearing Time

Recall that in Section 8.2 we discussed the notion of stability in a two-station (Kumar–Seidman) fluid network under a priority service discipline. Here, we define stability for a general multiclass fluid network as follows:

Definition 12.4 The multiclass fluid network specified in Section 12.1 is termed *stable* if there exists an admissible control (allocation) $\{T(t)\}$ such that for any initial state $q := Q(0)$, the fluid levels can all be driven down to zero at some time epoch $\tau < \infty$, i.e., $Q(t) = 0$ for all $t \geq \tau$.

Below, we show that in particular, if $h > 0$, the myopic procedure spelled out in the last section yields one such admissible control that guarantees the stability of the fluid network, under the traffic condition

$$\rho := CA^{-1}\lambda = D^{-1}(I - P')^{-1}\lambda < 1,$$

i.e., $\rho_j < 1$ for all $j \in \mathcal{J}$.

Theorem 12.5 Suppose $\rho < e$. Then, the fluid network is stable under the admissible control generated by the myopic procedure with $h > 0$ (i.e., $h_k > 0$ for all $k \in \mathcal{K}$) and with the regulation in (12.5) in force. In fact, if the myopic procedure terminates at iteration n (which must happen at $n \leq 2^K$), then $Q(t) = 0$ for all $t \geq t_n$.

Then we have the following corollary.

Corollary 12.6 The fluid network is stable if and only if $\rho < 1$.

The sufficiency part of the corollary follows clearly from Theorem 12.5, while the necessity part can be proved by choosing $Q_j(0) > 0$ for a $j \in \mathcal{J}$ with $\rho_j \geq 1$.

Proof (of Theorem 12.5). By Theorem 12.3, the myopic procedure terminates at some step $n \leq 2^K$ such that $t_{n+1} = \infty$. Let $q^n = Q(t_n)$ and $x^{\Pi(q^n)}$ be the optimal solution to the LP in (12.4) corresponding to $q = q^n$.

First we show that $x^{\Pi(q^n)} = A^{-1}\lambda$. From the feasibility of $x^{\Pi(q^n)}$, we have $(Ax^{\Pi(q^n)} - \lambda)_{\Pi(q^n)} \leq 0$; and from the myopic procedure and $t_{n+1} = \infty$, we have $(Ax^{\Pi(q^n)} - \lambda)_k \leq 0$ for all $k \notin \Pi(q^n)$. Hence,

$$Ax^{\Pi(q^n)} - \lambda \leq 0. \tag{12.10}$$

Now we claim that this inequality must hold as an equality and hence $x^{\Pi(q^n)} = A^{-1}\lambda$. For otherwise, let $a \neq \emptyset$ be such that $(Ax^{\Pi(q^n)} - \lambda)_a < 0$. Let $x^\epsilon = x^{\Pi(q^n)} + \epsilon A^{-1}e_a$, where $\epsilon > 0$ is a scalar and e_a is a vector whose kth element equals 1 if $k \in a$ and equals 0 otherwise. Then, for $\epsilon > 0$ small enough, x^ϵ is a feasible solution to the LP. Clearly, $x^\epsilon \geq 0$ (since $A^{-1} \geq 0$) and $Ax^\epsilon - \lambda \leq 0$ for $\epsilon > 0$ small enough. The inequality in (12.10) implies $x^{\Pi(q^n)} \leq A^{-1}\lambda$ (since $A^{-1} \geq 0$, A being an M-matrix), and hence

$$Cx^{\Pi(q^n)} \leq CA^{-1}\lambda = \rho < e,$$

which implies that $Cx^\epsilon \leq e$ for ϵ small enough. On the other hand,

$$c'x^\epsilon = h'Ax^\epsilon = h'Ax^{\Pi(q^n)} + \epsilon h'e_a > h'Ax^{\Pi(q^n)} = c'x^{\Pi(q^n)}$$

for $\epsilon > 0$, which contradicts $x^{\Pi(q^n)}$ being the optimal solution to the LP. Hence, $x^{\Pi(q^n)} = A^{-1}\lambda$.

Next, we show that $q^n = 0$. If $q^n \neq 0$, let $a = \mathcal{K} \setminus \Pi(q^n)$ (which is nonempty). Then $x^\epsilon = A^{-1}[\lambda + \epsilon e_a]$ is feasible to the LP in (12.4) with $q = q^n$ for $\epsilon > 0$ small enough. But

$$c'x^\epsilon = h'(\lambda + \epsilon e_a) > h'\lambda = c'x^{\Pi(q^n)}$$

for any $\epsilon > 0$, again contradicting $x^{\Pi(q^n)}$ being the optimal solution to the LP in (12.4).

Finally, with $Q(t_n) = q^n = 0$ and

$$T(t) = T(t_n) + x^{\Pi(q^n)}(t - t_n) = T(t_n) + A^{-1}\lambda(t - t_n),$$

for all $t \geq t_n$, we clearly have

$$Q(t) = Q(t_n) + \lambda(t - t_n) - A[T(t) - T(t_n)] = 0,$$

for all $t \geq t_n$.   □

Next, we define the *clearing time*, associated with a control T, as the time to drive all fluid levels down to zero. Formally, consider a fluid network, with the parameters (λ, μ, P, C) and $Q(0)$ given. For an admissible control T, the clearing time associated with T is

$$\tau_T := \inf\{t \geq 0 : \Psi(T)(t) = 0\},$$

where $\tau_T = \infty$ if $\Psi(T)(t) \neq 0$ for all $t \geq 0$. The *minimal* clearing time is defined as

$$\tau^* = \inf_{T \in \mathcal{A}} \tau_T.$$

The following theorem shows that the minimal clearing time is achievable under the traffic condition $\rho < e$.

Theorem 12.7 Suppose that $\rho < e$. Then, the minimal clearing time is

$$\tau^* := \max_{i \in \mathcal{K}} \left\{ \frac{(CA^{-1}Q(0))_i}{1 - \rho_i} \right\}, \tag{12.11}$$

which is achieved by an admissible control $T^*(t) = x^* t$ over the time interval $[0, \tau^*)$, where

$$x^* = \frac{A^{-1}Q(0)}{\tau^*} + A^{-1}\lambda.$$

Furthermore, suppose i^* is the station that achieves τ^* in (12.11). Then, $(Cx^*)_{i^*} = 1$

Proof. Set $q = Q(0)$. Suppose some admissible control $T(t)$ achieves a clear time τ. Then,

$$Q(\tau) = q + \lambda\tau - AT(\tau) = 0,$$

and hence,

$$T(\tau) = A^{-1}(q + \lambda\tau).$$

Since T is admissible, we must have $CT(\tau) \leq e\tau$, which, in turn, implies

$$CA^{-1}q \leq (e - CA^{-1}\lambda)\tau = (e - \rho)\tau.$$

Since the above holds for all components $i \in \mathcal{K}$ of the vectors on both sides, we have

$$\tau \geq \max_{i \in \mathcal{K}} \left\{ \frac{(CA^{-1}q)_i}{1 - \rho_i} \right\} := \tau^*.$$

It is left as an exercise (Exercise 2) to show that T^* is an admissible control. To verify that T^* indeed achieves the clearing time τ^*, observe the following:

$$Q(\tau^*) = q + \lambda\tau^* - A\left(\frac{A^{-1}q}{\tau^*} + A^{-1}\lambda\right)\tau^*$$
$$= q + \lambda\tau^* - q - \lambda\tau^*$$
$$= 0.$$

Finally, we also have

$$(Cx^*)_{i\cdot} = \left(\frac{CA^{-1}q}{\tau^*}\right)_{i\cdot} + (CA^{-1}\alpha)_{i\cdot}$$
$$= (1 - \rho_{i\cdot}) + \rho_{i\cdot}$$
$$= 1.$$

$\square$

12.4 The Single-Station Model

In this section we apply the myopic procedure to a single-station model. In the literature, this model is sometimes also referred to as a single-server network, based on the following equivalent view of the system: Transactions—continuous fluid or discrete jobs—are circulated among a set of stations according to the routing matrix P; and there is a single server that attends to all the stations and processes the transactions.

We start in the first of two subsections below with a step-by-step description of the myopic procedure as applied to this single-station model. In the second subsection, we formalize the algorithm, and set the stage for the optimality proof in the next section.

Throughout, we shall assume that there is enough capacity to process all the fluid circulating in the system; specifically, the traffic intensity satisfies $\rho = e'A^{-1}\lambda \leq 1$. We note that following our open network assumption (that P has a spectral radius less than one), the elements of A satisfy

$$a_{kk} > 0, \quad \forall k, \quad \text{and} \quad a_{kj} \leq 0, \quad \forall k \neq j,$$

and all principal minors of A have inverses, which are all nonnegative matrices. (Refer to Lemma 7.1.)

12.4.1 Description of the Solution Procedure

To start with, assume $\Pi(q) = \emptyset$. That is, each fluid class has a positive initial inventory. Then, in (12.4) $Cx \leq e$ is the only constraint, which in the single-station case specializes to $e'x \leq 1$ (recall that e is a vector of all 1's). We have the following LP:

$$\max \quad c'x \qquad (12.12)$$
$$\text{s.t.} \quad e'x \leq 1$$
$$x \geq 0$$

Suppose that $c_1 = \max_{k \in \mathcal{K}} \{c_k\}$, then the solution is

$$x_1^1 = 1; \qquad x_k^k = 0, \quad k \neq 1. \qquad (12.13)$$

Denote this solution by $x^1 = (x_k^1)_{k \in \mathcal{K}}$. The corresponding policy is to devote the entire capacity (100%) to class 1. The time interval $[0, t_1]$ over which this policy is in force is determined by

$$t_1 = \frac{q_1}{a_{11} - \lambda_1}. \qquad (12.14)$$

That is, the solution x^1 applies until class 1 fluid level drops to zero. (Note that $a_{ii} > \lambda_i$ for all $i \in \mathcal{K}$, following from the traffic intensity condition: $\rho = e'A^{-1}\lambda \leq 1$. Also note that $a_{ki} < 0$ for all $i \neq k$.)

Next, we update $\Pi(q) = \{1\}$, and proceed to a second LP

$$\max \quad c'x$$
$$\text{s.t.} \quad a_{11}x_1 + \sum_{j \neq 1} a_{1j}x_j \leq \lambda_1, \qquad (12.15)$$
$$e'x \leq 1,$$
$$x \geq 0.$$

To solve the above problem, first observe that there must exist an optimal solution that makes (12.15) binding. This is because starting with any optimal solution (which must exist, since the above LP is feasible and bounded) in which (12.15) is not binding, one can always increase x_1 and decrease another component, say x_j, by the same amount (sufficiently small to keep x_j nonnegative). Since $a_{11} > 0$, but $a_{1j} \leq 0$ for all $j \neq 1$, this adjustment will increase the value of $(a_{11}x_1 + \sum_{j \neq 1} a_{1j}x_j)$, so we can keep doing this until the constraint becomes tight. Meanwhile, due to the fact that $c_1 = \max_{i \in \mathcal{K}} c_i$, the objective value will not decrease during this procedure. Hence, we can effectively treat (12.15) as an equality, which gives us

$$x_1 = \frac{\lambda_1}{a_{11}} - \sum_{j \neq 1} \frac{a_{1j}}{a_{11}} x_j.$$

Now the problem becomes another LP with one fewer variable:

$$\max \quad \sum_{j=2}^{K} \left(c_j - c_1 \frac{a_{1j}}{a_{11}} \right) x_j$$

s.t.

$$\sum_{j=2}^{K} \left(1 - \frac{a_{1j}}{a_{11}} \right) x_j \leq 1 - \frac{\lambda_1}{a_{11}},$$

$$x \geq 0$$

Suppose, without loss of generality, that

$$2 = \arg\max_{j \neq 1} \left\{ \frac{c_j - c_1 \dfrac{a_{1j}}{a_{11}}}{1 - \dfrac{a_{1j}}{a_{11}}} \right\}.$$

Then, we have obtained the optimal solution, denoted by x^2,

$$x_1^2 = \frac{\lambda_1 - a_{12}}{a_{11} - a_{12}}, \quad x_2^2 = \frac{a_{11} - \lambda_1}{a_{11} - a_{12}}; \quad x_j = 0, \ j \neq 1, 2.$$

Specifically, this second LP identifies the second highest priority class, class 2; and the allocation is to give enough capacity to class 1 so as to maintain its level at zero, and give all the remaining capacity to class 2. This allocation applies up to t_2, with t_2 determined in the same fashion as t_1 in (12.14), with $q_1/(a_{11} - \lambda_1)$ replaced by $q_2/(a_{22} - \lambda_2)$, and with q_2 denoting the class 2 fluid level at t_1 (and likewise for q_k, $k \neq 2$).

At the third iteration, we have $\Pi(q) = \{1, 2\}$, and the corresponding LP is

$$\max \quad c'x$$
$$\text{s.t.} \quad a_{11}x_1 + \sum_{j \neq 1} a_{1j}x_j \leq \lambda_1, \qquad (12.16)$$

$$a_{22}x_2 + \sum_{j \neq 2} a_{2j}x_j \leq \lambda_2, \qquad (12.17)$$

$$e'x \leq 1,$$
$$x \geq 0.$$

Reasoning as before, if x^2 satisfies the new constraint in (12.17), then this is equivalent to the index set that defines $\tau^{\Pi(q)}$ being empty, and hence, the myopic procedure ends. Otherwise, we can again claim that there must exist an optimal solution to the above LP that makes *both* (12.16) and (12.17) binding. First, starting with any optimal solution, by increasing x_1 and decreasing x_j ($j \neq 1$; if (12.17) is already binding, set $j = 2$), we can augment the left-hand side of (12.16) while preserving the feasibility of both (12.16) and (12.17), and maintain the optimality (since $c_1 = \max_{k \in \mathcal{K}} \{c_k\}$). With (12.16) binding, solve for x_1 and suppress it from the problem. We then go back to the LP of the second iteration but with an additional constraint [from (12.17)]:

$$\sum_{j=2}^{K} \left(a_{2j} - a_{21} \frac{a_{1j}}{a_{11}} \right) x_j \leq \lambda_2 - a_{21} \frac{\lambda_1}{a_{11}}. \tag{12.18}$$

Note that here the coefficient of x_2 is positive (since $a_{11}a_{22} - a_{21}a_{12} > 0$, due to A being an M-matrix), while the coefficient of x_j, for $j \neq 2$, is nonpositive (since $a_{11}a_{2j} - a_{21}a_{1j} \leq 0$, again due to A being an M-matrix). Hence, the earlier argument can be repeated. That is, starting from any optimal solution, one can increase x_2 and decrease x_j ($j \neq 1, 2$) so that the left hand side of (12.18) is increased, while the objective value is not reduced.

12.4.2 Summary of the Algorithm

To summarize the above discussion, we now formalize the algorithm that solves the single-station model. We shall present the algorithm in matrix form. Recall that for any matrix (or vector) $G = (g_{ij})$, and any subsets of its row and column indices, α and β, with $\alpha \cap \beta = \emptyset$, $G_{\alpha\beta} = (g_{ij})_{i \in \alpha, j \in \beta}$ (the α-β block of G) and $G_\alpha = G_{\alpha\alpha}$. In particular, since $A = (I - P')D$,

$$A_\alpha := (I_\alpha - P'_\alpha)D_\alpha, \quad A_{\alpha\beta} := -P'_{\beta\alpha}D_\beta.$$

To initialize ($n = 1$), set

$$\alpha^1 = \emptyset \quad \text{and} \quad \beta^1 = \mathcal{K}.$$

Then for $n = 1, \ldots, K$,

- Compute

$$c^n := c'_{\beta^n} - c'_{\alpha^n} A_{\alpha^n}^{-1} A_{\alpha^n \beta^n}, \tag{12.19}$$

$$r^n := e'_{\beta^n} - e'_{\alpha^n} A_{\alpha^n}^{-1} A_{\alpha^n \beta^n}, \tag{12.20}$$

and identify

$$k_n := \arg \max_{k \in \beta^n} \left\{ \frac{c_k^n}{r_k^n} \right\}. \tag{12.21}$$

- Set

$$\alpha^{n+1} := \alpha^n \cup \{k_n\} \quad \text{and} \quad \beta^{n+1} := \mathcal{K} \setminus \alpha^{n+1}. \tag{12.22}$$

For simplicity, assume that k_n in (12.21) is unique (otherwise, break ties arbitrarily). Some explanation is in order. At the nth iteration, the LP in question takes the form

$$\max \quad c'_{\alpha^n} x_{\alpha^n} + c'_{\beta^n} x_{\beta^n} \tag{12.23}$$

$$\text{s.t.} \quad A_{\alpha^n} x_{\alpha^n} + A_{\alpha^n \beta^n} x_{\beta^n} \le \lambda_{\alpha^n}, \tag{12.24}$$

$$e'_{\alpha^n} x_{\alpha^n} + e'_{\beta^n} x_{\beta^n} \le 1, \tag{12.25}$$

$$x \ge 0,$$

and its optimal solution takes the form

$$x^n_{\alpha^n} = A^{-1}_{\alpha^n}(\lambda_{\alpha^n} - A_{\alpha^n \beta^n} x_{\beta^n}) \tag{12.26}$$

$$x^n_{k_n} = (1 - e'_{\alpha^n} A^{-1}_{\alpha^n} \lambda_{\alpha^n})/r^n_{k_n}, \tag{12.27}$$

$$x^n_j = 0, \qquad j \in \beta^n \setminus \{k_n\}. \tag{12.28}$$

To see this, suppose $\beta^n \ne \emptyset$. Then, as discussed earlier, we can first make the inequalities in (12.24) binding by decreasing the value of x_j, $j \in \beta^n$, and increasing x_ℓ, $\ell \in \alpha^n$, while ensuring that the objective value is not reduced. Then, with (12.24) being an equality, we can solve the matrix equation and obtain x_{α^n} in (12.26). Substituting this solution into (12.25) reduces the LP to the following, with c^n and r^n following (12.19) and (12.20), respectively:

$$\max \quad c^n x_{\beta^n}$$

$$\text{s.t.} \quad r^n x_{\beta^n} \le 1 - e'_{\alpha^n} A^{-1}_{\alpha^n} \lambda_{\alpha^n},$$

$$x_{\beta^n} \ge 0.$$

Since there is only a single constraint in the above LP, it is clear that with the possible exception of one variable, all other variables should be zero; and this exception is identified as x_{k_n} in (12.27), with k_n following (12.21).

To argue that the solutions in (12.26) and (12.27) satisfy the nonnegativity constraint, note the following: For (12.26), nonnegativity follows from A being an M-matrix, which implies that $A^{-1}_{\alpha^n}$ is an M-matrix and hence $A^{-1}_{\alpha^n}$ is nonnegative, and $A_{\alpha^n \beta^n}$ is nonpositive. These properties also guarantee that $r^n \ge 0$, which, along with the traffic intensity condition $\rho = e' A^{-1} \lambda \le 1$, ensures the nonnegativity of (12.27).

Suppose the class k_n fluid level is $q^{n-1}_{k_n}$ when the control generated by x is first used. Then this control should be followed for a duration of

$$\tau^n = \frac{q^{n-1}_{k_n}}{a_{k_n k_n} x_{k_n} + \sum_{j \in \alpha^n} a_{k_n j} x_j - \alpha_{k_n}}. \tag{12.29}$$

This is the time needed for class k_n fluid to drop to the zero level. At this point, it would not be feasible to follow this solution, and this explains the update of α^{n+1} and β^{n+1} in (12.22) and the need for the next iteration.

After a certain number of iterations, which is no more than K, we will reach $\beta^n = \emptyset$, and the algorithm ends.

12.4.3 Remarks

The above algorithm generates an allocation that works as follows. Initially, when all fluid classes have positive inventories, the class k_1 is identified, and the entire 100% capacity is allocated to this class. This allocation is continued until t_1, when Q_{k_1}, the inventory level of k_1, drops to zero. At t_1, a second class k_2 is identified, and the allocation is modified such that Q_{k_1} is maintained at zero while the remaining capacity is given entirely to k_2. This allocation then continues until t_2, when Q_{k_2} drops to zero. At t_2, a third class k_3 is identified, and the allocation becomes such that both Q_{k_1} and Q_{k_2} are maintained at zero while the remaining capacity is given to k_3; and so forth.

It is important to note that once a class is driven down to the zero level, it will be maintained at zero, and this corresponds to the notion of a priority class, for the obvious reason that sufficient capacity is allocated to it to keep it at zero before any remaining capacity is allocated to other classes.

Interestingly, the set of indices $(k_n)_{n=1}^K$ generated by the algorithm is consistent with the index set that solves the scheduling control of a multiclass single-station *stochastic* queueing model, known as Klimov's problem (under long-run average cost objective); refer to Chapter 11, in particular Section 11.3.2 and Algorithm 11.30. In Klimov's problem, the optimal control is identified as a priority rule following the indices: Class k_1 jobs have priority over class k_2 jobs, which, in turn, have priority over class k_3 jobs, and so forth. The processing (service) of jobs then follows this priority scheme.

If we relax the traffic condition, i.e., to allow $e'A^{-1}\lambda > 1$, then it is possible that the algorithm will stop before it reaches the Kth iteration. As in the network case, if the algorithm stops at step n, then the optimal solution to the nth LP remains optimal to the $(n+1)$st LP. It can be verified that this corresponds to the situation that the total processing requirement of the classes in α^n has already saturated the station, so that no additional class can be processed.

Although the algorithm identifies the (priority) index set by assuming that all classes start from a positive fluid level, this does not mean that the priority of any class is in any way affected by whether or not its initial fluid level is positive. Once the priority ordering is set, it will be followed, regardless of the initial state: If a higher-priority class is initially at zero level while a lower-priority class is at a positive level, we still need to devote enough capacity to maintain the higher-priority class at zero level, as well as to drive other high-priority classes that have positive fluid levels down to zero, and only process the lower-priority class if there is residual capacity.

This is easily justified by examining the algorithm as well. Suppose the algorithm has identified the index set as $1, 2, \ldots, n, \ldots$, with 1 having the highest priority. Let n_0 be the smallest class index such that the initial inventory q_{n_0} is positive, i.e., $q_n = 0$, for all $n < n_0$. In this case, we can start the algorithm from the n_0th iteration. In other words, for those (lower-priority) classes $n > n_0$, even if $q_n = 0$ (so that the corresponding constraints should be included into the LP), we can effectively treat these as if their initial fluid levels were positive, i.e., ignore the corresponding constraints. This is because the optimal solution of the n_0th LP in (4.5) automatically satisfies these constraints. (In particular, note that the coefficients of $x_k^{n_0}$ ($k = 1, \ldots, n_0$) in these constraints are all nonpositive.)

12.5   Proof of Global Optimality: Single-Station Model

In this section we establish the global optimality of the myopic control in the single-station model.

To begin with, let us formally construct the myopic control T^* corresponding to the algorithm in the last section. Let Q^* denote the corresponding fluid level process. Let x^n be the solution of the LP corresponding to the nth iteration of the myopic algorithm, as specified in (12.26)–(12.28). Let τ^n follow (12.29), with q^n specified below. Let $\{t_n\}$ be an increasing sequence of time epochs:

$$t_0 = 0, \qquad t_n = \sum_{i=1}^{n} \tau^i, \quad 1 \le n \le K.$$

Then we can write

$$T^*(t) = x^1 t, \quad t \in [0, t_1),$$

$$T^*(t) = \sum_{i=1}^{n-1} x^i \cdot (t_i - t_{i-1}) + x^n \cdot (t - t_{n-1}), \quad t \in [t_{n-1}, t_n), \qquad (12.30)$$

for $1 \le n \le K$, and

$$Q^*(t) = q^{n-1} + (\lambda - Ax^n)(t - t_{n-1}), \quad t \in [t_{n-1}, t_n),$$

for $1 \le n \le K$, where

$$q^0 := Q(0),$$
$$q^n := Q^*(t_n) = q^{n-1} + (\lambda - Ax^n)\tau^n, \quad 1 \le n \le K.$$

(Note that it is possible that $\tau^n = 0$ for some $n < K$; this corresponds to the class k_n fluid level being always zero.)

We want to show that for any given $t \in (0, t_K)$, the allocation $T^*(t)$ specified above is the optimal solution to the following LP:

$$\max \quad c'T(t) \tag{12.31}$$
$$\text{s.t.} \quad AT(t) \le q + \lambda t,$$
$$e'T(t) \le t,$$
$$T(t) \ge 0,$$

where $q = q^0 = Q(0)$. Note that from the way T^* is constructed, along with the constraints in the above LP, we know that T^* is an admissible control, in particular, both $T^*(t)$ and $t - e'T^*(t)$ are nondecreasing in t. Also note that $T^*(t)$ is clearly optimal for $t > t_K$ since $Q^*(t) \equiv 0$ for $t \ge t_K$.

Therefore, what remains is to show that T^* maximizes the objective function in (12.31) for all t. To this end, consider the dual LP

$$\min \quad y'(q + \lambda t) + zt \tag{12.32}$$
$$\text{s.t.} \quad y'A + ze \ge c,$$
$$y \ge 0.$$

Here y (a vector) and z (a scalar) are the dual variables.

Consider a fixed $t \in [t_{n-1}, t_n)$. Without loss of generality, suppose

$$\alpha^n = \{1, 2, \ldots, n-1\}, \quad k_n = n.$$

To simplify notation, write

$$\alpha := \alpha^n, \quad \beta := \mathcal{K} \setminus \alpha, \quad \beta^- = \beta \setminus \{n\}.$$

From the algorithm of the last section, we know that $T^*(t)$ of (12.30) satisfies

$$T_\alpha^*(t) > 0, \quad T_n^*(t) > 0, \quad T_{\beta^-}^*(t) = 0. \tag{12.33}$$

We want to show that the following constitutes a feasible solution to the dual LP in (12.32), and that it also satisfies the complementary slackness, and is hence the optimal (dual) solution

$$y_\alpha' = (c_\alpha' - ze_\alpha')A_\alpha^{-1}, \tag{12.34}$$
$$y_\beta = 0, \tag{12.35}$$

and

$$z = \frac{c_n - c_\alpha' A_\alpha^{-1} A_{\alpha n}}{1 - e_\alpha' A_\alpha^{-1} A_{\alpha n}}. \tag{12.36}$$

Combining (12.34) and (12.36) yields

$$y_\alpha' A_\alpha + ze_\alpha' = c_\alpha', \tag{12.37}$$
$$y_\alpha' A_{\alpha n} + z = c_n. \tag{12.38}$$

And these two, where compared against $T_\alpha^*(t) > 0$ and $T_n^*(t) > 0$ in (12.33), do comply with complementary slackness. Furthermore, from (12.37), we have

$$y_\alpha' = (c_\alpha' - ze_\alpha')A_\alpha^{-1}, \qquad (12.39)$$

which, upon substituting into (12.38), yields (12.36).

Next, the dual feasibility requires the following inequality:

$$y_\alpha A_{\alpha\beta-} + ze_{\beta-} \geq c_{\beta-}.$$

Substituting (12.39) into the above, we can write it as follows:

$$z \geq \max_{i \in \beta^-} \frac{(c_{\beta-} - c_\alpha A_\alpha^{-1}A_{\alpha\beta-})_i}{(e_{\beta-} - e_\alpha A_\alpha^{-1}A_{\alpha\beta-})_i} = \frac{c_i^n}{r_i^n}. \qquad (12.40)$$

Note that the equality above follows from the specification of the algorithm in the last section, which also enables us to reexpress (12.36) as $z = c_n^n/r_n^n$. Since $k_n = n$ as stipulated earlier, the inequality in (12.40) follows immediately from the maximality of k_n.

As a by-product of the above discussion, we also know that $z \geq 0$, since c^n and r^n, for all n, are both nonnegative. This is due to (a) A being an M-matrix, and (b) the traffic condition $\rho = e'A^{-1}\alpha \leq 1$.

So what remains is to show the nonnegativity of the y components in (12.34). We do this in the lemma below.

Lemma 12.8 The dual solution in (12.34) is nonnegative: $y_\alpha \geq 0$. Specifically, we have

$$(c_\alpha A_\alpha^{-1})_j/(e_\alpha A_\alpha^{-1})_j \geq z,$$

for all $j \in \alpha$.

Proof. Use induction on n. Adding back the n index so far omitted, we want to show the following, with u^j denoting the jth unit vector (i.e., a vector with the jth component equal to 1 and all other components equal to 0):

$$\frac{c_{\alpha^n}A_{\alpha^n}^{-1}u^j}{e_{\alpha^n}A_{\alpha^n}^{-1}u^j} \geq z^n, \qquad (12.41)$$

for $j \in \alpha$. Recall that $z^n = c_{k_n}^n/r_{k_n}^n$.

When $n = 1$, the left hand side of (12.41) is reduced to c_1, whereas

$$z = \max_{j \neq 1} \left\{ \frac{c_j - c_1\dfrac{a_{1j}}{a_{11}}}{1 - \dfrac{a_{1j}}{a_{11}}} \right\}.$$

The right hand side above is obviously dominated by c_1, since $c_1 \geq c_j$. Hence, (12.41) holds for $n = 1$.

Suppose (12.41) holds for $n - 1$; we want to show it also holds for n. As before, let k_{n-1} denote the last index that was included into α. For any given $j \in \alpha$, express $A_{\alpha^n}^{-1} u^j$ as follows:

$$A_{\alpha^n}^{-1} u^j = \begin{pmatrix} v\delta + A_{\alpha^{n-1}}^{-1} u_-^j \\ v \end{pmatrix} \tag{12.42}$$

where v is a scalar, δ a vector of dimension $n - 1$, and u_-^j the jth unit vector of dimension one less than u^j. Also note that $\alpha^n = \alpha^{n-1} \cup k_{n-1}$. Multiplying both sides of (12.42) by

$$A_{\alpha^n} = \begin{pmatrix} A_{\alpha^{n-1}} & A_{\alpha^{n-1} k_{n-1}} \\ A_{k_{n-1} \alpha^{n-1}} & a_{k_{n-1} k_{n-1}} \end{pmatrix},$$

we can solve for v and δ to obtain

$$\delta = -A_{\alpha^{n-1}} A_{\alpha^{n-1} k_{n-1}} u_-^j,$$

$$v = -\frac{A_{k_{n-1} \alpha^{n-1}} A_{\alpha^{n-1}}^{-1} u_-^j}{a_{k_{n-1} k_{n-1}} + A_{k_{n-1} \alpha^{n-1}} \delta}.$$

The above, combined with (12.42), yields

$$\frac{c_{\alpha^n} A_{\alpha^n}^{-1} u^j}{e_{\alpha^n} A_{\alpha^n}^{-1} u^j} = \frac{v(c_{k_{n-1}} - c_{\alpha^{n-1}} A_{\alpha^{n-1}}^{-1} A_{\alpha^{n-1} k_{n-1}}) + c_{\alpha^{n-1}} A_{\alpha^{n-1}}^{-1} u_-^j}{v(e_{k_{n-1}} - e_{\alpha^{n-1}} A_{\alpha^{n-1}}^{-1} A_{\alpha^{n-1} k_{n-1}}) + e_{\alpha^{n-1}} A_{\alpha^{n-1}}^{-1} u_-^j}. \tag{12.43}$$

Now recall that

$$
\begin{aligned}
z^{n-1} &= \frac{c_{k_{n-1}}^{n-1}}{r_{k_{n-1}}^{n-1}} \\
&= \frac{c_{k_{n-1}} - c_{\alpha^{n-1}} A_{\alpha^{n-1}}^{-1} A_{\alpha^{n-1} k_{n-1}}}{e_{k_{n-1}} - e_{\alpha^{n-1}} A_{\alpha^{n-1}}^{-1} A_{\alpha^{n-1} k_{n-1}}} \\
&\leq \frac{c_{\alpha^{n-1}} A_{\alpha^{n-1}}^{-1} u_-^j}{e_{\alpha^{n-1}} A_{\alpha^{n-1}}^{-1} u_-^j},
\end{aligned}
\tag{12.44}
$$

where the inequality is due to the induction hypothesis. Combining this inequality with (12.43), we have

$$\frac{c_{\alpha^n} A_{\alpha^n}^{-1} u^j}{e_{\alpha^n} A_{\alpha^n}^{-1} u^j} \geq z^{n-1}, \tag{12.45}$$

taking into account that both the numerator and the denominator of the expression in (12.44) are nonnegative, and so is v (all due to A being an M-matrix).

We can now repeat the above manipulation, but with $A_{\alpha^n i}$ (for some $i \in \beta^n$) replacing u^j_-, to obtain the following

$$c_{\alpha^n} A_{\alpha^n}^{-1} A_{\alpha^n i} = v(c_{k_{n-1}} - c_{\alpha^{n-1}} A_{\alpha^{n-1}}^{-1} A_{\alpha^{n-1} k_{n-1}}) + c_{\alpha^{n-1}} A_{\alpha^{n-1}}^{-1} A_{\alpha^{n-1} i}$$

$$e_{\alpha^n} A_{\alpha^n}^{-1} A_{\alpha^n i} = v(e_{k_{n-1}} - e_{\alpha^{n-1}} A_{\alpha^{n-1}}^{-1} A_{\alpha^{n-1} k_{n-1}}) + e_{\alpha^{n-1}} A_{\alpha^{n-1}}^{-1} A_{\alpha^{n-1} i}.$$

Hence, we have

$$\frac{c_i^n}{r_i^n} = \frac{c_i - c_{\alpha^n} A_{\alpha^n}^{-1} A_{\alpha^n i}}{e_i - e_{\alpha^n} A_{\alpha^n}^{-1} A_{\alpha^n i}}$$

$$= \frac{-v(c_{k_{n-1}} - c_{\alpha^{n-1}} A_{\alpha^{n-1}}^{-1} A_{\alpha^{n-1} k_{n-1}}) + (c_i - c_{\alpha^{n-1}} A_{\alpha^{n-1}}^{-1} A_{\alpha^{n-1} i})}{-v(e_{k_{n-1}} - e_{\alpha^{n-1}} A_{\alpha^{n-1}}^{-1} A_{\alpha^{n-1} k_{n-1}}) + (e_i - e_{\alpha^{n-1}} A_{\alpha^{n-1}}^{-1} A_{\alpha^{n-1} i})}.$$

Since

$$\frac{c_i - c_{\alpha^{n-1}} A_{\alpha^{n-1}}^{-1} A_{\alpha^{n-1} i}}{e_i - e_{\alpha^{n-1}} A_{\alpha^{n-1}}^{-1} A_{\alpha^{n-1} i}} \leq \frac{c_{k_{n-1}} - c_{\alpha^{n-1}} A_{\alpha^{n-1}}^{-1} A_{\alpha^{n-1} k_{n-1}}}{e_{k_{n-1}} - e_{\alpha^{n-1}} A_{\alpha^{n-1}}^{-1} A_{\alpha^{n-1} k_{n-1}}} = z^{n-1}$$

following the definition of k_{n-1} (analogous to k_{n+1}, being the arg max), we have, from combining the above, $c_i^n / r_i^n \leq z^{n-1}$ for all $i \in \beta^n$. Hence

$$z^n = \max_{i \in \beta^n} \left\{ \frac{c_i^n}{e_i^n} \right\} \leq z^{n-1},$$

which, combined with (12.45), yields the desired inequality in (12.41). $\square$

In summary, we have shown that (a) T^* is primal feasible, (b) (y, z) is dual feasible, and (c) the complementary slackness condition is fulfilled; hence the claimed global optimality of T^*.

Theorem 12.9 The allocation T^* in (12.30), with the x^n's generated by the algorithm in Section 12.4.2, is globally optimal.

Remark 12.10 Note that any allocation, say $\tilde{T}(t)$, that is a feasible solution to the LP in (12.31) need not be an admissible control, as there is no guarantee that $\tilde{T}(t)$ and $t - e'\tilde{T}(t)$ are necessarily nondecreasing. (That $T^*(t)$ satisfies these monotone properties has to do with its special form as specified in (12.30).) Therefore, by showing that $c'T^*(t) \geq c'\tilde{T}(t)$ for all t, the optimality that we have established for T^* is in fact stronger than what is required of the global optimality.

12.6   Notes

For each multiclass stochastic network there is a corresponding deterministic fluid network, which takes only the first-order data (means and rates) from the stochastic model, and further assumes that the transactions circulating in the network are continuous flows instead of discrete jobs. It is also known that with appropriate scaling, the fluid network is a limit of the stochastic network, in the sense of a functional strong law of large numbers or a functional law of the iterated logarithm; refer to Section 6.3, Section 7.8, Section 7.6, and Section 9.5.

In recent years, the fluid model has also played a central role in studying the stability of stochastic networks. As demonstrated in Chapter 8, the stability of a fluid network implies the stability of the corresponding queueing network. Refer to Section 8.7 for recent research in this area.

That the fluid model is a powerful tool in studying dynamic scheduling of multiclass stochastic networks has been demonstrated in Chen and Yao [5], which is also the basis of Section 12.1 through Section 12.5. The proof of Lemma 12.8 in Section 12.5 is adapted from the proof of Lemma 4.5 of Lu [10], where a more general fluid model is studied, one with additional constraints (upper limits) on the fluid levels. Other related studies on optimal scheduling of fluid networks include Atkins and Chen [1], Avram, et al. [2], Bäuerle and Rieder [3], Chen and Yao [6], Dai and Weiss [9], Meyn [11], Sethuraman [12], and Weiss [13].

Definition 12.4 and Theorem 12.7 are drawn from Connors et al. [8], where the myopic procedure via solving a sequence of LPs is extended and applied to generate "what-next" real-time schedules for a semiconductor fabrication facility. Exercises 3-5 are also drawn from [8].

12.7 Exercises

1. Formulate the problem of finding $\tau(q)$ in (12.5) as a linear program problem.

2. Show that T^* in Theorem 12.7 is admissible.

3. Show that if a fluid network is stable, then for each station i in the network, we must have

$$(MA^{-1}Q(0))_i > 0 \quad \Rightarrow \quad \rho_i < 1, \qquad (12.46)$$
$$(MA^{-1}Q(0))_i = 0 \quad \Rightarrow \quad \rho_i \leq 1.$$

4. Show that if $\rho \leq e$, and the inequality in (12.46) holds, then

$$\tau := \max_{i:\rho_i<1} \frac{(MA^{-1}Q(0))_i}{1 - \rho_i}$$

is a clearing time, and hence the network is stable under the allocation

$$T(t) = \frac{A^{-1}Q(0)}{\tau}(t \wedge \tau) + A^{-1}\lambda,$$

which is admissible.

5. Prove the necessity part of Corollary 12.6.

References

[1] ATKINS, D. AND CHEN, H., Performance Evaluation of Scheduling Control of Queueing Networks: Fluid Model and Heuristics. *Queueing Systems*, **21** (1995), 391–413.

[2] Avram, F., D. Bertsimas, and M. Ricard, Fluid models of sequencing problems in open queueing networks: an optimal control approach. In: *Stochastic Networks, volume 71 of Proceedings of the International Mathematics Association*, F.P. Kelly and R.J. Williams (ed.), Springer-Verlag, New York, 1995, 199–234,

[3] BÄUERLE, N. AND RIEDER, U., Optimal Control of Single-Server Fluid Networks. Preprint, 1998.

[4] CHEN, H. AND MANDELBAUM, A., Hierarchical Modeling of Stochastic Networks, Part I: Fluid Models. In: *Stochastic Modeling and Analysis of Manufacturing Systems*, D.D. Yao (ed.), Springer-Verlag, New York, 1994, 47–105.

[5] CHEN, H. AND YAO, D.D., Dynamic Scheduling Control of a Multiclass Fluid Network. *Operations Research*, **41** (1993), 1104–1115.

[6] CHEN, H. AND YAO, D.D. (1996). Stable Priority Disciplines for Multiclass Networks. In: *Stochastic Networks: Stability and Rare Events*, P. Glasserman, K. Sigman, and D.D. Yao (eds.), Springer-Verlag, New York, 1996, 27–39

[7] CHVÁTAL, V., *Linear Programming*. W.H. Freeman, New York, 1983.

[8] CONNORS, D., FEIGIN, G. AND YAO, D.D., Scheduling Semiconductor Lines Using a Fluid Network Model. *IEEE Trans. Robotics and Automation,* **10** (1994), 88–98.

[9] DAI, J.G. AND G. WEISS, A fluid heuristic for minimizing makespan in a job-shop. Technical report, 1999.

[10] LU, Y., *Dynamic Scheduling of Stochastic Networks with Side Constraints.* Ph.D. thesis, Columbia University, 1998.

[11] Meyn, S. (2000). Feedback regulation for sequencing and routing in multiclass queueing networks. Preprint.

[12] SETHURAMAN, J. *Scheduling job shops and multiclass queueing networks using fluid and semidefinite relaxations.* PhD. thesis, Operations Research Center, MIT, August 1999.

[13] Weiss, G. (1995). On optimal draining of fluid re-entrant lines. In: *Stochastic Networks, volume 71 of Proceedings of the International Mathematics Association,* F.P. Kelly and R.J. Williams (ed.), Springer-Verlag, New York, 1995, 91–103.

Index

Aggregation
  nodes, 53–55
  servers, 51–53
Almost sure convergence, *98*, 99,
  100
Arrangement ordering, *56*, 62
Arrival theorems, 88–91
ASTA (arrivals see time averages),
  92

Balance equations
  detailed, 6, 7, 24, 32, 71
  full, 6, 16, 71, 72
  partial, 71
Birth-death process, 2–4, 6, 7, 12,
  16, 212, 225
  birth rate, 1, 2, 6
  death rate, 1, 6
Birth-death queue, 1–10
BNA/FM, 322
BNA/FM algorithm, 177
Bottleneck
  balanced, *167*
  non-bottleneck, *167*
  strict, *167*

Branching bandit process, 343
Breakdowns
  autonomous, 299
  operational, 300
Brownian approximation, 148, 173,
  183, *see* SRBM approx-
  imation
Brownian motion, 102–106

Clearing time, *376*
$c\mu$-rule, 340
Completely-S matrix, *202*, 305
Concavity/Convexity, 45–48
Conservation laws
  definition, 335
  examples, 336–339
  polymatroid structure, 334–
  336
Continuous Mapping Theorem, *101*
Convergence in distribution, *see*
  Weak convergence
Convergence in probability, *98*, 99,
  100

Convergence with probability one,
    see Almost sure conver-
    gence
Convergence-Together Theorem, *101*
Convolution algorithm, 25–26
Coupling, 37, 59

Diffusion approximation, *125*
    feedforward network, 284
    G/G/1 queue, 126, 138–142,
        148, 153
    generalized closed Jackson net-
        work, 199–201
    generalized Jackson network,
        158, 185–193
    Kumar-Seidman network, 232–
        238, 243, 246
    multiclass network, 292, 306,
        308, 310
    multiclass queue, 287
Donsker's theorem, 105, 110

Effective inflow rate, *166*
Effective outflow rate, 166
Equal in distribution, *100*
Equilibrium distribution, see sta-
    tionary distribution
    birth-death process, 2, 3
    Jackson network, 16, 17, 20–
        22, 30–32, 38
    Kelly network, 69, 70
    Markov chain, 20, 28, 29, 70
Equilibrium rate, *8*, 37–40, 42, 45,
        46, 48, 52, 60, 61
Ergodicity, *4*
Exit process, 29, 31, 69, 84, 87, 88
Extended polymatroid
    connections to generalized con-
        servation laws, 351
    connections to sub/supermodularity,
        347–351
    equivalent definitions, 345–347
    optimization, 351–356

Flow-balance equation, see balance
    equation

Fluid approximation
    G/G/1 queue, 135–138, 141,
        145, 150
    generalized Jackson network,
        184–185
    Kumar-Seidman network, 229–
        232
    multiclass network, 243
Fluid model, 306
    heterogeneous network, 243,
        367
    homogeneous network, 157, 158,
        165–170, 194
    Kumar-Seidman network, 211,
        214–223, 225, 231, 239,
        242
    multiclass fluid network, 367,
        367, 374, 375, 387, 388
    single station, 125, 137, 145
Functional central limit theorem,
        109
    G/G/1 queue, see diffusion
        approximation
    i.i.d. sequence, 186, 234
    i.i.d. squence, 110–111
    simulation plot, 109
Functional law-of-iterated logarithm
    feedforward network, 272
    G/G/1 queue, 125, 144–146
    generalized Jackson network,
        158, 177–181, 193
    i.i.d. sequence, 111–112
Functional strong approximation,
        see Strong approximation
Functional strong law-of-large-numbers
    G/G/1 queue, see fluid ap-
        proximation
    generalized Jackson network,
        see fluid approximation
    i.i.d. sequence, 109
    simulation plot, 109

Generalized conservation laws
    definition, 342
    examples, 342–345

extended polymatroid structure, 345–351
Gittins index, 358
Global optimality, *369*, 383–387

Heavy traffic approximation, *see* Diffusion approximation
Heavy traffic condition, 141, 192, 215, 306
Heavy traffic limit theorem, *see* Diffusion approximation, 305

In equilibrium, 4, 7, 8, 17, 19, 25, 57, 69, 75
In steady state, *see* In equilibrium
Irreducible, *19*
Irreducible closed network, 158, 193–201
Ito's formula, *104*, 171

Jackson network
    closed network, 19–21
    irreducible closed network, *see* closed network
    open network, 16–19
    semi-open network, 21–23, 25, 32, 33
    time reversal, 28–32

Klimov's problem, 341, 342, 367, 382
Kumar-Seidman network
    description, 211
    diffusion approximation, 232–238
    fluid approximation, 229–232
    fluid model, 215–223
    quasi-reversibility, 93
    simulation results, 212
    stability, 223–229

Least element, *163*
Likelihood ratio ordering, 8, *8*, 9–12, 37, 39–42, 52, 53, 55, 60–62

Linear programming
    dual, 332
    over a polymatroid, 332
    over an extended polymatroid, 351
    scheduling applications, 339
    vertices, 333
Lipschitz continuity, 128

Majorization ordering, *56*, 62
Markov chain, 1, 2, 4–7, 16, 20, 28, 29, 32, 69, 70, 74, 76, 78, 82–84, 91, 212, 225
Martingale, *102*, 303
Mean value analysis, 26–28
Metric space, *100*
Minimal clearing time, *376*
M-matrix, 16, *162*
Monotonicity, 38–45
Multiple server queue, 49–51
Myopic procedure, 370, *372*, 373

Non-anticipation, *163*

Oblique reflection mapping, *see* reflection mapping
Optimal control/dynamic control
    fluid networks, 367–387
    multiclass queueing networks, 367–387

PASTA (Poisson arrivals see time averages), 89
Performance polytope, 329, 351
PF_2 property, 38–39, 43
PH-distribution, 76–81, 91
Phase-type distribution, *see* PH-distribution
P-matrix, *203*
Poisson event epochs, 3, 10, 43, 46, 53
Poisson flows, 7, 11, 21, 29, 31, 69, 74, 79, 81–84, 87–89, 92, 93
    non-loop criterion, 92

Poisson process, 3, 16, 21, 29, 31,
33, 70, 71, 74, 75, 77, 78,
84, 86, 87, 157, 342
Poisson-in-Poisson-out, 69, 74, 75,
91
Polymatroid
connections to conservation
laws, 334–336
equivalent definitions, 330–331
optimization, 339–341
rank function, 330

Quasi-reversible queue, 32, 84, 87–
92
definition, 74
network of, 81–87
product form solution, 84
properties, 75
Queue length
feedforward network, 271
G/G/1 queue, 127
generalized Jackson network,
160
Kumar-Seidman network, 229
multiclass network, 296
multiclass queues, 256

Random Time-Change Theorem,
101
RBM, *see* reflected Brownian motion
Reflected Brownian motion, 125,
126, *129*, 132, 147, 157,
158, 170–173, 233, 275,
321
on a simplex, 195
simulation plot, 129
Reflection mapping, 125, 127–134,
136, 157, 161–165, 167,
203, 205, 231, 257, 303,
306
on a simplex, 194

Scale property, *163*
Schur convex (concave) functions,
56, 63

Semi-martingale, *303*
Semi-martingale reflected Brownian motion, *see* SRBM
Shifted likelihood ratio ordering,
42–45, 61
Skorohod problem, *162*, 202, 204,
310, 311
Skorohod Representation Theorem,
101, 116, 139, 187, 234
Skorohod topologies, 99, 101, 116,
202
S-matrix, *202*
Sojourn time
feedforward network, 272
multiclass network, 296
multiclass queue, 256
SRBM, 291, 305, 310–312
SRBM approximation, 292, 300–
307
Stability
fluid network, *374*
fluid network under a priority discipline, *217*
Kelly-type network, 240
Kumar-Seidman network, 212,
224
multiclass queueing network,
212
SPT network, 239
three-station variation of Bramson network, 241
three-station variation of Kumar-
Seidman network, 238
variation of Bramson network,
240
Stationary distribution, 229
Jackson network, 16
Markov chain, 28
queue length, 126, 170, 198,
212, 279
reflected Brownian motion, 126,
129, *172*, 280
reflected Brownian motion on
a simplex, 196
sojourn time, 279

SRBM, 292, 305, 310–312
  workload, 126, 278
Stochastic ordering, 1, 8–12, 39,
    40, 43, 45, 62
Stochastic scheduling, 329–356
Strong approximation, 306
  feedforward network, 272–278,
    282
  G/G/1 queue, 125, 146–150
  generalized Jackson network,
    158, 181–184
  i.i.d. sequence, 112–114
  multiclass queue, 261–267
Strong Markov process, 105
Strong Markov property, 105
Subcritical buffer, 169
Submartingale, 102
Submodularity, 63, 330, 347–351
Supercritical buffer, 169
Supermartingale, 102
Supermodularity, 63, 330, 347–351
Symmetric queue, 76–82, 91

Throughput function, 33, 48, 51,
    59–62
  as equilibrium rate, 23, 24
  computation, 25–28
  concavity, 45–48
  convexity, 45–48
  monotonicity, 38–45
Time-reversal, 5, 7, 15, 28, 30–
    32, 69, 70, 72–75, 77–79,
    82–89, 91
Time-reversibility, 4–7
Time-reversible, see time-reversibility
Traffic equation
  generalized closed Jackson net-
    work, 195
  generalized Jackson network,
    166
  Jackson network, 16

u.o.c., 100
Uniform norm, 99
Uniform topology, 99

Uniformization, 3, 10–12, 43, 45
Uniformly on all compact sets, see
    u.o.c.
Upper-semi-continuous, 164

Weak convergence, 99, 99, 100
Weak stability
  fluid network under a prior-
    ity discipline, 218
Workload
  feedforward network, 271
  G/G/1 queue, 127
  generalized Jackson network,
    160
  multiclass network, 296
  multiclass queues, 256

Applications of Mathematics

(continued from page ii)

37 Yin/Zhang, **Continuous-Time Markov Chains and Applications** (1998)

38 Dembo/Zeitouni, **Large Deviations Techniques and Applications**, Second Ed. (1998)

39 Karatzas/Shreve, **Methods of Mathematical Finance** (1998)

40 Fayolle/Iasnogorodski/Malyshev, **Random Walks in the Quarter Plane** (1999)

41 Aven/Jensen, **Stochastic Models in Reliability** (1999)

42 Hernández-Lerma/Lasserre, **Further Topics on Discrete-Time Markov Control Processes** (1999)

43 Yong/Zhou, **Stochastic Controls: Hamiltonian Systems and HJB Equations** (1999)

44 Serfozo, **Introduction to Stochastic Networks** (1999)

45 Steele, **Stochastic Calculus and Financial Applications** (2000)

46 Chen/Yao, **Fundamentals of Queueing Networks: Performance, Asymptotics, and Optimization** (2001)

47 Kushner, **Heavy Traffic Analysis of Controlled Queueing and Communication Networks** (2001)